Horticulture: Plants for People and Places, Volume 2

Geoffrey R. Dixon • David E. Aldous
Editors

Horticulture: Plants for People and Places, Volume 2

Environmental Horticulture

Springer

Editors
Geoffrey R. Dixon
GreenGene International
Hill Rising, Sherborne, Dorset
United Kingdom

David E. Aldous
School of Agriculture and Food Sciences
The University of Queensland, Lawes
Queensland
Australia

ISBN 978-94-017-8580-8 ISBN 978-94-017-8581-5 (eBook)
ISBN 978-94-017-8640-9 (set)
DOI 10.1007/978-94-017-8581-5
Springer Dordrecht Heidelberg New York London

Library of Congress Control Number: 2014936300

Printed on acid-free paper

Springer is part of Springer Science+Business Media (www.springer.com)

We dedicate these Books to our wives;
Mrs. Kathy Dixon and Mrs. Kaye Aldous
in gratitude for their lifetimes of unstinting
support, forbearance and understanding

Professor David E. Aldous – deceased 1st November 2013.

The concepts underlying the Trilogy Horticulture—Plants People and Places were developed by David Aldous and me during the International Horticultural Congress 2010 in Lisbon, Portugal. These Books celebrate our common views of the scholastic and intellectual depth and breadth of our discipline and the manner by which it is evolving in response to the economic, environmental and social challenges of the

21st Century. Jointly, over more than two years we enlisted international authorities as lead authors, reviewed drafts and edited final texts. Despite there being half the world between us we formed a deep rapport. His sudden and wholly unexpected death from a brain aneurysm came as a shattering blow leaving me with the tasks of seeing our work through to completion. This Trilogy stands as a legacy to horticulture as "the first of all the Arts and Sciences" from an internationally acclaimed, respected and much loved: scientist, educator and author. It is appropriate that the Trilogy should be dedicated to David united with our original intention of paying tribute to our respective wives.

Professor Geoffrey R. Dixon

Preface

"This Trilogy of books answers the question "What is Horticulture?". Their contents span from tropical plantations growing exotics crops such as cocoa, pineapples and rubber through to the interior landscaping of high-rise office tower blocks and to other landscape applications which encourage physical and mental health. The common thread uniting this Discipline is the identification, breeding, manipulation of growth and stimulation of flowering and fruiting in plants either for food or environmental improvement. Understanding the scientific principles of why plant productivity increases following physical, chemical and biological stimuli has fascinated horticulturists for several millennia.

Epicurus (341BC–270BC) the Athenian philosopher of the 3rd century BC believed that plants achieved "the highest good was calmness of mind". Calmness comes to some Horticulturists with the satisfaction of entering vast hectarages of bountiful orchards, to others from well designed and carefully maintained landscapes, while others are entranced by participation in conserving components of the Earth's fragile biodiversity. Horticulture while being a scientific discipline has much wider and deeper dimensions. There are historic, artistic and cultural facets which are shared with the Humanities and these aspects are included within this Trilogy. Wherever Horticulturists gather together they share a common language which interprets useful scientific knowledge and cultural understanding for the common benefit of mankind. For while Horticulture is about achieving an intensity of growth and development, flowering and fruiting, it is wholly conscious that this must be achieved sustainably such that the resources used are matched by those passed on for use by future generations.

The structure of this Trilogy is such that it traces the evolution in emphasis which has developed in Horticultural philosophy across the second half of the 20th and into the 21st century. Following the worldwide conflicts of the 1940s the key desires were the achievement of food sufficiency and the eradication of hunger. In the increasingly affluent developed world there is food sufficiency par excellence obtained from the planet. Never before has such an array of plenty been made available year round. This plenty is nowhere more evident that in the fresh fruit and vegetable aisles of our supermarkets. Horticulture has given retail shoppers the gift of high quality and diversity of produce by manipulating plant growth, reproduction and postharvest care across the globe.

This second volume illustrates in considerable depth the scientific, managerial and technological concepts which underpin Environmental Horticulture. It covers considerations of: Horticulture and the Environment, Woody Ornamentals, Herbs and Pharmaceuticals, Urban Greening, Rural Trees, Urban Trees, Turfgrass Science, Interior and External Landscaping, Biodiversity, Climate Change and Organic Production. These subjects are united by consideration of the need for the sustainable use of resources and careful conservation applied to all points where Horticulture, natural flora and fauna, and the environment coincide. Horticulture plays an enormous role in aiding environmental care and support. Indeed this discipline could be considered as having founded much of the basis for what is now considered to be ecological, environmental science and the analysis and understanding of eco-system services. This is illustrated by the current enthusiasm for developing green "eco-towns" and "garden cities". Such concepts are rooted in Horticulture as illustrated for example by Dame Sylvia Crowe's town and country planning in the period 1930–1950. Recent revivals reflect the need for rejuvenating cities and communities which have exhibited declining economies, poor employment opportunities, and returning poverty. Initiatives have provided employment opportunities that link living networks of green plants, in residential, peri-urban and rural areas, with business, leisure and tourism development, and also address environmental issues associated with ensuring air and water quality, habitat conservation and sustainable recreation development.

Research shows that green open space provides plant communities that offer shade and release oxygen, act as fuel sources and sinks for the sequestration of atmospheric carbon dioxide, control soil erosion, stabilize dust particles, reduce glare, noise and visual pollution, conserve surface water resources, and act as wildlife or biodiversity corridors, that protect and conserve diverse populations of flora and fauna. The presence of healthy plant communities provides significant economic benefits in the form of increased tourism, employment opportunities, and commercial real estate values. Considerable cost-benefits are added, such as carbon sequestration, providing carbon credits and stormwater attenuation. There is also evidence that green open space provides energy savings in terms of reducing air conditioning and cooling costs, pollution and lowering health care expenses.

Particular attention is given to the threats posed by climate change. This is the single largest challenge which faces scientists of all disciplines. Horticulture has much to offer by way of mitigation. Such measures will then permit nations to deal with the consequences of an increasing world population and increasing urbanisation. Unless mankind is able to limit the rate of population expansion there will be continuing damaging effects on natural green spaces. This will further degrade the World's natural resources, such as wetlands, grasslands, wild areas, coastal systems and temperate and tropical forests. Losing the biological functions, and the ecosystem services of these natural green spaces, places mankind in even greater jeopardy. Horticulturists along with other ecologists are beginning to develop understandings of the means by which mitigation of this damage may be achieved.

The first volume in this Trilogy covers Crop Production Horticulture (volume 1) and the final volume is devoted to Social Horticulture (volume 3). Volume 1 illustrates in considerable depth the science and technology which underpin the continuous production of Horticultural Produce. Firstly there is a consideration of the aspects of industrial development based on basic scientific discoveries. This is followed by chapters written by acknowledged world experts covering the production of: Field Vegetables, Temperate Fruit, Tropical Fruit, Citrus, Plantation Crops, Berry Crops, Viticulture, Protected Crops, Flower Crops, Developing New Crops, Post-harvest Handling, Supply Chain Management and the Environmental Impact of Production. Production Horticulture may now be found supporting the economies of less developed nations and consequently the final Chapter focuses especially on the impact of a changing environment on Crop Production Horticulture in Africa.

Volume 3, Social Horticulture brings the evolution of the Discipline firmly into the 21st Century. It breaks new ground by providing a detailed analysis of the value of Horticulture as a force for enhancing society in the form of social welfare, health and well-being, how this knowledge is transferred within and between generations, and the place of Horticulture in the Arts and Humanities. Volume 3 covers considerations of: Horticulture and Society, Diet and Health, Psychological Health, Wildlife, Horticulture and Public Welfare, Education, Extension, Economics, Exports and Biosecurity, Scholarship and Art, Scholarship and Literature, Scholarship and History, and the relationship between Horticulture and Gardening.

The value of Horticulture for human development was emphasised by Jorge Sampaio (United Nations High Representative for the Alliance of Civilisations and previously the President of the Republic of Portugal) in his opening address to the 28th International Horticultural Congress in Lisbon, 2010. He stated that Horticulture can achieve "a lot to overcome hunger and ensure food security". This is in the face of estimates that the world's population, particularly in developing counties, will reach 9.1 billion by 2050. Horticulture has especially important role in easing this burden. Intensive plant production has much to offer as urbanization continues at an accelerating pace. Shortly about 70 % of the world's population will choose to live in urban and peri-urban areas of many countries. Despite affluence in parts of the World many millions of the world's population still continue to be undernourished and in poor health. Climatic change, over population, soil degradation, water and energy shortages, pollution, and crippling destruction of biodiversity are the challenges facing all of humanity. Horticulture in its Production, Environmental and Social facets offers important knowledge and expertise that will help minimise these changes. This is well explained in "Harvesting the Sun" in a digest recently published by the International Society for Horticultural Science. In summary form the international interactions between horticultural science, technology, business and management are displayed. This offers suggestions as to how, over the early part of the 21st Century, world food production must rise by at least some 110 % to meet the demands of expanding populations in countries such as China, India, parts of Asia and in South America.

Considerable breadth and depth of intellect are demanded of those who seek an understanding of horticulture. This is not a discipline for the faint hearted since the true disciple needs a considerable base in the physical, chemical and the biological sciences together with a knowledge of natural resources, linked with an understanding of the application of economics, engineering and the social sciences. Added to this should also come an appreciation of the artistic, historic and cultural dimensions of the Discipline. The teaching of fully comprehensive horticultural science courses in higher educational institutions has regrettably diminished worldwide. It is to be hoped that this Trilogy may go some small way in providing an insight into the scale, scope, and excitement of the Discipline as well as the intellectual rigour demanded of those who seek a properly proportioned understanding of it.

Enormous thanks go to all those who have contributed to these three volumes. Their devotion, hard work and understanding of the Editors' requests are greatly appreciated. Thanks are also due to our colleagues in Springer for all their continuing help, guidance and understanding. In particular we would like to thank Dr. Maryse Walsh, Commissioning Editor and Ir. Melanie Van Overbeek, Senior Publishing Assistant.

Professor Geoffrey R. Dixon affectionately records his thanks to his mentor Professor Herbert Miles, then Head of the Horticulture Department of Wye College, University of London (now Imperial College, London) who challenged him to "define Horticulture". Regrettably, it has taken half a century of enquiry to respond effectively.

Sherborne, Dorset, United Kingdom — Geoffrey R. Dixon
Queensland, Australia — David E. Aldous
August 2013

Contents

Volume 1 Production Horticulture

1 **An Introductory Perspective to Horticulture: Plants for People and Places** ... 1
Geoffrey R. Dixon and David E. Aldous

2 **Science Drives Horticulture's Progress and Profit** ... 27
Geoffrey R. Dixon, Ian J. Warrington, R. Drew and G. Buck-Sorlin

3 **Vegetable Crops: Linking Production, Breeding and Marketing** ... 75
Daniel I. Leskovar, Kevin M. Crosby, Marco A. Palma and Menahem Edelstein

4 **Temperate Fruit Species** ... 97
Guglielmo Costa and Angelo Ramina

5 **Tropical and Subtropical Fruits** ... 123
Victor Galán Saúco, Maria Herrero and Jose I. Hormaza

6 **Citrus Production** ... 159
Manuel Agustí, Carlos Mesejo, Carmina Reig and Amparo Martínez-Fuentes

7 **Viticulture and Wine Science** ... 197
Yann Guisard, John Blackman, Andrew Clark, Bruno Holzapfel, Andrew Rawson, Suzy Rogiers, Leigh Schmidtke, Jason Smith and Christopher Steel

8 **Plantation Crops** ... 263
Yan Diczbalis, Jeff Daniells, Smilja Lambert and Chris Searle

9 **Berry Crops** ... 301
R. M. Brennan, P. D. S. Caligari, J. R. Clark, P. N. Brás de Oliveira, C. E. Finn, J. F. Hancock, D. Jarret, G. A. Lobos, S. Raffle and D. Simpson

10 **Protected Crops** ... 327
Nazim Gruda and Josef Tanny

11 **The Role of Ornamentals in Human Life** ... 407
Jaap M. van Tuyl, Paul Arens, William B. Miller and Neil O. Anderson

12 **New Ornamental Plants for Horticulture** ... 435
Kevin Seaton, Andreas Bettin and Heiner Grüneberg

13 **Postharvest Care and the Treatment of Fruits and Vegetables** ... 465
Peter M. A. Toivonen, Elizabeth J. Mitcham and Leon A. Terry

14 **Designing New Supply Chain Networks: Tomato and Mango Case Studies** ... 485
Jack G. A. J. van der Vorst, Rob E. Schouten, Pieternel A. Luning and Olaf van Kooten

15 **Environmental Impact of Production Horticulture** ... 503
Henry Wainwright, Charlotte Jordan and Harry Day

Index ... 523

Volume 2 Environmental Horticulture

16 **Horticulture and The Environment** ... 603
Robert Lillywhite

17 **Woody Ornamentals** ... 619
Paul E. Read and Christina M. Bavougian

18 **Medicinal and Aromatic Plants—Uses and Functions** ... 645
Maiko Inoue and Lyle E. Craker

19 **Urban Greening—Macro-Scale Landscaping** ... 671
Gert Groening and Stefanie Hennecke

20 **Urban Trees** ... 693
Mark Johnston and Andrew Hirons

21 **Trees in the Rural Landscape** ... 713
Glynn Percival, Emma Schaffert and Luke Hailey

22 Management of Sports Turf and Amenity Grasslands 731
David E. Aldous, Alan Hunter, Peter M. Martin, Panayiotis A. Nektarios and Keith W. McAuliffe

23 Interior Landscapes 763
Ross W. F. Cameron

24 Biodiversity and Green Open Space 787
Ghillean T. Prance, Geoffrey R. Dixon and David E. Aldous

25 An Assessment of the Effects of Climate Change on Horticulture 817
Geoffrey R. Dixon, Rosemary H. Collier and Indrabrata Bhattacharya

26 Concepts and Philosophy Underpinning Organic Horticulture 859
David Pearson and Pia Rowe

Index 873

Volume 3 Social Horticulture

27 Horticulture and Society 953
Tony Kendle and Jane Stoneham

28 Fruit and Vegetables and Health: An Overview 965
Yves Desjardins

29 Health and Well-Being 1001
Ross W. F. Cameron

30 Human Dimensions of Wildlife Gardening: Its Development, Controversies and Psychological Benefits 1025
Susanna Curtin and Dorothy Fox

31 Horticultural Science's Role in Meeting the Need of Urban Populations 1047
Virginia I. Lohr and P. Diane Relf

32 Education and Training Futures in Horticulture and Horticultural Science 1087
David E. Aldous, Geoffrey R. Dixon, Rebecca L. Darnell and James E. Pratley

33 Extension Approaches for Horticultural Innovation 1117
Peter F. McSweeney, Chris C. Williams, Ruth A. Nettle, John P. Rayner and Robin G. Brumfield

34 Increasing the Economic Role for Smallholder Farmers in the World Market for Horticultural Food ... 1139
Roy Murray-Prior, Peter Batt, Luis Hualda, Sylvia Concepcion and Maria Fay Rola-Rubzen

35 International Plant Trade and Biosecurity ... 1171
Aaron Maxwell, Anna Maria Vettraino, René Eschen and Vera Andjic

36 Horticulture and Art ... 1197
Jules Janick

37 Scholarship and Literature in Horticulture ... 1225
Ian Warrington and Jules Janick

38 A Short History of Scholarship in Horticulture and Pomology ... 1255
Silviero Sansavini

39 Gardening and Horticulture ... 1307
David Rae

Index ... 1339

Contributors

Manuel Agustí Instituto Agroforestal Mediterráneo, Universitat Politècnica de València, València, Spain

David E. Aldous School of Land, Crop and Food Science, The University of Queensland, Gatton Campus, Lawes, Queensland, Australia

Dipartimento Colture Arboree, University of Bologna, Bologna, Italy

Neil O. Anderson Flower Breeding & Genetics Department of Horticultural Science, University of Minnesota, Saint Paul, MN, USA

Vera Andjic Department of Agriculture, Perth, Australia

Paul Arens Wageningen University and Research Centre, Plant Breeding, Wageningen, The Netherlands

Peter Batt School of Management, Curtin University, Perth, WA, Australia

Christina M. Bavougian Department of Agronomy and Horticulture, University of Nebraska, Lincoln, NE, USA

Andreas Bettin Faculty of Agricultural Sciences and Landscape Architecture, University of Applied Sciences Osnabruck, Osnabrück, Germany

Indrabrata Bhattacharya Department of Plant Pathology, Bidhan Chandra Krishi Viswavidyalaya, Nadia, West Bengal, India

John Blackman National Wine and Grape Industry Centre, Wagga Wagga, NSW, Australia

P. N. Brás de Oliveira Instituto Nacional de Investigacao Agraria e Veterinaria, Oeiras, Portugal

R. M. Brennan Fruit Breeding and Genetics Group, James Hutton Institute, Invergowrie, Dundee, Scotland, UK

Robin G. Brumfield Department of Agricultural, Food and Resource Economics, Rutgers, The State University of New Jersey, New Brunswick, NJ, USA

G. Buck-Sorlin AGROCAMPUS WEST Angers Centre, National Institute of Horticulture and Landscape, UMR1345 Research Institute of Horticulture and Seeds (IRHS), Angers, France

P. D. S. Caligari Instituto de Biología Vegetal y Biotecnología, Universidad de Talca, Talca, Chile

Ross W. F. Cameron Department of Landscape, The University of Sheffield, Sheffield, South Yorkshire, UK

Andrew Clark National Wine and Grape Industry Centre, Wagga Wagga, NSW, Australia

J. R. Clark University of Arkansas, Fayetteville, AR, USA

Rosemary H. Collier Warwick Crop Centre, The University of Warwick, Warwick, UK

Sylvia Concepcion School of Management, University of the Philippines Mindanao, Mintal, Davao, The Philippines

Guglielmo Costa Department of Agricultural Sciences—DipSA, Alma Mater Studiorum, University of Bologna, Bologna, Italy

Lyle E. Craker Medical Plant Program, University of Massachusetts, Amherst, MA, USA

Kevin M. Crosby Department of Horticultural Sciences, Vegetable and Fruit Improvement Center, Texas A&M University, College Station, US

Susanna Curtin School of Tourism, Bournemouth University, Poole, Dorset, UK

Jeff Daniells Department of Agriculture, Fisheries and Forestry, Centre for Wet Tropics Agriculture, South Johnstone, Queensland, Australia

Rebecca L. Darnell Horticultural Sciences Department, Gainsville, FL, USA

Harry Day Juneau, USA

Yves Desjardins Institute of Nutrition and Functional Foods/Horticulture Research Center, Laval University, Québec City, QC, Canada

Yan Diczbalis Department of Agriculture, Fisheries and Forestry, Centre for Wet Tropics Agriculture, South Johnstone, Queensland, Australia

Geoffrey R. Dixon School of Agriculture, Earley Gate, University of Reading, Reading, UK

GreenGene International, Hill Rising, Sherborne, Dorset, UK

R. Drew School of Biomolecular and Physical Sciences, Griffith University, Nathan, Queensland, Australia

Menahem Edelstein Department of Vegetable Crops, Agricultural Research Organization, Newe Ya'ar Research Center, Ramat Yishay, Israel

René Eschen CABI, Delémont, Switzerland

C. E. Finn USDA-ARS, HCRL, Corvallis, OR, USA

Dorothy Fox School of Tourism, Bournemouth University, Poole, Dorset, UK

Victor Galán Saúco Departamento de Fruticultura Tropical, Instituto Canario de Investigaciones Agrarias, La Laguna, Tenerife, Spain

Gert Groening Forschungsstelle Gartenkultur und Freiraumentwicklung, Institut für Geschichte und Theorie der Gestaltung, Universität der Künste, Berlin, Germany

Nazim Gruda Division of Horticultural Sciences, University of Bonn, Bonn, Germany

Department for Innovation Promotion, Federal Office for Agriculture and Food, Bonn, Germany

Heiner Grüneberg Department of Horticultural Plant Systems, Humboldt-Universität zu Berlin, Berlin, Germany

Yann Guisard National Wine and Grape Industry Centre, Orange, NSW, Australia

Luke Hailey Bartlett Tree Research Laboratory, Bartlett Tree Research Laboratory, Reading, UK

J. F. Hancock Department of Horticulture, Michigan State University, East Lansing, MI, USA

Stefanie Hennecke Fachgebiet Freiraumplanung, Universität Kassel, Kassel, Germany

Maria Herrero Department of Pomology, Estación Experimental de Aula Dei—CSIC, Zaragoza, Spain

Andrew Hirons Lecturer in Arboriculture, Myerscough College, Lancashire, UK

Bruno Holzapfel National Wine and Grape Industry Centre, Wagga Wagga, NSW, Australia

Jose I. Hormaza Instituto de Hortofruticultura Subtropical y Mediterránea La Mayora (IHSM-CSIC-UMA), Algarrobo-Costa, Malaga, Spain

Luis Hualda School of Management, Curtin University, Perth, WA, Australia

Alan Hunter College of Life Sciences, School of Agriculture and Food Science, Agriculture & Food Science Centre, University College Dublin, Belfield, Dublin 4, Ireland

Maiko Inoue Medical Plant Program, University of Massachusetts, Amherst, MA, USA

Jules Janick Department of Horticulture and Landscape Architecture, Purdue University, West Lafayette, IN, USA

D. Jarret Fruit Breeding and Genetics Group, James Hutton Institute, Invergowrie, Dundee, Scotland, UK

Mark Johnston Arboriculture and Urban Forestry, Myerscough College, Lancashire, UK

Charlotte Jordan Redhood City, USA

Tony Kendle Eden Project, Cornwall, UK

Olaf van Kooten Horticultural Production Chains Group, Wageningen University, Wageningen, The Netherlands

Smilja Lambert Cocoa Sustainability Research Manager Asia/Pacific, Mars Chocolate—Cocoa Sustainability, Cairns, Queensland, Australia

Daniel I. Leskovar Texas A&M AgriLife Research Center, Texas A&M University, Uvalde, TX, US

Robert Lillywhite Warwick Crop Centre, University of Warwick, Warwickshire, UK

G. A. Lobos Plant Breeding and Phenomic Center, Faculty of Agricultural Sciences, Universidad de Talca, Talca, Chile

Virginia I. Lohr Department of Horticulture, Washington State University, Pullman, WA, United States of America

Pieternel A. Luning Food Quality and Design Group, Wageningen University, Wageningen, The Netherlands

Peter M. Martin Amenity Horticulture Research Unit, University of Sydney Plant Breeding Institute, Cobbitty, NSW, Australia

Amparo Martínez-Fuentes Instituto Agroforestal Mediterráneo, Universitat Politècnica de València, València, Spain

Aaron Maxwell School of Veterinary and Life Sciences, Murdoch University, Perth, Australia

Keith W. McAuliffe Sports Turf Research Institute, Ormiston, QLD, Australia

Peter F. McSweeney Department of Agriculture and Food Systems, Melbourne School of Land and Environment, The University of Melbourne, Melbourne, VIC, Australia

Carlos Mesejo Instituto Agroforestal Mediterráneo, Universitat Politècnica de València, València, Spain

William B. Miller Department of Horticulture, Cornell University, Ithaca, NY, USA

Elizabeth J. Mitcham Horticulture Collaborative Research Support Program, University of California, Davis, CA, USA

Roy Murray-Prior School of Management, Curtin University, Perth, WA, Australia

Panayiotis A. Nektarios Department of Crop Science, Lab. of Floriculture and Landscape Architecture, Agricultural University of Athens, Athens, Greece

Ruth A. Nettle Department of Agriculture and Food Systems, Melbourne School of Land and Environment, The University of Melbourne, Melbourne, VIC, Australia

Marco A. Palma Department of Agricultural Economics, Texas A&M University, College Station, US

David Pearson Faculty of Arts and Design, University of Canberra, Australian Capital Territory, Australia

Glynn Percival Bartlett Tree Research Laboratory, Bartlett Tree Research Laboratory, Reading, UK

Ghillean T. Prance The Old Vicarage, Lyme Regis, UK

James E. Pratley Charles Sturt University, School of Agricultural and Wine Sciences, Wagga Wagga, NSW, Australia

David Rae Royal Botanic Garden Edinburgh, Edinburgh, Scotland

S. Raffle Horticultural Development Company, Agriculture and Horticulture Development Board, Kenilworth, Warwickshire, UK

Angelo Ramina Department of Agronomy, Food, Natural resources, Animals and Environment—DAFNAE, University of Padova, Padova, Legnaro, Italy

Andrew Rawson School of Agricultural and wine Sciences, Orange, NSW, Australia

John P. Rayner Department of Resource Management and Geography, Melbourne School of Land and Environment, University of Melbourne, Richmond, VIC, Australia

Paul E. Read Department of Agronomy and Horticulture, University of Nebraska, Lincoln, NE, USA

Carmina Reig Instituto Agroforestal Mediterráneo, Universitat Politècnica de València, València, Spain

P. Diane Relf Department of Horticulture, Virginia Tech, VPI & SU, Blacksburg, VA, United States of America

Suzy Rogiers National Wine and Grape Industry Centre, Wagga Wagga, NSW, Australia

Maria Fay Rola-Rubzen CBS—Research & Development, Curtin University, Perth, WA, Australia

Pia Rowe Faculty of Arts and Design, University of Canberra, Australian Capital Territory, Australia

Silviero Sansavini Dipartimento di Scienze Agrarie, University of Bologna, Bologna, Italy

Emma Schaffert Bartlett Tree Research Laboratory, Bartlett Tree Research Laboratory, Reading, UK

Leigh Schmidtke National Wine and Grape Industry Centre, Wagga Wagga, NSW, Australia

Rob E. Schouten Horticultural Production Chains Group, Wageningen University, Wageningen, The Netherlands

Chris Searle Suncoast Gold Macadamias, Bundaberg, Queensland, Australia

Kevin Seaton Department of Agriculture and Food Western Australia, South Perth, WA, Australia

D. Simpson East Malling Research, East Malling, Kent, UK

Jason Smith National Wine and Grape Industry Centre, Wagga Wagga, NSW, Australia

Christopher Steel National Wine and Grape Industry Centre, Wagga Wagga, NSW, Australia

Jane Stoneham Sensory Trust, Cornwall, UK

Josef Tanny Institute of Soil, Water & Environmental Sciences, Agricultural Research Organization, Bet Dagan, Israel

Leon A. Terry Plant Science Laboratory, Vincent Building, Cranfield University, Bedfordshire, UK

Peter M. A. Toivonen Agriculture and Agri-Food Canada, Pacific Agri-Food Research Centre, Summerland, Canada

Jaap M. van Tuyl Wageningen University and Research Center, Plant Breeding, Wageningen, The Netherlands

Anna Maria Vettraino DIBAF, University of Tuscia-Viterbo, Viterbo, Italy

Jack G. A. J. van der Vorst Operations Research and Logistics Group, Wageningen University, Wageningen, The Netherlands

Henry Wainwright The Real IPM Company (K) Ltd, Thika, Kenya

Ian Warrington Massey University, Palmerston North, New Zealand

Chris C. Williams Department of Resource Management and Geography, Melbourne School of Land and Environment, University of Melbourne, Richmond, VIC, Australia

Chapter 16
Horticulture and The Environment

Robert Lillywhite

Abstract Horticulture interacts with the environment in many different ways. This chapter discusses three aspects: firstly the exploitation and degradation of natural resources (land, water); secondly, the use of, and generation of pollution from, synthetic resources (energy, fertilizer and pesticides); and thirdly, the visual and cultural impact of horticultural infrastructure. Horticultural production, in comparison to other agricultural sectors, occupies a small land area but due to its intensity of production, its environmental impact can be relatively greater. Access to, and management, of water resources is the aspect of greatest concern since loss of water resources, deterioration of water quality and pollution of surface water with pesticide residues, nitrates and phosphates are detrimental to both horticultural production and the wider environment. Land use and acidification are important locally but unlike the emission of greenhouse gases, are less important at the global scale. Intensive horticultural production can be responsible for both environmental and visual pollution but they are generally small-scale in global terms and any adverse impacts should be viewed in the context of the benefits that the industry provides in terms of economics, social employment and improved diets and health.

Keywords Horticulture · Environment · Water · Greenhouse gas · Land use · Eutrophication · Acidification · Pesticides

Horticulture encompasses a huge range of sectors and crops and some form of production occurs in almost every country of the world, from the cold of Northern Scandinavia to the tropics. This geographical variation has allowed a large range of crops to evolve and humans have been quick to exploit the food potential. Almost every crop within every location has evolved its own production system and all of them interact with the environment at some point. Mostly, those interactions are negative since humans are still driven by the need to produce more food but there are examples of positive interactions, e.g. carbon sequestration. This chapter is predominantly devoted to a description of the negative interactions although some positive interactions will also be briefly discussed.

R. Lillywhite (✉)
Warwick Crop Centre, University of Warwick, Wellesbourne,
Warwickshire CV35 9EF, UK
e-mail: robert.lillywhite@warwick.ac.uk

G. R. Dixon, D. E. Aldous (eds.), *Horticulture: Plants for People and Places, Volume 2*,
DOI 10.1007/978-94-017-8581-5_16, © Springer Science+Business Media Dordrecht 2014

The huge range of crops makes it difficult to define exactly what horticulture is, and this chapter makes no apology for not attempting to do so. Rather it concentrates on environmental impacts and illustrates these using case studies from the academic literature. It is impossible within a single chapter to discuss every horticultural crop, and may even be pointless to do so, as most will share the same issues.

We live in a time of increasing population, with a correspondingly greater demand for food, but with the slow realization that the production of food crops is having real and very serious detrimental environmental impacts. This chapter concentrates on describing the main environmental burdens, how they are quantified and what their impact is. However, it should be noted that the same environmental impacts arise from all agricultural production, be it for food, fibre or fuel, so in defence of horticulture it is important to remember that horticulture is practiced at a much smaller scale than the arable and livestock industries and that its environmental impacts should be viewed in this context. If you need further reassurance on this point, remember that the consumption of horticultural products is an essential part of a healthy and balanced diet.

The interaction between horticultural production and the environment can be condensed into two categories: exploitation and degradation of natural resources (land, water) and the use of, and generation of pollution from, synthetic resources (energy, fertilizer and pesticides). To this a third category may be added, which is the visual and cultural impact of the infrastructure that supports horticultural production.

This chapter will focus on the following environmental burdens and impacts:

- Land use and soil
- Water
- Pesticide residues
- Eutrophication
- Greenhouse gas emissions
- Acidification
- Visual impact

Land

Land is the first of two critical resources (the other being water) than are essential for successful horticultural production. The importance of land should be considered from a number of different perspectives. Firstly, the area of land required in order to support horticultural production; secondly, the type, or quality, of that land and finally the environmental impacts of using land. The latter category encompasses soil quality and erosion, and how that land might otherwise be used.

Land is a hugely valuable resource but in reality little is used to support horticultural production. Accurate and robust data on the land area used in horticulture is difficult to obtain so the UK, where good land use data is available, is used as a proxy to illustrate the position. In the UK, horticulture occupies 3 % of UK agricultural land which increases to 5% if potatoes are included (Defra 2012), and it's unlikely that this percentage varies much by country. However, while the area of land used might be small, the high intensity of production of many horticultural crops means that its environmental impact is likely to be disproportionately high to the area of land under cultivation.

Horticultural crops tend to be high value in comparison to other field based crops and their production has mostly migrated to areas containing the best soils. What is best, of course, depends on the crop being grown and the quality requirements of the customer. In the UK, brassicas are mostly grown on water-retentive silts but light sands soils are preferred for onions. This preference for particular soil types does lead to clustering of crops in suitable areas, e.g. in the UK, brassicas in coastal areas of Lancashire and Lincolnshire. This is obviously beneficial to the structure of the industry but can result in the long-term build-up of pest and disease pressures. It can also cause in-direct environmental concerns as a result of increased labour and infrastructure to support the harvesting, processing and logistics of crops. The costs to landscapes and local facilities can be illustrated by looking at protected soft fruit protection in Europe where the proliferation of polytunnels and imported labour has caused both environmental and social problems in some countries (Lillywhite et al. 2007).

Clustering also occurs in response to other variables, principally water and light. Abundant water attracts crops that require irrigation and high light levels are always preferred to maximise crop growth rates. This can often result in very densely populated pockets of horticultural activity and can result in exploitation of land for building and cultivation and lead to a loss of soil quality. Loss of soil quality has become an issue around the world. Erosion, loss of soil organic matter and carbon, decreases in soil fertility and loss of soil biodiversity are all common problems with heavily worked soils. Inappropriate water management only adds to the pressure on soils; repeated irrigation can cause salinization and both too much and too little water can affect soil structure. The best land is scarce, and therefore expensive, and farmers and growers need to produce as intensively as possible to generate the maximum economic return to justify their investments. This type of exploitative production is exasperated by global competition for fresh produce and can result in permanent damage to the best soils.

Horticulture differs greatly in its impact on soil. Intensive double, or even triple, cropping of short duration cash crops, i.e. salad leaves, can place a considerable burden on soil, both in terms of loss of nutrients and organic matter, and in further reducing soil quality through increased compaction. In complete contrast, plantation and orchard crops, once established, may have a positive environmental impact on soil depending on production strategies. Long-term crops are known to sequester carbon in both above- and below-ground biomass.

Sloping land is particularly prone to soil loss via water runoff. García-Ruiz (2010) reported that an extreme rainfall event in Spain was responsible for the loss of 282 t of soil per hectare from a vineyard and while this is an extreme example it does illustrate that using sloping land for vineyards and orchard crops (apples, pears, olives, almonds etc.) can promote severe losses and that permanent tree crops do not always provide as much soil stability as expected.

Horticultural production is generally more intensive, and makes a greater demand on its soil, than comparable arable or livestock systems. This has both pros and cons. It is true than horticulture tends to be more environmentally demanding of its soils and therefore has the potential to degrade soils quicker than other systems but horticultural soils tend to be of higher value than other soils, and consequently tend to be managed more sympathetically. However, overall we conclude that the quality of horticultural soils is still declining and despite many initiatives, the trend looks unlikely to be reversed in the near future. In terms of the environmental challenges facing the horticultural industry, we rank the loss of soil and soil quality at number one. In the long-term, we must protect, actually enhance, the quality of our soils if we wish to continue to eat.

Water

Water is the single most precious resource on the planet; without it life would cease to exist. Globally, agricultural production consumes approximately 90% of all freshwater resources (Allan 2011) with approximately 60% of production being rain-fed and the remaining 40% irrigated. A majority of that irrigated production is horticultural crops. So, while horticultural production may occupy only a small land area and be responsible for limited greenhouse gas emissions, it is in the area of water use and the consequences that arise from water use that horticulture probably faces its biggest issues. We consider that increased water use, especially in semi-arid areas, and pollution of water resources from pollutants and pesticide residues will be the biggest challenge facing horticulture in the future.

Until quite recently, management of water for horticulture focused on ensuring that supply met demand, however, the result of that approach is an ever increasing demand for water; a demand that cannot always be met. Allan (2011) suggested that irrigators, in the end, always run out of water since once the initial step to irrigate for increased yield and improved quality has been taken, there is no turning back. This unfortunate scenario has already been demonstrated in the area around the Aral Sea where irrigation for cotton production has resulted in dramatic lowering of water levels (down to 17% compared to previous years). It is now being repeated across the non-renewable aquifers of North Africa and even the renewable aquifers that support asparagus production in Peru are coming under severe pressure. Water use for commercial large scale horticultural production is currently following the slash and burn techniques used for clearing virgin forests. Except in this case, it's a matter of exhausting scarce water supplies and then moving on to the next area. Southern Europe is a good example. Twenty to thirty years ago, Spain started to supply Northern Europe with early and late season fruits and vegetables but the intensity of production reduced water resources so production migrated to North Africa in search of a more secure supply of water (and cheaper labour). Production is now moving southwards to West Africa having started to exhaust existing water supplies in North Africa.

Access to water is likely to remain the biggest issue for mainstream horticulture for many years. Identifying good horticultural land with access to water remains the primary goal for many horticultural businesses since every other aspect of business is mobile. Modern and sustainable horticulture requires that water demand is managed and various approaches are possible. Scientific advances in plant breeding for water-use efficiency, advanced irrigation scheduling and partial root zone drying are driving reductions in water use but more work is still required to understand the actual volumes of water used in crop production and how to reduce them.

Water scientists partition water into three colour-coded components to help them understand where water comes from and how water is used: Green water is soil water derived from rainfall and supports dryland or non-irrigated production. Green water is normally assumed to be the volume of water that is lost through evapotranspiration during crop growth so location (as a proxy for temperature, relative humidity and wind effect) has a major impact on the volumetric water footprint. Green water is often seen as 'free' and its use unproblematic but this is not necessarily always the case. In the majority of situations, interception of rainfall by the plant canopy of cash crops is less than that of natural vegetation; this leads to the scenario where replacing natural vegetation by cash crops can increase ground water and river flows (Scanlon et al. 2007). However, there may be situations where natural vegetation, for example, scrub, is replaced by forestry, either for biomass or oil production, which captures a greater volume of rainfall with subsequent detrimental impact on the natural water environment. Globally, 60 % of the world's food is produced under rain-fed conditions on 83 % of the cultivated land so it is important to realise that green water productivity is the key to global food security (FAO 2002).

Blue water is the water abstracted from surface (river, lake and reservoir) and groundwater resources for agricultural, industrial and domestic uses. This is the water used for crop irrigation. Blue water has to be paid for. Whether it is abstracted from ground or surface water or taken from the public supply, its supply incurs a cost. That cost is always economic but sometimes environmental and social as well. Blue water is used globally to irrigate agricultural crops although the sustainability of that use varies greatly from region to region. Blue water can be used sustainably, and without long-term detrimental impact, where ground water abstracted during the growing season is fully replenished through rainfall during the following wet season, however, this simple and obvious water balance approach is too often ignored by the weight of economic argument. Pressure on blue water resources can be alleviated by the efficient use of water in agriculture, for example, scheduling and accurate irrigation, however, it should be remembered that efficient use is not necessarily sustainable use. In most countries, the biggest demand for blue water is for the public supply and commercial use and they always have priority over agricultural use.

Grey water is conceptually interesting but difficult to deal with in practice. Grey water can be either polluted water (the common definition) or the volume of freshwater required to dilute polluted water so that it meets the requirement of the local discharge license (the definition that the Water Footprint Network use). Grey water is more theoretical than either green or blue since it is not necessarily a true volume (depending on definition) and can therefore be very confusing where water management is being discussed. It will not be discussed further.

Various organisations work on understanding water use in horticultural products with the best known being the Water Footprint Network. They define the volumetric water footprint as the virtual water content of a product; that is, the total volume of freshwater consumed within all stages of the supply chain (Hoekstra et al. 2011). The adjective 'virtual' refers to the fact that most of the water consumed is not contained in the final product but during the production process. The term 'water footprint' was chosen to be analogous to the ecological footprint and represents the appropriation of resources; its use can therefore be more "user focused" as opposed to virtual water, which is an inherent property of the commodity.

The real-water content of products is generally negligible if compared to the virtual-water content. For example, 1 t of lettuce may contain 800 to 900 L of water but the supply chain (predominantly the production stage) consumed 161,000 L. On average the total comprises 133,000 L of green water and 28,000 L of blue water.

Water in Context

The use of water should always be considered in the context of location. The production of 'thirsty' crops in regions where rainfall is plentiful and there are no competing demands for the water is not a problem; indeed, in some cases the ability of a crop to use up water that would otherwise cause flooding may be actively beneficial. Similarly a country may import a great deal of virtual water but it may be sourcing its supplies from regions where water is in abundance—or vice versa. Table 16.1 shows the average value (and the complete range) for a number of horticultural crops. Under optimum production conditions, water use for many crops is relatively low but values can increase dramatically when crops are grown outside of their natural range and irrigation is required to supplement scarce rainfall. This issue is made worse when crops are grown in hot temperatures with greatly increased evapotranspiration.

The volumetric water footprint is a useful metric and a helpful auditing tool but on its own it cannot be used to assess the environmental impacts of water consumption. Additional information on water availability and (micro) climate are required to fully understand the impacts of production. As a consequence when considering the water footprint of a product it is necessary to assess whether the country or region of production is water scarce, water self-sufficient or water dependent.

Horticulture doesn't use a huge quantity of water but it does use more than its fair share. This can often be justified due to its output of high quality specialized crops and water use itself may not be a problem in a location with sufficient rainfall. However there are examples, Southern Spain, North Africa, California and the Ica valley in Peru, to name but a few, where horticultural use of water has had long-term detrimental impacts on the environment. The horticultural industry has both an image and actual problem with sustainable water use and water is likely to remain the number two issue for the foreseeable future.

Table 16.1 The volumetric green and blue water footprint of selected horticultural crops

Crop	Average water footprint (L/kg^{-1}) product	Range
Apple	69	34–6,979
Asparagus	1,613	148–6,348
Banana	757	22–11,904
Beans (green)	405	38–2,087
Carrot	134	11–4,297
Cauliflower	210	6–5,854
Coconut	2,671	61–25,814
Cucumber	248	n/a–5,122
Grapes	522	5–30,306
Lettuce	161	4–2,412
Pineapple	224	61–11,584
Potato	224	7–2,749
Orange	511	18–14,037
Tomato	171	n/a–5,011
Water melon	175	n/a–4,324

Compiled the WaterStat database, Water Footprint Network, NB Enschede, The Netherlands

Pesticide Residues

In many, if not the majority of markets, horticultural produce is sold as much on appearance as it is on taste. While there are different levels of acceptance of quality issues across the world, better looking produce always sells first. It can often command a price premium as well so growers strive to produce the best looking product that they can. In some countries, notably developed countries in the west, this situation is enforced by the retailer's quality standards and assurance schemes. The consequence is that, on average, fresh produce receive more applications of pesticides than other categories of crop. Other temptations also occur; over application is still common, poor application leading to direct pesticide losses to water courses occurs, illegal pesticides are still used and counterfeit pesticides of unknown quality are available. To some extent, all of these practices are part of modern horticulture but both short- and long term effects are mostly unexplored.

Of all the environmental impacts generated by horticulture, that of pesticides is the best studied and regulated. All active ingredients go through stringent tests before commercial use and their direct detrimental impact on various living organisms is mostly well understood. Various methods are available to quantify environmental impact, from a basic amount per area to more sophisticated approaches including computer modelling (Kovach et al. 1992; Reus and Leenderste 2000; Lewis et al. 2003). Forward thinking producers are now able to specify pesticides that reduce non-target damage and will degrade more quickly after contact with soil. In combination with the use of spatially-variable application within an integrated programme of control now allows the best producers to limit releases of residues to the environment.

Pesticide residues affect both flora and fauna and can accumulate in soil. The intended effects of herbicides on selected flora are very well understood but the non-target consequences of application can be severe if limited in expanse. The effects of pesticides on fauna are long-term and potentially disastrous. Accumulation of active ingredients in birds and fish has been regularly reported for many years and, at the time of writing, concerns are being expressed that insecticides are having far reaching and detrimental effects on non-target pollinators. For the horticultural industry in particular, and fruit especially, a lack of pollinating insects could have very serious consequences for the industry. While accumulation of pesticides in above-ground biomass is very worrying, it has the advantage of being fairly easily observed and examined; the same cannot be said for soil and water. Below-ground accumulation in soil could potentially be more damaging in the long-term as the effects on the soil microbial community is harder to investigate and far more difficult to remediate.

Fruit production in particular is known for its high level of pesticide use. Pesticide residues from banana plantations were found to exceed a calculated chronic risk ratio and pose a threat to aquatic organisms and the wetland areas in the Suerte River basin (Castillo LE et al. 2000). Residues of the herbicide bromacil are now prevalent in groundwater and may present a cancer risk in the long term. Echeverria-Saenz et al. (2012) showed a clear relationship between high levels of herbicides, ecological quality of macro-invertebrates and habitat deterioration. As a consequence, much effort is being currently being expended in developing new approaches to pest and disease control. Both chemical (new active ingredients and bio-pesticides) and non-chemical (natural predators and barrier solutions for pest control) solutions are emerging onto the market and should hopefully enable growers to protect the yield and quality of their crops without adding to the pesticide residues already in the environment.

A note of caution here since in many ways crop protection is becoming polarised. The availability of environment friendly new products is being matched by a rise in illegal and counterfeit products which pose a risk to both the environment and human health (EU Parliament 2013). The effect of pesticide residues is mostly hidden in our soils and water so it is wise to pay attention to any terrestrial impacts since they can forewarn of potential problems. However, the rise in consumption of organically grown produce is a clear indicator that consumers want food that is free of pesticide residues and that is the route that a modern and sophisticated horticultural industry should continue to take.

Eutrophication Potential

Eutrophication is an environmental response to excessive plant nutrients. The normal consequence is increased biomass growth, especially algal, in water bodies, leading to hypoxia. Eutrophication can be both anthropogenic and natural but this section deals exclusively with man-made causes, predominantly the loss of nitrogen and phosphorus fertilizers through leachate and run-off.

In concept, the estimation of potential eutrophication is relatively simple. The contribution of four fertiliser-derived compounds (phosphate, nitrate, ammonia and oxides of nitrogen) is assessed, based on their potential contribution, and the results are summed to provide a value for potential eutrophication. This is expressed in terms of phosphate equivalents. Consistency within the estimation process is maintained by using known leachate coefficients, weighting and normalisation values. The fraction of nitrogen leached as nitrate is commonly taken to be 15 % (Silgram et al. 2001) and the loss of phosphorus as phosphate as 6.5 % (Johnes 1996). Weighting of the different contributor compounds is normally based on the work of Azapagic et al., 2004 where phosphate, nitrate, ammonia and oxides of nitrogen are weighted at 1.00, 0.42, 0.33 and 0.13 respectively.

In practice, potential eutrophication is influenced by many different factors, including nutrient (fertilizer) type, applied water (both rainfall and irrigation events), proximity of production sites to water bodies, soil type (and therefore percolation rates) and topography. These factors combine to make estimation of potential eutrophication difficult and like the carbon footprint, any values should be taken as a rough estimate only. In many instances where there is inadequate rainfall or irrigation to cause leaching or where fields are flat and do not readily drain into watercourses, potential eutrophication remains just that, potential, and in reality while nutrients may be lost below the root zone they do not cause eutrophication.

Does horticultural production contribute to eutrophication? The answer to this question is obviously yes, but like horticulture's contribution to greenhouse gases, the contribution is small in comparison to other agricultural sectors. The largest contribution within horticulture are where large amounts of fertilizer are used, for example, brassica crops in Europe, and especially where large amounts of nitrogen and phosphorus fertilizer are used together with high irrigation rates or in areas of high rainfall. Potential eutrophication can become actual eutrophication when these scenarios are true in close proximity to water courses. Fertilizer use can be used as a proxy to establish horticulture's contribution to eutrophication (and acidification). Horticultural production tends to be more intensive and use higher rates of fertilizer in comparison to other agricultural sectors, so we assume that its contribution is greater than the pro-rata land use of 3 % and could be as high as 5 % of total fertilizer use. However, this level of use is still low.

The effects of eutrophication can be severe but it is difficult to attribute blame to any one sector when most environments are comprised of a mixture of arable, horticultural and livestock enterprises, and may often also contain human settlements which have their own discrete sources of excess nutrients. However, some good examples of eutrophication do exist. The most prominent is probably Lake Naivasha in Kenya where intensive horticultural production has affected both the water level and phosphorus concentration in the lake's water but there are others. Cranberry production was identified by Garrison and Fitzgerald (2005) as the likely source of nutrients that resulted in serious eutrophication in a Wisconsin lake.

In general terms, eutrophication from horticultural production is rarely a problem in mixed agricultural systems, especially those in an active hydrological environment. However, problems can occur in areas of intensive horticulture production, especially close to water bodies with limited flows and outlets.

Greenhouse Gas Emissions

Horticultural production is responsible for the emission of carbon dioxide and nitrous oxide; two of the principal greenhouse gases. Carbon dioxide is produced at every stage of resource production and use (especially fertilizer) and small amounts are emitted at every stage of the production process from energy use, however it is the manufacture and construction of materials for use within protected cropping that dominates overall carbon dioxide emissions. Nitrous oxide is produced by the application of nitrogen fertilizers, tillage of agricultural land and emission from manures/composts used for plant nutrition. The production of nitrous oxide from field horticulture is proportional to the amount of nitrogen fertilizer applied, so crops like potato and cauliflower which have high nitrogen requirements tend to emit more nitrous oxide in comparison to crops like carrot and onion which have a lower requirement for nitrogen.

In common with many areas of environmental accounting, accurate assessment of greenhouse gas emissions is difficult and considerable variation is possible for the same crop. Crop location (and therefore soil type, soil fertility and water availability) influences all aspects of crop agronomy and therefore emissions of greenhouse gases. It is entirely possible, and should be expected, that the carbon footprint of a product can vary two-fold or more so all values should be treated as an estimate only. The last ten years has seen considerable research in this area and the calculation of carbon footprints for multiple crops. A selection is presented in Table 16.2.

What contribution does horticulture make to global greenhouse gas emissions? In the UK, the agricultural sector is responsible for 8 % of total greenhouse gas emissions and the horticultural sector represents just 3 % (by area) of the agricultural sector. Therefore a rough calculation suggests that the horticulture sector might be responsible for 0.24 % of UK greenhouse gas emissions which is 1.4 million t carbon dioxide equivalents. Given the huge range of horticultural crops, it is not possible to assume that the UK is typical of all countries but the over-riding conclusion must be that horticulture contributes very little to global greenhouse gas emissions.

Although horticultural production is generally viewed as having a negative effect on the environment, there are some situations where it may offset greenhouse gas emissions and increase soil organic matter contents. It is accepted that forests sequester carbon, both in the living biomass and their soils, so the same principle can be applied to long term plantation and orchard crops. The size of any benefit is very much dependant on crop type, duration of growth, management and land use

Table 16.2 The carbon footprint (normalised emissions of carbon dioxide, methane and nitrous oxide) of selected horticultural crops. (Compiled from Audsley et al. 2009; Edward-Jones et al. 2009; Berners-Lee 2010; Lillywhite 2010; Gunady et al. 2012; Ingwersen 2012; WRAP 2013)

Crop	Typical carbon footprint (CO2e) kg/kg product	Range (CO_2e)
Apple	0.55	0.32–0.88
Asparagus	1.94	1.00–3.50
Banana	0.48	0.25–1.33
Beans (green)	1.55	1.40–10.70
Carrot	0.30	0.50–5.50
Cauliflower	2.22	1.94–2.39
Coconut	1.78	
Cucumber	3.30	1.30–3.79
Grapes	0.59	0.42–0.75
Lettuce	0.45	0.23–3.75
Pineapple	1.78	0.19
Potato	0.12	0.09–0.14
Orange	0.50	
Mushroom	2.75	2.75–4.10
Strawberry	0.60	0.60–7.20
Tomato	1.49	0.60–1.39
Water melon	1.33	1.30–3.79

change status and while it would be churlish to suggest that horticultural crops can deliver the same environmental benefits as virgin forest, they can help to reduce the environmental impact compared to annual crops. The principal of partially undisturbed production systems unpins the concept of agroecology and can be equally applied in mainstream perennial horticulture.

Wu et al. (2012) reported that apple orchards in China sequestered carbon until they were 18 years old and had reached a new equilibrium level but that production could not be considered carbon neutral since the greenhouse emissions associated with production were still greater than had been sequestered. This work demonstrates that for perennial crops, the concept of carbon neutrality is achievable and further work on the management side will undoubtedly bring it closer to reality.

In summary, the majority of horticultural production emits greenhouse gases, principally carbon dioxide and nitrous oxide and while it is important to reduce emissions as far is economically possible, the level of emissions in comparison to other food crops is small and given the health benefits of horticultural products, easy enough to justify. Annual crops will always generate higher emissions through soil turnover and a requirement for higher nutrition compared to long-term crops so environmental impacts can be reduced through perennial crops. The greatest benefit will come from orchard and plantation crops with life-spans of 25 years or more but even shorter term crops like strawberry and asparagus may reduce environmental impacts.

Acidification Potential

Acidification is a consequence of acids (and other compounds which can be transformed into acids) being emitted to the atmosphere and subsequently deposited in surface soils and water. Increased acidity of these environments can result in negative consequences for crop growth and the surrounding natural environment. The acidification potential is driven by three inputs: sulphur dioxide, oxides of nitrogen and ammonia, and is quantified in terms of sulphate equivalents (Azapagic et al. 2004).

Acidification was a serious issue in Europe 20 years ago when sulphur released from the combustion of coal for electricity was subsequently transferred to land and water and caused major ecological damage. However, current generation technologies have eliminated this problem and acidification caused primarily by sulphur dioxides is no longer an issue. The main cause of acidification is now excessive application of nitrogen fertilizers.

Acidification of soil is now a major issue in intensive production systems in many countries. Guo et al. (2010) report significant acidification of topsoils in China which they attribute to over fertilization. The authors identified the production of glasshouse vegetables to be a serious concern with fertilizer rates above 4,000 kg N ha year; this is far in excess of potential plant uptakes and at a rate which causes serious eutrophication as well.

There is no doubt that horticultural crops receive higher rates of nitrogen fertilizer compared to other crops, and the continuing pressure to increase crop yields is likely to exasperate the problem. Therefore it is likely that acidification will continue to be a problem but it is one that can be resolved through improved understanding of crop nutrient requirements, better extension services and farmer education.

Visual Impacts

The aesthetics of field, orchard or plantation horticulture differs little from arable cropping but there is one aspect of horticultural production that attracts increasing concern and that is the proliferation of glasshouses, polytunnels, polythene and fleece. While small-scale protected cropping has always been part of the horticultural landscape, the increase in temperate zones of protection, be it for improved quality, extending the season or for pest and disease prevention, has been considerable. The impact can be seen at different scales, from an individual farmers field covered in polythene to advance the crop, through clusters of glasshouses in sensitive landscapes to regional dominance, for example, Almería in Southern Spain, where the impact can be seen from space.

The expansion of protected cropping is a major horticultural success story of recent years. It has maintained (or saved in some cases) the economic viability of the industry, reduced pesticide applications through improved IPM, improved the quality of many crops and expanded the market for others, notably soft fruits in

Europe, quite considerably. All of this has been welcomed by consumers and will have contributed to health policy objectives. Despite these advantages, it should be recognised that many people consider the visual impact to be a blot on, and an industrialisation of, the landscape and part of the unacceptable face of farming. The fact that it is a visual rather than an invisible impact, like the other issues considered here, is probably the most important fact. In terms of environmental impact, protected crops have higher greenhouse gas emissions due to the manufacture of the protection, and may increase water and leachate run-off but at a national or global level, this is a small price to pay for the advantages that they bring.

None the less, it is important that the industry recognises the issue and brings about workable solutions. Location is always going to be the most important aspect of the argument so care should be taken to site new structures in areas with less dense populations. Thanet Earth in the UK has demonstrated that this is quite possible by constructing the largest glasshouse in the UK on the Isle of Thanet in Kent. Polytunnels should be sited as to not change the character of the landscape and early season polythene should be removed as soon as is practical. A common sense approach that is considerate of local opinion while at the same time promoting the benefits should be employed. While none want to witness a landscape spoilt, everything should be viewed in context; the combined area of all the permanently protected cropping in the UK is probably less than 6,000 ha which represents less than 0.1 % of the total croppable area. We suggest that the visual impact of this is a small price to pay for the huge benefits that flow from protected horticulture.

Summary

Horticulture's relationship with the environment is quite complicated and can perhaps be considered in three categories. First are the hidden impacts that have little consumer awareness: greenhouse gases, eutrophication and acidification. Of these, eutrophication is probably the worst but it is only a problem where horticultural production is concentrated into small areas, e.g. Lake Naivasha in Kenya. The second category contains the hidden impacts that consumers are aware of, namely water use and pesticide residues. These should be of concern to everyone since water resources should be preserved and pesticide residues reduced to maintain the integrity of the industry and the environment in which it operates. We would argue that water and pesticides are likely to dominate industry thinking over the next 10 to 20 years as producers strive to become more environmentally friendly and produce more sustainability. Thirdly, are the visible impacts, land use, especially where land is covered in fleece, polythene, polytunnels or glasshouses. This is perhaps the most difficult of the three categories to deal with since their use is a reflection of technological advancement and great gains in productivity and quality have been achieved through physically protecting crops. However, it is still quite easy to have sympathy with its detractors since large groupings of glasshouses and polytunnels are undeniably unattractive. This is well captured in the infamous images of Almería from space where every piece of land is covered in glass or plastic. The detrimental aspects of production should be considered in the context of the benefits that horticulture brings: mass employment, prosperity and foods that taste good and are essential in a balanced diet. On reflection, the latter out-weigh the negative aspects of production.

References

Allan JA (2011) Virtual water: tackling the threat to our planet's most precious resource. Tauris, London

Audsley E, Brander M, Chatterton J, Murphy-Bokern D, Webster C, Williams A (2009) How low can we go? An assessment of greenhouse gas emissions from the UK food system and the scope to reduce them by 2050. WWF–UK

Azapagic A, Roland C, Perdan S (eds) (2004) Sustainable development in practice: case studies for engineers and scientists. Wiley, Hoboken

Berners-Lee M (2010) How bad are bananas: the carbon footprint of everything. Profile Books, London

Castillo LE, Ruepert C, Solis E (2000) Pesticide residues in the aquatic environment of banana plantation areas in the north Atlantic zone of Costa Rica. Environ Toxicol Chem 19(8):1942–1950

Defra (2012) Agriculture in the United Kingdom 2011. Defra, London

Echeverria-Saenz S, Mena F, Pinnock M, Ruepert C, Solano K, de la CE, Campos B, Sanchez-Avila J, Lacorte S, Barata C (2012) Environmental hazards of pesticides from pineapple crop production in the Rio Jimenez watershed (Caribbean Coast, Costa Rica). Sci Total Environ 440:106–114

Edward-Jones G, Plassmann K, York EH, Hounsome B, Jones DL, Milà i Canals L (2009) Vulnerability of exporting nations to the development of a carbon label in the United Kingdom. Environ Sci Policy 12:479–490

EU Parliament (2013) http://www.theparliament.com/latest-news/article/newsarticle/eu-urged-to-step-up-efforts-on-illegal-and-counterfeit-pesticides/. Accessed April 2013

FAO (2002) Crops and drops: Making the best use of water for agriculture. Food and Agriculture Organization of the United Nations, Rome

García-Ruiz JM (2010) The effects of land uses on soil erosion in Spain: a review. Catena 81:1–11

Garrison PJ, Fitzgerald SA (2005) The role of shoreland development and commercial cranberry farming in a lake in Wisconsin, USA. J Paleolimnol 33(2):169–188

Gunady MGA, Biswas W, Solah VA, James AP (2012) Evaluating the global warming potential of the fresh produce supply chain for strawberries, romaine/cos lettuces and button mushrooms in Western Australia using life cycle assessment (LCA). J Cleaner Prod 28:81–87

Guo JH, Liu XJ, Zhang Y, Shen JL, Han WX, Zhang WF, Christie P, Goulding KWT, Vitousek PM, Zhang FS (2010) Significant acidification in major Chinese croplands. Science 327:1008–1010

Hoekstra AY, Chapagain AK, Aldaya MM, Mekonnen MM (2011) The water footprint assessment manual: setting the global standard. Earthscan, New York

Ingwersen WW (2012) Life cycle assessment of fresh pineapple from Costa Rica. J Cleaner Prod 35:152–163

Johnes PJ (1996) Evaluation and management of the impact of land use change on the nitrogen and phosphorus load delivered to surface waters: the export coefficient modelling approach. J Hydrol 183:323–349

Kovach J, Petzoldt C, Degnil J, Tette J (1992) A method to measure the environmental impact of pesticides. NY Food Life Sci Bull 139:1–8

Lewis KA, Brown CD, Hart A, Tzilivakis J (2003) p-EMA (III): overview and application of a software system designed to assess the environmental risk of agricultural pesticides. Agronomie 23(1):85–96

Lillywhite RD (2010) The use of environmental footprints in horticulture: case studies. Final report of Defra project WU0114. University of Warwick & Defra, London

Lillywhite RD, Chandler D, Grant W, Lewis K, Firth C, Schmutz U, Halpin D (2007) Environmental footprint and sustainability of horticulture (including potatoes): a comparison with other agricultural sectors. Final report of Defra project WQ0101. University of Warwick & Defra, London

Reus JA, Leendertse PC (2000) The environmental yardstick for pesticides: a practical indicator used in the Netherlands. Crop Prot 19:637–641

Scanlon BR, Jolly I, Sophocleous M, Zhang L (2007) Global impacts of conversions from natural to agricultural ecosystems on water resources: quantity versus quality. Water resources research 43

Silgram M, Waring R, Anthony S, Webb J (2001) Intercomparison of national & IPCC methods for estimating N loss from agricultural land. Nutr Cycling Agroecosyst 60:189–195

WRAP (2013) An initial assessment of the environmental impact of grocery products. Product Sustainability Forum & WRAP, Banbury

Wu T, Wang Y, Yu CJ, Chiarawipa R, Zhang XZ, Han ZH, Wu LH (2012) Carbon sequestration by fruit trees—Chinese apple orchards as an example. Plos One 7(6):e38883

Chapter 17
Woody Ornamentals

Paul E. Read and Christina M. Bavougian

Abstract Woody ornamental plants are the backbone of massive horticultural industries. They provide the basis for the retail selling of plants through nurseries, garden centres and increasingly supermarkets and multiple retailers. Equally they provide the foundation for all urban, peri-urban and rural landscape designs. These are the horticultural products which bring pleasure, leisure and relaxation to populations worldwide. Here is a sector where impulse buying results in considerable price elasticity. Purchasing plants in the retail sector is a pleasurable event and not governed by cost. This chapter provides a major review of the science and technology which supports this sector of horticulture. It has developed very substantially over the last 50 years. Applied science and technology has provided whole new avenues for propagation, husbandry, nutrition, pest and pathogen control and vastly expanded the range of plants available. Realisation is now developing that some plants despite their attractiveness can become invasive, even noxious weeds. As a result, researchers have devised ways to create seedless woody ornamentals from plants that are weedy, invasive or produce toxic fruits, as was the case with the recently publicized fruitless 'Anna Bela' tung tree (*Aleurites fordii*) (Rinehart et al., Hort Sci 48:123–125, 2013). The future for woody ornamentals looks bright. Although there are challenges to providing plant materials that will meet the needs of customers, the woody ornamental industry will continue to grow and flourish into the foreseeable future, enriching the lives of future generations.

Keywords Woody ornamental · Leisure · Pleasure · Propagation · Multiplication · Pest and pathogen control · New crops

P. E. Read (✉) · C. M. Bavougian
Department of Agronomy and Horticulture, University of Nebraska, Lincoln, NE 68583-0724, USA
e-mail: pread@unl.edu

C. M. Bavougian
e-mail: christina.huck@huskers.unl.edu

G. R. Dixon, D. E. Aldous (eds.), *Horticulture: Plants for People and Places, Volume 2*,
DOI 10.1007/978-94-017-8581-5_17,

Fig. 17.1 American redbud (*Cercis canadensis*) provides an attractive contrast to the ornamental textures and stature of Norway spruce (*Picea abies*). (Photo courtesy of Nebraska Statewide Arboretum, Lincoln, NE, USA)

Introduction

Nearly all plants that have cellulosic and lignified cells could be used as ornamentals. However, this chapter will focus on trees and shrubs that are not covered elsewhere in this Trilogy. Such plants termed as "woody ornamentals" may be considered for landscape uses to take advantage of their unique characteristics, including presence or absence of flowers and flower color; foliage color and morphology; plant form, including stature, shape and size; and textural features. Figs. 17.1 and 17.2 illustrate how flowering trees can interact with evergreens and can be used to frame attractive vistas.

Scope of the Industry

The United States is the world's largest producer of (and market for) nursery crops with total gross sales of US$ 4.65 billion in 2006 (Anon 2007). The nursery and greenhouse industry makes up the fastest-growing sector of U.S. agriculture; gross grower sales are increasing at approximately US$ 500 million annually (Mizell et al. 2012). In 2002, nursery production contributed US$ 18.1 billion to the U.S.

Fig. 17.2 The pendulous branches and flowers of *Cladrastis* beckon the observer to travel along the path in this interesting landscape. (Photo courtesy of Nebraska Statewide Arboretum, Lincoln, NE, USA)

economy and provided almost 2 million jobs (Behe et al. 2008). Of total gross sales, 76 % was production of woody ornamentals including broadleaf evergreens, deciduous shade trees, deciduous shrubs, deciduous flowering trees, coniferous evergreens, palms, and other woody ornamentals, vines and ground covers (Anon 2007). An additional 7 % of total sales was attributed to propagative nursery materials, a portion of which should also be included here because much of the root-stock, lining-out stock, seedlings, tissue culture, and other products were undoubtedly woody ornamentals.

According to Mizell et al. (2012), "ten states account for more than two-thirds of all nursery-crop output in the U.S.: CA (20 %), FL (11 %), NC (8 %), TX (8 %), OH (5 %), MI (2–4 %), PA (2–4 %), and NY (2–4 %)." In 2006, the states with the largest area in nursery production were OR: 94,250 acres; NC: 48,454 acres, PA: 46,839 acres, MI: 45,886 acres, and FL: 40,706 acres (USDA 2007). California was not among the top five states for production area (even though it has the highest gross nursery crop sales for any state) because of what is grown there, i.e., high-value container-produced broadleaf evergreen plants such as azalea, boxwood (*Buxus* spp), *Euonymus*, and holly (*Ilex*)(Anon 2007).

Container production of nursery crops has increased in popularity in recent decades; currently well over 50 % of woody ornamentals produced in the U.S. are grown in containers (Hall et al. 2005). In 2006, 88 % of broadleaf evergreens, 77 % of deciduous shrubs, 59 % of coniferous evergreens, 50 % of deciduous flowering trees, and 47 % of deciduous shade trees were container-grown (Anon 2007). Containers provide many advantages over field (in-ground) production of nursery crops, including convenience of marketing and transportation, consumer appeal, product longevity, and establishment success after transplanting (Ingram et al. 1993; Gilman 2001). When field-grown plants are harvested, approximately 30 % of root area is lost or damaged, whereas containerized plants are sold with their root systems intact (Harris and Gilman 1991, 1993; Thomas 2000). Container production requires considerably less land than field production. Relative to field-produced ornamentals, containerized plants are easier to transport and suffer less injury during shipping and handling. Container-grown ornamentals therefore have a larger "window of marketability" for saleable plants, due to their improved longevity in the retail market (Mathers et al. 2007).

Field nurseries incur costs associated with labor and equipment needed to dig plants for sale that are not needed in container nurseries (Whitcomb 1984); however, container nurseries require more intense management. One reason for this is that their root systems, confined in containers with high surface area to volume ratio, are subject to greater temperature extremes and fluctuations than in-ground plants (Ingram et al. 1993).

Container production is overwhelmingly popular for certain types of woody ornamentals (most notably broadleaf evergreens and deciduous shrubs), but other categories still rely on field production (deciduous flowering trees and, especially, deciduous shade trees). Many of these trees are grown for the landscape industry, which often requires larger plant material than could be produced efficiently in containers (LeBude and Bilderback 2007).

Propagation

For the practical use of woody ornamentals as specimen plants or utilization in landscape plantings, economical multiplication of sufficient numbers to meet planting demands is a critical part of woody ornamental production and use. As noted in 2, huge numbers of woody ornamental plants must be generated in the U.S. alone, with an obvious concomitant need for similarly large numbers in other developed countries. Propagation methods may include, but are not limited to, the use of seeds, cuttings, grafting, layering and micropropagation (in vitro, or tissue culture) (Soyler and Khamer 2006). When asexual methods are employed instead of sexual (usually seeds), many advantages are accrued, including speed of production, circumvention or elimination of the juvenile phase, propagation of plants that do not produce seeds or that have seeds that are not viable, and taking advantage of characteristics that can be achieved by use of rootstocks. Efficiency of production and high product quality, including uniformity, are parameters requisite to choice of propagation method (Read et al 1986).

Seed Propagation

Seed propagation is the most ancient and often is the simplest of methods for propagation of woody ornamentals. Seed provenance, location of geographic origin, must be considered if collected from naturally occurring stands; source of seed must be considered in order to assure trueness to type (genetic fidelity) and freedom from contamination from pollen of undesirable genotypes. In many species, freshly harvested seed can be planted immediately following harvest. However, seed dormancy may prevent such an approach and special techniques must be employed to encourage satisfactory germination capability. Seeds such as *Ginkgo* may have immature embryos at harvest time, but the embryos of most seeds are mature when harvested. Simple cleaning practices can assist the propagator in realizing good propagation success.

Seed dormancy consists of several types. Germination may be inhibited by seed physiology, anatomy or both. If seeds are dormant immediately after harvest, they are considered to exhibit primary dormancy, whereas if external environmental factors induce dormancy, it is referred to as secondary dormancy. Whether seeds exhibit primary or secondary dormancy, if physiologically dormant, special treatments may be necessary. Seeds with under-developed embryos (e.g., *Fraxinus excelsior*) require an after-ripening treatment of 20–30°C for 3 months. High temperature after-ripening treatments are necessary for some palm seeds (38–40°C) for 30 to 90 days. Other types of physiological dormancy may be overcome by a moist chilling treatment referred to as stratification (Soyler and Khawar 2006). Examples include *Berberis thunbergii,* which requires 90 days at 0–5°C and *Malus* seeds need to be exposed to temperatures of 3–5°C for up to 60 days, depending upon species and cultivar.

Many woody ornamentals have hard seed coats that prevent imbibition of water, obstruct gaseous exchange or may physically prevent the embryo from exiting the seed coat. Such seed coats must be penetrated by physical or chemical means, a process referred to as *scarification.* Many woody ornamental members of the *Fabaceae*, such as *Cercis, Gymnocladus* and *Gleditsia*, have hard seed coats and require scarification. *Cercis canadensis* also exhibits double dormancy, that is, it requires both scarification and stratification. Another type of double dormancy is found in some *Fraxinus* species because they have immature embryos and also need to be stratified. Once seed dormancy requirements are met or if seeds are not dormant, seed sowing can take place in a suitable medium. Medium characteristics for container production are discussed under that topic, while seeds destined for field production need to be planted into an appropriate seed bed. For seed bed preparation, see Sect. 5.

Cutting Propagation

Propagation by cuttings involves severing a part of the plant from the parent plant (stock plant) and providing conditions that encourage regeneration of the missing parts of the plant. For example, when propagating by stem cuttings, stimulation of adventitious roots is necessary, while root cuttings require generation of a shoot. A critical component for successful propagation by cuttings is the condition of the stock plant; that is, cuttings should only be taken from healthy, disease-free parent plants that are true to the desired plant phenotype to be cloned. Types of cuttings that are used to propagate woody ornamentals include leafy stem cuttings, softwood cuttings, semi-hardwood cuttings, hardwood deciduous cuttings, hardwood evergreen cuttings and root cuttings. For successful propagation by cuttings,

Fig. 17.3 Close-up of a mist nozzle employed to mist leafy cuttings. (Photo courtesy of John Preece, National Clonal Germplasm Repository, USDA)

Fig. 17.4 Cuttings of a variety of woody landscape plants being rooted under mist in a commercial nursery. (Photo courtesy of John Preece, National Clonal Germplasm Repository, USDA)

environmental conditions must be provided that are conducive to production of the requisite missing plant parts.

For leafy stem cuttings such as softwood cuttings and semi-hardwood cuttings, because the leaves will be losing moisture through transpiration, high relative humidity must be maintained near the leaf lamina to reduce moisture loss related to transpiration. To provide the necessary high humidity, several methods may be employed. Perhaps the most common method is the use of intermittent mist. Such systems employ a time clock to control a fine mist spray of water applied intermittently, e.g., 15 seconds of spray every 30 minutes during the daylight hours (see Figs. 17.3 and 17.4). This timing may vary, depending upon species, condition of the cuttings, growth regulator treatment (if any), day-length, porosity of the rooting medium and ambient and root-zone temperature. Systems that provide fog, an atmosphere that is nearly 100 % relative humidity, are also commonly employed.

Fog may be introduced into an entire greenhouse or under a tent over a greenhouse bench. Some propagators use hand spraying of the cuttings, but this approach is labor intensive and difficult to maintain consistent humidity control. On a small scale, such as for home propagators, a pot with cuttings in its medium may be enclosed in a plastic bag. Another technique to reduce transpiration losses is to remove some of the leaves, or parts of the leaves, to reduce the amount of surface that will be transpiring. Hardwood cuttings usually do not require humidity modification until new leafy growth emerges from the buds on the cuttings.

Regardless of the type of stem cutting, application of an auxin-type plant growth regulator (hormone) to the base of the cutting may be useful. Although indole acetic acid (IAA) is the naturally occurring hormone in nearly all higher plants, because it is light labile, analogs of IAA are generally preferred for application to stimulate adventitious roots. Examples include indole butyric acid (IBA) and naphthalene acetic acid (NAA); either IBA or NAA, or combinations, are often found as the active ingredients in commercially available rooting powders or root stimulating solutions. Effective concentrations and methods of application vary with species, cultivar, type of cutting and environmental conditions under which the cuttings are being rooted. Some cuttings will root faster than those not treated, while others can be stimulated to (King et al 2012) root that are difficult or impossible to root without growth regulator applications. Most root cuttings do not respond to growth regulator applications (Blythe and Sibley 2012).

Examples of woody ornamentals that can be propagated by softwood cuttings include *Syringa* (lilac) and various maples (*Acer* spp.), while many *Rhodendron* and *Citrus* species can be rooted as semi-hardwood cuttings. Many species are propagated by deciduous hardwood cuttings (e.g., *Vitis, Cornus, Salix*) and evergreen hardwood propagation is commonly employed for *Taxus, Juniperus* and *Thuja*. Sumac (*Rhus* spp.) types and *Chaenomeles japonica* are easily propagated by root cuttings, as are many members of the *Populus* and *Rubus* genera.

Grafting and Budding

Propagation by grafting and budding (a form of grafting) is accomplished by the joining together of two plants or plant parts so that they will ultimately grow as one. Although grafting occurs occasionally in nature, grafting requires specialized skill on the part of the propagator and is generally both more labor intensive and expensive than propagation by cuttings or seeds. Therefore, grafting is employed for plants that are difficult to propagate by other means or to take advantage of characteristics imparted by the rootstock. The desirable upper portion of a graft is called the scion, while the lower part is referred to as the rootstock. When another piece of stem is included between the scion and the rootstock, it is referred to as an interstock or interstem. Interstocks are sometimes used to impart growth regulating effects on the scion or the rootstock, but often are used to overcome graft incompatibility of scion and rootstock as in the use of 'Old Home' pear interstem to overcome incompatibility between scions of 'Bartlett' (also known as 'Williams') pear and quince rootstocks.It is nearly impossible to mention all types of grafting, but a few worthy of note include the cleft graft, whip and tongue graft, approach graft, saddle graft, wedge graft and various types of bud grafts, such as T-budding and chip budding. Bud grafting is a more economical use of plant material and may be chosen if propagating material of the desired scion is in limited supply. Each technique requires different skills and may be the appropriate choice, depending on species, cultivar and even season of the year. Regardless of grafting method or type, it is critical for the cambium of the scion to be aligned with the cambium of the rootstock and that the union not be allowed to become dried out or attacked by insects or disease organisms. Many unique woody ornamentals are grafted in order to capture their unique traits, such as color sports and shrubs and trees with variegated leaf patterns. Details of grafting are covered more completely in other resources (Hartmann et al. 2010).

Propagation by Layering; Layerage

Layering, or layerage, involves propagation of a cutting while it is still attached to the mother plant. Mound layering is a commonly employed practice, although layerage is rapidly becoming supplanted by improved cutting propagation methods or by micropropagation. Mound layering has been historically used for propagation of clonal apple rootstocks, quince and *Ribes* spp. such as currants and gooseberries, but can also be useful for propagating some ornamental members of the *Rosaceae*. The method begins by planting a healthy stock plant in well-drained soil a year before propagation is to commence. In the spring of the second year, the mother plant is cut back to a few centimeters above the ground, where a few shoots will begin growth. After they have reached about 8–12 cm in length, sawdust, loose soil or other suitable medium is mounded to about half the height of the shoots, followed by a second mounding a few weeks later. A third mounding will usually take place in mid to late summer to about one-half the height of the shoots. The layered shoots usually will have produced adventitious roots by the end of the summer. They can then be removed from the parent plant and lined out in the nursery row, i.e., treated as transplants.

Micropropagation

As plant tissue culture technology has grown from its humble beginnings when Haberlandt (1902) reported on his attempts to produce plants in vitro, through the identification of indole acetic acid (IAA) a naturally occurring auxin (Thimann and Went 1934), and the discovery of cytokinins by Skoog's group (Skoog and Miller 1957), it soon became apparent that micropropagation, or propagation of plants in vitro, has become another powerful tool available to the plant propagator. The

Fig. 17.5 Proliferating shoot culture of *Hibiscus Moscheutos*. (Photo courtesy of John Preece, National Clonal Germplasm Repository, USDA)

process of micropropagation involves placement of small plant pieces (explants), somewhat like a miniature cutting, aseptically onto a medium designed to encourage axillary bud growth in vitro. Typically, the medium consists of various materials dissolved or suspended in water, including macronutrient salts, micronutrients, amino acids, vitamins, an energy source such as sucrose, and plant growth regulators. The medium is often gelled with a gelling agent such as agar, but some micropropagation takes place in a liquid medium. A great deal of research reports have been generated since the landmark publication describing what is today a standard medium, MS medium (Murashige and Skoog 1962). The primary goal is to enhance the growth and multiplication of existing meristems, usually axillary buds that are part of the explant. The balance of cytokinin to auxin content of the medium determines whether bud and shoot growth or root growth will occur. A high cytokinin to auxin ratio favors shoot growth, with high auxin stimulating root growth (Murashige 1974).

Commercial nurseries utilize micropropagation to produce large numbers of woody ornamentals. Examples include *Cercis*, *Hibiscus*, *Rhododendron* and *Berberis*, while research has continued on numerous additional species (e.g., *Acanthopanax*, Yang and Read 1997; *Acer*, Preece et al. 1991; and *Quercus*, Fishel et al. 2003; Fig. 17.5). In addition to manipulating the medium constituents and environmental conditions (Read 1990, 1992), methodology has been developed to encourage softwood out-growth from dormant stems placed in a forcing solution (Yang et al 1986; Yang and Read 1989; Fig. 17.6), while forced large branch segments of mature woody species such as *Quercus* and *Acer* have produced shoots from epicormic buds suitable for use as explant material (Henry and Preece 1997). An interesting recent development in studies of invasive plants has stimulated use of in vitro systems to create sterile (seedless) plants, thus making them non-invasive. Examples include 'Lilac Chip' *Buddleia* spp. (Zampini 2013 and *Euonymus alatus* (Thamina et al. 2011), the latter resulting from regeneration of triploid plants from endosperm tissues (Knapp et al 2001). For further reading on micropropagation,

Fig. 17.6 Woody stems of *Viburnum* x *rhytidophylloides* '*Willowwood*' being forced to promote softwood growth for use as explant material for micropropagation.

consider Cloning Plants—Micropropagation/Tissue Culture, a chapter soon to be published in the Encyclopedia of Agriculture and Food Systems (Read and Preece 2013).

Site Selection

Poor choice of the location where woody ornamentals are to be grown, whether commercially or for home landscape use, is perhaps the most common cause of failure. A suitable site must be in a location that has a climate conducive to healthy growth and development of the plant for its desired purpose. It must also provide appropriate levels of light (full sun for many species), water availability with good internal soil drainage and protection from environmental stresses such as extremes of cold or heat, drying winds or potential animal depredation (see Sect. 6). Low lying areas that can become frost pockets are undesirable and should be avoided. Since cold air is heavier than warm air, it will flow to the low areas where it is trapped and can cause damage because of the cold temperatures. The properties of the existing soil for woody ornamentals to be grown in the ground can be important, as is the composition of the medium for container-grown ornamentals. Water-holding ability; cation exchange capacity (the ability to hold and provide nutrients); soil reaction (pH); freedom from pathogens, insects and their eggs, weed seeds and toxic substances are also critical elements to consider when selecting a site for production of woody ornamentals. Studying the characteristics of the location to which the particular species is endemic can be instructive in determining the appropriate site conditions for that plant. Ease of access to the site for conducting management practices is also important.

Site Preparation

Important site preparation activities include soil amendment with organic matter and establishment of cover crops. These best management practices are the most effective ways to improve soil quality and reduce the potential for erosion. LeBude and Bilderback recommend applying organic amendments in four-foot-wide strips along planting rows, rather than covering the whole field (LeBude and Bilderback 2007).

If lime or other soil amendments will be used, they should also be incorporated before planting. Cover crops can add organic matter and provide weed control before planting field nursery crops. They can be disked under before planting or killed with herbicide before seedling emergence. Establishment of permanent filter strips (usually cool-season grasses) will help to slow the movement of surface runoff and reduce the risk of chemicals and nutrients moving off-site. Site preparation for container nurseries includes such activities as weed control and mulching the container yard, and preparation of potting media.

Pre-plant weed control is primarily relevant to field nurseries, although certain aspects also pertain to container nurseries. Preventing the introduction of weed seeds is a major concern for both types of nurseries. Weed seeds can enter nursery production areas from weeds in or near the growing area, stock plants, irrigation water, and from soil and organic material such as manures, composts, and mulches. Measures such as installing screens on irrigation water inflow sources and proper composting of manures and yard wastes can help mitigate the risk of weed seed introduction (Wilen and Elmore 2009).

Before planting field nursery crops, weeds must be removed. The most effective method is to irrigate the area so that weed seeds germinate, and then cultivate or apply a nonspecific/broad-spectrum herbicide. Perennial weeds may require chemical treatment as they sometimes are not controlled by cultivation alone (Altland et al. 2003). In some cases, soil fumigation is used to control weeds, pests, and/or pathogens. Ideally, soil sterilants would only be used in extreme cases because the chemicals are highly toxic. Two examples are metham-sodium, a liquid, and dazomet, a granular formulation. Both of these products work by releasing methyl isothiocyanate gas. Successful fumigation requires intensive soil preparation, specific moisture and temperature conditions, and at least several weeks' delay before planting (LeBude and Bilderback 2007). Soil solarization using clear polyethylene can be employed to kill weed seeds, and in some cases may control pests and pathogens.

Planting/Lining Out

Field and container nurseries have very different concerns regarding planting/lining out. For field nurseries, planting density must account for machinery movement in the field, as well as space needed by plants depending on how long they will be grown (which is entirely dependent on marketing plan). Wide spacing is necessary for trees that are to be sold to professional landscapers or for municipal use (LeBude and Bilderback 2007). Field nursery crops can be planted in late fall, winter, or early spring, depending on location and climate (Harris and Gilman 1993).

Container nurseries must consider pot size and type of substrate. Ingram et al (1993) provide a comprehensive discussion of container nursery planting considerations, including substrate properties, container size, and response to environmental conditions Traditionally, container stock has been grown in soil-less media, because field soil must be treated to kill weed seeds and is expensive to transport. Rideout (2012) describes a method whereby steam is used to "clean" and kill weed seeds in used potting soil and containers (also see Sect. 6). This method allows potting mix and other materials to be reused. It is important to prevent weed seed contamination of container mixes by covering piles if they are outdoors (Wilen and Elmore 2009). It is also a good idea to conduct germination tests to check for weed seeds in container mixes before planting.

Container substrates are vital components of the production system, providing physical support, oxygen, water and nutrients. One important consideration in selecting materials for container media is the distribution of pore spaces; there must be adequate water holding capacity while maintaining aerobic conditions for root respiration. Many different substrates are commonly used (including hardwood bark, composted yard waste or manure, cotton gin wastes, clay minerals, rice or peanut hulls, peat, coconut coir, vermiculite, perlite, sawdust, and many others), although in the southeastern United States the industry standard is a mixture of pine bark and sand at a ratio of 8:1. Mathers et al. (2007) provide a review of the physical and chemical properties of various container medium components.

For both types of nurseries, planting depth is vitally important. Liners must not be planted too deeply. The root ball should be covered by approximately 2.5 cm of soil, leaving the root collar just above the soil surface. Watering-in well will ensure sufficient root-soil contact, by eliminating air pockets and firming the soil (LeBude and Bilderback 2007). When planting woody ornamentals in landscapes and field nurseries, it is important to dig sufficiently large holes in order to avoid problems such as girdling roots as the plants grow.

Weed Management

Weed management in container nurseries.

Container nurseries present some unique weed management challenges. Weed seeds may be transported along with ornamental liners as they are shipped to various locations. The economic threshold for some ornamentals is less than one weed per container, partly due to consumer expectations and also because resources are limited within the container. One of the most important measures in container nursery weed prevention is to ensure proper sanitation and to minimize wind dispersal of weed seeds from adjacent areas. Cultural and mechanical weed control methods can be used in container nurseries, as well as chemicals. Case et al (2005) provide a useful review of container weed control practices.

Cultural weed control methods for container nurseries include mulches, sub-surface irrigation, and large-porous substrates. In general, mulches have not been used extensively as weed suppressants in container production; however, many materials have been proposed. Organic mulches, such as tree bark in various forms, shredded newspaper, rice hulls, and pelletized wool, have received the most attention. Additional products include fiber or textile discs, polyethylene sleeves, biodegradable films, and dolomitic lime and other minerals. Organic mulches with a high C:N (carbon:nitrogen) ratio reduce the amount of available nitrogen because of microbial activity and therefore may require an additional nitrogen source for good plant growth.

Because hand-weeding is so time-consuming, most commercial container nurseries utilize chemical weed control. This can be challenging since relatively few herbicides are labeled for use in ornamental production, and almost all of them are selective. Therefore, accurate weed identification is critical. Scouting should be performed at least several times per year.

The most common type of herbicide formulation utilized by container nurseries is granular pre-emergent. Because these chemicals are selective [controlling either grasses, broadleaf weeds, or sedges (*Cyperus* spp.)], they are often applied in combinations. According to Mathers 2002), the most common pre-emergent herbicides used in container production are oryzalin, prodiamine, pendimethalin, trifluralin, oryzalin + oxyfluorfen, isoxaben + trifluralin, pendimethalin + oxyfluorfen, and oxadiazon + prodiamine. Most of these are dinitroanaline (DNA) herbicides which may damage or inhibit roots of container stock if applied too soon after potting. Container nursery plants are most vulnerable to herbicide damage at that time, because large macropores can allow surface-applied chemicals to reach the roots (Altland 2002; Mathers et al. 2007). It would be impractical to provide specific herbicide recommendations here, because there are hundreds of woody ornamental species with a wide range of herbicide tolerances.

A chemical barrier is maintained on the container substrate surface by 3–5 applications of pre-emergent herbicide per year. The frequency of these applications exposes nursery workers to potential health risks, while leaching and off-site run-off are additional concerns. Chemical and cultural methods can be combined in order to reduce herbicide use. Several researchers have demonstrated the weed suppression potential of organic mulches treated with herbicides (Ferguson et al. 2008; Mathers et al. 2004; Samtani et al. 2007). The chemicals adsorb to the organic material, in some cases reducing leaching and decreasing the amount of herbicide needed.

Although post-emergent herbicides are usually not used in container production, glyphosate is often applied in container yards around fences, structures and between pots. Occasionally it is spot-applied inside containers, if particularly difficult weeds become established.

Weed management in field nurseries.

Because the root zones of in-ground plants are less constricted than containerized plants, the introduction of weed seeds in field nurseries is somewhat less of a concern. Therefore, use of organic materials (as mulch and/or soil amendments contributing organic matter and nutrients) is more practical in field nurseries than in container nurseries. Proper composting of wheat straw, pine straw, lawn waste should mitigate the risk of weed seeds being introduced to the nursery area. Additional weed management strategies for field nurseries include cultivation, cover crops, soil solarization, organic and synthetic mulching materials, pre- and post-emergence herbicides, and soil fumigation. Weed pressure may also be avoided by transplanting rather than direct-seeding woody ornamental field nursery beds.

Cultivation in field nurseries is often preceded by irrigation to germinate weed seeds. Hoeing or hand-weeding are also generally necessary because cultivation does not remove 100 % of weeds. Mulching can effectively reduce weed seed germination by shading the soil surface. Optimal mulch depth ranges from 5 to 15 cm, depending on the texture of the material. Cover crops may be used to achieve various weed management goals (see Wilen and Elmore 2009). They may be mowed and left in place as a living mulch or killed with herbicide, frost or senescence for a non-living mulch. Cover crops also provide benefits such as erosion control, soil improvement, habitat for beneficial organisms, and improved working conditions in the field. Recommendations for appropriate cover crop species vary depending on location and specific management goals. Care must be taken to avoid excessive competition of cover crops, especially during the establishment of nursery crop seedlings or transplants.

Because hand-weeding requires so much labor, chemical weed control methods are used almost ubiquitously in field nurseries. A number of selective herbicides (both pre- and post-emergence) are labeled for use in field-grown woody ornamentals. The selectivity of these can be based on physiological or morphological differences between crop plants and target weeds (South and Carey 2005). When nonselective herbicides are used, they must be applied in such a way as to avoid contact with seedlings (using spray shields). Although granular formulations pose the least risk of injury to crop plants, liquid formulations are most commonly used in field nurseries because they are relatively inexpensive. Timing of herbicide application is of critical importance, both to avoid damaging the ornamentals and to ensure successful weed control. For additional information, see Altland et al. (2003) and South and Carey (2005).

Irrigation

Plant physiologists sometimes joke that plants can live without water…, but not for very long! Because water is required for nearly all plant processes, it is important to recognize the need for application of water (irrigation) when it is not available in supplies sufficient for normal plant growth. Photosynthesis, respiration, transpiration and translocation are all plant processes for which water is critical. The properties of water make it the ideal solvent both for movement of solutes within the plant and for uptake of nutrients by the roots and transport of mineral nutrients into and throughout the vascular system (translocation). Transpiration is described as the loss of water in vapor form, mostly via evaporation through the stoma. Transpiration takes advantage of the heat of vaporization of water—for one gram of water to change from liquid form to vapor form requires a loss of 540 calories—resulting in cooling of the plant. Horticulturists capitalize on this phenomenon by using mist irrigation to cool plants growing in excessively warm conditions. Also, as noted in section 3, misting of plants can assist with propagation of leafy cuttings.Another application of the heat of vaporization is use of fan and pad systems for cooling greenhouses.

In growing woody ornamentals, as with most other horticultural plants, an adequate supply of contamination-free, high quality water is necessary. If water is inadequate, incipient wilting may occur. Plants experiencing incipient wilting will recover turgidity following re-watering. In extremes of water deficiency, the permanent wilting point may be reached, about −15 bars (1.5 Mpa), and re-watered plants will not recover, will remain wilted and die. There are many methods of applying water (irrigation) which are described in detail in a number of sources (see Preece and Read 2005). Anti-transpirants are compounds sprayed on the foliage—such as wax emulsions or other compounds designed to block or close the stoma—thus reducing the amount of water lost by transpiration. Research has been inconclusive regarding the efficacy of such treatments.

Water's unique properties are also brought to bear when frost danger is imminent. By application of sprinkler irrigation as the ambient temperature near the plants approaches 0°C, the heat of fusion effectively keeps the plant from freezing. This is because for water to change to the solid form (ice), 80 calories per gram of water must be given up by the water before it can become solid. This is known as the heat of fusion and this property of water is often taken advantage of during cold temperature inversion events, usually in the spring of the year.

Fertilization

Fertilizer practices for woody ornamentals vary considerably with species, cultivar, stage of growth, desired growth rate in the landscape, availability of nutrients in the soil or growing medium and many other factors. Because there are so many variables, including the huge number of species and cultivars of woody ornamentals, no attempt will be made to make specific recommendations for kinds and amounts of fertilizers to apply. For meeting the nutritional needs of specific woody ornamentals, it is prudent to solicit advice from local horticulture professionals such as university extension advisors and educators, local producers and experts from garden centers, arboreta and botanical gardens.

All plants require nutrients in order to grow in a healthy manner and for the horticulturist to achieve this desired growth, attention must be given to meeting the needs of the species or cultivar being grown. Plant physiologists have determined that the following nutrients are required for normal plant growth: carbon (C), hydrogen (H), oxygen (O), nitrogen (N), phosphorous (P), potassium (K), calcium (Ca), magnesium (Mg), sulfur (S), iron (Fe), manganese (Mn), zinc (Zn), copper

(Cu), molybdenum (Mo), boron (B), chlorine (Cl) and nickel (Ni). The first three (C, H and O) are provided by the atmosphere and water, while most of the others listed are commonly found in most field and garden soils. The latter two (Cl and Ni) are questioned by some authorities as to the universality of their need by higher plants, along with silicon (Si), which is found in the structure of a few plants such as horsetails (*Equisetum* spp).

Nitrogen, phosphorous and potassium are required in the greatest amounts and are the primary components of most commonly available fertilizers. N, P, K, Ca, Mg and S are referred to as macronutrients, while Fe, Mn, Zn, Cu, Mo, B, Cl and Ni are considered to be micronutrients because they are needed by the plant in very small amounts. Nutrient availability is influenced by soil reaction (pH) and application of amendments may be necessary to adjust the soil pH to a desired level. Most woody ornamentals grow best at a pH of 6.2–7.0. A pH of 7.0 is considered neutral, while pH levels below 7.0 are considered acidic and those above 7.0 are termed alkaline. If soils are too acid, application of lime, usually ground limestone, will help adjust the soil reaction to be less acidic. Reducing the pH of a high pH soil can be accomplished by adding sulfur or sulfur-containing compounds. However, this is generally not practical for field soils because of cost.

There are a number of woody species that are obligate for more acidic soils (4.5–5.5), including members of the *Ericaceae* such as rhododendrons and azaleas (*Rhododendron* spp), *Kalmia*, heather (*Calluna*) and heaths (*Erica* spp). Choosing an acid site for field production of such species or use of a container medium that is acidic is recommended. Addition of acidic peats, e.g. peats derived from *Sphagnum* bogs, to a soil mix is one method to obtain an appropriate pH for use as a container medium.

Soil sampling and testing is generally recommended. It can not only determine soil reaction, but also the results of a soil test will provide information regarding sufficiency of nutrients needed by the plants that will be produced. This will be instructive for the grower—the soil testing service often makes recommendations—but horticulture professionals can be consulted for advice on what fertilizers to apply and in what amounts. Commercial fertilizers may be used to supply the plant's needs or various organic materials such as manures, composts and other organic sources may be employed for ensuring healthy plant growth.

Pest Control

As noted in 5 for fertilizers, specific recommendations will not be made here regarding control or management of specific woody ornamental pests because of the vast number of species and cultivars in the woody ornamental category. Use of information from reliable sources will provide potentially up-to-date advice for controlling diseases and the numerous pests that may attack specific woody ornamental plants. Also, because problems that fall into the category of "non-infectious diseases" are generally physiological problems, this section will only discuss infectious pathogens, insects and other arthropods, and other specific pests that may cause damage to woody ornamental plants. In many cases, diseases to be discussed in this section may be pathogens that attack other types of plants, such as fruit crops and herbaceous ornamentals.

Infectious diseases are often caused by specific organisms that are referred to as plant pathogens. They typically include fungi (singular fungus) and fungal-like organisms, bacteria, viruses and viroids, and mollicutes. Attacks by nematodes may also be considered as pests. A number of organisms fall into the category of water molds. These include members of the *Pythium* and *Phytophthora* genera. The latter is perhaps most infamously known for being the causal agent of Late Blight in potatoes, a disease that resulted in the Irish potato famine that took place in the 19th Century. However, many of the water molds (*Phytophthora* included) cause seedling damage referred to as damping-off. Symptoms may be the classic collapse of cells near the soil line following germination leading to the seedling falling over and dying (post-emergence damping-off). Pre-emergence damping-off is evidenced by the pathogen attacking the seed after emergence of the radicle, but before the seedling grows above the soil line. The water molds flourish and are most pathogenic under conditions of high moisture (hence the name water mold), but *Rhizoctonia* is a damping-off pathogen that functions under soil conditions conducive to optimum growth of higher plants, yet is a frequent causal agent of damping-off (Agrios 2005).

The woody ornamentals grower's best weapon for combating many diseases, including damping-off, is hygiene. Use of clean tools, healthy, pathogen–free plants, a growing medium free of pathogens, and employing clean management practices are examples of hygiene. To achieve a pathogen-free growing or propagation medium, the medium must be composed of materials that are already free of pathogens or the medium constituents must be treated to eliminate the pathogens. A common practice employed to produce a pathogen free medium is to treat the soil and other medium ingredients with high heat for a prescribed length of time at a specific temperature. The usual practice is to treat the soil mass with steam heat at 60–70°C for 30 min. This practice will also kill insects, insect eggs, nematodes and most weed seeds. Various chemical fumigants can also be employed to obtain the same results.

Additional problems that can be caused by R*hizoctonia*, *Phytophthora* and *Armillaria* pathogens causing root and stem rots. Fungal pathogens known as rusts are also serious problems with certain woody ornamentals. One example is caused by *Gymnosporangium* spp, the pathogen that causes Cedar-Apple Rust. Cedar-apple rust is an example of a disease that attacks two different host plants and is obligate for both host plant species in order to complete its life cycle. The practical horticulturist employs this knowledge to combat this pathogen by eliminating the alternate host species. For example, if the desired species is ornamental crabapples (*Malus* spp.) or hawhorn (*Crataegus* spp), removal of all cedar trees and shrubs within the nearby landscape will help to eliminate the problem.

Bacteria are microscopic, single-celled organisms that contain no chlorophyll. Examples of bacterial pathogens important to growers of woody ornamentals include crown gall, caused by *Agrobacterium tumefaciens,* a pathogen that attacks a large number of woody plant species. It is a soil borne pathogen that is almost ubiquitous, but usually requires a wound to facilitate entry and pathogenicity, so preventing injuries to the plant, especially to the trunk near the soil line can be helpful in combating this pathogen. Fire Blight is another bacterial pathogen that is a serious problem in many members of the Rosaceae such as ornamental crabapples, *Cotoneaster*, hawthorns(*Crataegus* spp) and mountain ash (*Sorbus* spp). It is caused by the bacterium *Erwinia amylovora*. It is difficult to control, but as with fungal pathogens, prevention through hygiene and good management practices will aid in its control.

Viruses and viroids often exhibit symptoms of variegation in the foliage, making it difficult to determine whether the plant is diseased or genetically variegated. Typically, if the variegation is uniform within and among plants, it is probably genetic. However, if uniformity is lacking, the plant may be succumbing to a virus. Other symptoms of virus and viroid diseases are distortion of the plant's growth, stunting, strap-like leaves, chlorotic streaks, necrotic leaf spots and leaf rolling, among numerous other symptoms.

One method to overcome or escape the ravages of bacterial and virus pathogens is to combine heat treatment with meristem culture. The stock plant is grown under near-lethal high temperatures and the resulting rapid growth enables the meristem to elongate faster than invasion by the pathogenic organism occurs. The meristem is then excised and cultured in vitro (see again 3.5). The plantlet that develops from the meristem is potentially free of the disease organism and subsequently can be used as a stock plant. Such stock plants must be tested to ensure that they are indeed free of the pathogen in question.

Nematodes are microscopic organisms that are sometimes referred to as eelworms and are typically soil-borne. Plant diseases caused by nematodes result from their feeding on the roots of a range of plant species and produce symptoms such as root knots, cysts or galls, abnormal root branching or other abnormal growth. Such damage can lead to above ground symptoms, including death of the plant being attacked by nematodes. As with bacteria and fungi, sanitation and use of clean tools, along with use of a growing medium that has been heat treated or otherwise ensured to be pathogen free. Movement of soil from a nematode-infested field by equipment used in that field to a clean field can lead to inoculating the field with the undesirable organism. Again, fumigation or other chemical approaches may be necessary.

Insects are members of the Phylum Arthropoda, class Insecta, typified structurally by three body segments (head, thorax and abdomen), three pairs of segmented legs, two antennae and may have two pairs of wings in the adult stage (Cadré and Resh 2012; Resh and Cadré 2009). They have three different means of attacking plants, having chewing mouthparts, piercing-sucking mouthparts or rasping sucking mouthparts. Larval forms of insects that have chewing mouthparts are commonly worm-like in appearance, soil borne and may consume plant roots, e.g., the larvae of the Japanese beetle (*Popillia japonica*) or the June beetle. Other larval forms may consume leaves of trees and shrubs, while the adults may also consume leaves, as is the case with Japanese beetles.

Piercing sucking insects such as aphids, penetrate leafy or other soft plant tissue and the damage that they inflict is by removal of plant sap. In the case of aphids, the sap is consumed as a result of the plant's sap pressure, while other insects of this group can develop a strong pumping mechanism to consume the plant sap. A concomitant problem resulting from piercing of the plant tissue is that transfer of virus particles may occur. If the insect has been attacking a virus-infested plant and then flies to and penetrates a healthy plant, it transfers the virus pathogen to the healthy plant. Many virus and phytoplasma pathogens are transmitted in this way.

Insects with rasping sucking mouthparts essentially cause symptoms that appear as a though the plant surfaces were scraped away. Thrips are an example of rasping-sucking insects. They are not as serious or common a problem for woody plants as insects with chewing or piercing-sucking mouthparts.

Spider mites are members of the Arthropoda, class Arachidna, but differ from insects in that they have only two body parts and have four pairs of legs. The two body parts are an abdomen and a cephalothorax that has characteristics of both a head and a thorax. Although there are many spider mite species, many of them nearly microscopic, the two-spotted spider mite (*Tetranychus urticae*) is an example that frequently infests and causes damage to woody ornamentals. They are relatively small, about 0.5 mm in length and lay eggs that hatch and the offspring begin feeding immediately. They suck fluids from plant leaves, leaving a stippling appearance that sometimes is mistaken for nutrient deficiencies or virus diseases. Spider mites multiply at a very rapid rate, especially at high temperatures. As a result, a small infestation can quickly become a serious problem under warm conditions.

In some parts of the world, snails and slugs can be serious problems. They are of particular significance in sub-tropical and tropical regions. These organisms are members of the phylum Mollusca and are characterized by soft bodies with a head, eyes and sensory tentacles and a foot for locomotion. Some of these organisms are protected by a hard shell (snails), while those without a shell are referred to as slugs. Within the mouth is a tooth-covered radula, which enables these organisms to ingest the soft parts of woody plants, such as leaves and new softwood growth. Generally, slugs and snails are controlled by cultural practices or by chemical means.

Molluscicides are chemicals specifically developed to poison slugs and snails, while a pan of beer at ground level has been demonstrated to attract these creatures and they drown when they fall into the beer. Since they hide in moist dark places, eliminating hiding places such as clods of soil, pieces of wood or other plant material that is on the ground can help in their control.

Vertebrate pests are sometimes a serious problem for both commercial growers of woody ornamentals and for home gardeners. In North America, various types of wildlife such as deer, moose and elk can be a problem, since they are animals that browse on woody species. Other problem species include rabbits, mice, voles, beavers and many other animals, depending upon in what part of the world the woody ornamental plants are being grown. Although many of these problem animals are frightened easily by signs of human habitation, others are not. Exclusion by fencing seems to be the most effective approach for minimizing the negative effects of many vertebrate pests.

Pruning

Pruning has long been recognized as an important component of horticultural practice (Bailey 1919; Bradley 1733). Indeed, pruning may be used to influence a wide variety of plant functions and characteristics. It can be used to alter plants' growth rate and vigor, improve their structure or form, alter their canopy architecture, remove dead branches or those damaged by pests or pathogens, and to enhance flowering and/or fruiting. However, most ornamental plants do not require regular yearly pruning to the extent that many fruit crops do. Labor costs associated with pruning woody ornamentals in the landscape can be minimized by selecting appropriate and high quality plant material and by placing such materials far enough from existing structures and plantings.

Pruning Shoots

When and how to prune depend on species' growth habit (excurrent or decurrent) and flowering habit (whether the species flowers on current season's growth, previous season's wood, or spurs on older wood) (Brown et al. 2004; Gilman 2002). Other factors may also influence decisions on when and how to prune, such as management needs and objectives. However, the following are some general guidelines.

If a branch is to be entirely removed, prune close to the stem, being sure to leave the branch collar intact. If a branch is to be partially removed, prune just beyond a bud or lateral branch. Otherwise, the stem will die back to the bud or branch, leaving an unsightly stub and larger area for potential disease development. It is important to begin pruning ornamental trees early, particularly for species with decurrent growth habit. Multiple prunings per year may even be necessary during the first several years. In most cases it is unnecessary to prune at transplanting. Thereafter, light prunings removing less than 10% of foliage may generally be performed at any time of the year; however, most species respond the best if moderate or heavy prunings are performed during winter dormancy (Gilman 2002). See Preece and Read (2005) for additional general pruning information. For specific pruning needs, authoritative recommendations for "nearly 450 genera of trees, shrubs and conifers" may be found in Brown et al. (2004).

Espalier and topiary art are specialized techniques that take advantage of the horticulturist's knowledge of how plants respond to pruning.Espalier is a French system employed to train trees or shrubs within one plane, usually against a wall (a south –facing wall in the Northern Hemisphere). Although there can be several designs employed, one example is shown in Fig. 17.7. Topiary is an art form in which plants are pruned into shapes such as the scroll work in the landscape shown in Fig. 17.8. Plants can also be pruned into shapes of animals or other objects.

Fig. 17.7 *Left*, ornamental pear (*Pyrus* spp) trained against the wall in a classic espalier pruning system; *right*, close-up of an espalier-trained flowering quince (*Chaenomeles* spp)

Fig. 17.8 Scroll-like topiary training of boxwood (*Buxus* spp) in a formal garden in England

Historically, coppicing and pollarding are also specialized pruning techniques that were employed for timber and charcoal production, but now have uses for energy plantations and by horticulturists wishing to obtain young stems and rejuvenate plants that can tolerate such severe pruning. Coppicing involves cutting a tree or shrub back to near ground level to encourage multiple stem regrowth, while pollarding is used to severely restrict the growth of larger trees such as *Platanus*. Because many plants can be more successfully propagated from juvenile material, such rejuvenation can provide propagation material that responds well to rooting or as explants for micropropagation (Dickerson 2013; Read and Bavougian 2012). Generally severe pruning as used in pollarding is restricted to a limited number of species and best utilized with relatively young trees. *Catalpa* is a species that is often pollarded to encourage larger, more ornamental leaves, while hybrid willow (*Salix* spp), poplars (*Populus* spp) and other species employed for bioenergy plantations are coppiced repeatedly every few years. Coppicing is also practiced for obtaining stems that are ornamental, sometimes referred to as "woody florals" (Greer and Dole 2009). Examples of the latter include several *Cornus* species, *Taxus* and species of *Salix* that have ornamental value, such as *Salix* 'Golden Spiral'.

Pruning Roots

In general, root pruning reduces a plant's growth, producing a dwarfing effect. It is rarely used with woody ornamentals in this capacity, although trees and shrubs are often root-pruned in preparation for transplanting. This practice encourages fibrous root growth within the root ball area (Brown et al. 2004; Gilman 2002). Root pruning may also be employed if transplants' roots are excessively long, and in order to prevent circling roots from girdling the plant. See Preece and Read (2005) for additional information.

Harvesting

Digging Field-Grown Plants and Transplanting

Digging field-grown woody ornamentals requires care in the digging process and in subsequent transport and transplanting if required. It is easy to damage roots by rough or careless handling, thus causing a reduction to their capacity to absorb water and nutrients. Prior to digging the plants, the digging equipment must be inspected and adjusted to an optimum operating condition, which will help avoid damage to the plants. As plants are dug, it is imperative to avoid leaving them for a lengthy period in the field exposed to sun and wind. Handling them as quickly and gently as possible should result in a high percent success in the balling and burlapping process or when transplanting into containers. Mizell et al. (2012) describe a modified form of field production, where plants are grown in-ground, in bags made of porous fabric. In-ground bags have a solid plastic bottom which prevents tap root formation, and plant roots are confined within the bag allowing more roots (80 %+) to be harvested when the crop is dug. Theoretically, this will improve survival and decrease transplant shock. The ball and burlap approach is rapidly being replaced by growing in containers following field digging. Whether bare-root, balled and burlapped or container-grown, they should be stored under cool and humid conditions to prevent drying out and loss of weight because of respiration.

Handling and Transporting Balled and Burlapped and Container-Grown Plants

It is important to keep balled and burlapped and container-grown plants out of the sun and wind when they are delivered to the nursery or sales area. Initially, there is potential for significant water loss, so handling carefully is imperative. Watering frequency may be necessary considerably more often than when the plants were in the production field or structure. Balled and burlapped plants will have been freshly dug from the field and will have lost a portion of their water and nutrient absorbing roots. Container-grown plants generally are grown in protected areas, such as greenhouses, lath houses or high tunnels. In either case, the ability for the plants to absorb water and nutrients, especially if they have sizeable leaf area, will be impaired, necessitating careful monitoring of the water and fertilizer needs of the plants. Attention needs to be paid to practices that minimize water loss during transport, subsequent storage and movement to the sales or display area. It is imperative to keep the plants in prime condition, not only so that they will look appealing and sell readily, but also so that the customer will be likely to experience a high degree of success when transplanting into the landscape.

Marketing

It has often been said, "There is no money to be made in growing plants (including woody ornamentals)—not even in growing great plants—there is just profit in selling them!" The successful producer of woody ornamental plants is no exception to this old saying. Producing high quality plants, along with selection of products that will be in demand by customers are critical components of making a profit in the woody ornamental industry. In addition, attention must be paid to advertising and promotion and to providing the customer with an appropriate level of service. Today's customer is more and more interested in gaining information about their prospective purchases. Therefore, it is important to employ knowledgeable sales people who are good communicators and create a friendly atmosphere, thus enhancing the customer's shopping experience and potential success with their woody ornamental purchases. Once the salesperson understands the needs and desires of the customer, the knowledgeable salesperson can assist the customer to obtain the plants appropriate for their needs. It is always helpful to have a good knowledge of the environmental conditions in which the plants will be grown so that choices can be made that will lead to success in planting the woody ornamentals being purchased. Successful marketers must employ individuals who are educated in plant selection and culture, thus creating a satisfied customer who will tell their friends of the great service and will themselves be repeat customers, coming back again for more product. As is often the case, both professional horticulturists and leisure gardeners will continue to try to grow desirable plants that may not be recommended

Fig. 17.9 A specimen plant of bald cypress (*Taxodium distychum*) in a landscape near Lincoln, Nebraska, a location well beyond the native range for this species. (Photo courtesy of Nebraska Statewide Arboretum, Lincoln, NE, USA)

for a given location, but successes often can be accomplished (Fig. 17.9). This attitude ensures continuing development in the ornamental plant industry.

References

Agrios BN (2005) Plant pathology, 5th edn. Elsevier Academic Press, San Diego

Altland J, Gilliam C, Wehtje G (2003) Weed control in field nurseries. Hort Tech 13:9–14

Altland J (2002) Herbicide timing for container weed control. Digger 46:46–48

Anon (2007) US Department of Agriculture (2007) Nursery crops, 2006 summary. usda01.library.cornell.edu/usda/current/NursProd/NursProd-09-26-2007.pdf. Accessed 17 Sept 2012.

Bailey LH (1919) The pruning-manual, 18th edn. MacMillan, New York

Behe BK, Dennis JH, Hall CR, Hodges AW, Brumfield RG (2008) Regional marketing practices in US nursery production. Hort Sci 43:2070–2075

Bradley R (1733) The compleat fruit and flower gardener, 3rd edn. London

Brown GE, Kirkham T, Lancaster R (2004) The pruning of trees, shrubs and conifers, 2nd edn. Timber Press, Portland

Blythe EK, Sibley JL (2012) Winter stem cutting propagation of Heller's Japanese holly with and without use of a conventional auxin treatment. Hort Tech 22:771–773

Cardé RT, Resh VH (2012) A world of insects. Harvard University, Cambridge

Case L, Mathers H, Senesac A (2005) A review of weed control practices in container nurseries. Hort Tech 15:535–545

Dickerson T (2013) Garden practice: coppicing and pollarding. The Garden 138:51–54
Ferguson J, Rathinasabapathi B, Warren C (2008) Southern red cedar and southern magnolia wood chip mulches for weed suppression in containerized woody ornamentals. Hort Tech 18:266–270
Fishel DW, Zaczek JJ, Preece JE (2003) Positional influence on rooting of shoots forced from the main bole in swamp white oak and northern red oak. Cana J Forest Res 33:705–711
Gilman EF (2001) Effect of nursery production method, irrigation, and inoculation with mycorrhizae-forming fungi on establishment of *Quercus virginiana*. J Arboriculture 27:30–39
Gilman EF (2002) An illustrated guide to pruning. 2nd edn. Delmar Thomson Learning, Albany
Greer L, Dole JM (2009) Woody cut stems for growers and florists. Timber Press, Portland
Haberlandt G (1902) Kulturversuche mit isolierten Pflanzenzellen. Sber Akad Wiss Wien 111:69–92
Hall CR, Hodges AW, Haydu JJ (2005) Economic impacts of the green industry in the United States. http://hbin.tamu.edu/greenimpact.html. Accessed 17 Sept 2012
Harris JR, Gilman EF (1991) Production method affects growth and root regeneration of leyland cypress, laurel oak, and slash pine. J Arboriculture 17:64–69
Harris JR, Gilman EF (1993) Production method affects growth and post-transplant establishment of 'East Palatka' holly. J Am Soc Hort Sci 118:194–200
Hartmann HT, Kester DE, Davies FT Jr, Geneve RL (2010) Hartmann & Kester's plant propagation: principles and practices, 8th edn. Prentice Hall, Upper Saddle River, 915 p
Henry PH, Preece JE (1997) Production and rooting of shoots generated from dormant stem sections of maple species. Hort Sci 32:1274–1275
Ingram DL, Henley RW, Yeager TH (1993) Growth media for container grown ornamental plants. University of Florida Cooperative Extension Service Bulletin 241
King AR, Arnold MA, Welsh DF, Watson WT (2012) Developmental stage and growth regulator concentration differentially affect vegetative propagation of select clones of *Taxodium* Rich. Hort Sci 47:238–248
Knapp JE, Kausch AP, Auer C, Brand MH (2001) Transformation of Rhododendron through microprojectile bombardment. Plant Cell Rep 20:749–754
LeBude AV, Bilderback TE (2007) Field production of nursery stock: field preparation, planting, and planting density. In: Bilderback TE, LeBude AV, Neal J, Safely C, Adkins C, Feitshans T. Best management practices for field production of nursery stock. North Carolina Nursery and Landscape Association (NCNLA), Raleigh
Mathers HM (2002) Choosing the right herbicide, Part I. Buckeye, pp 14–16
Mathers HM, Pope J, Case LT (2004) Evaluation of micro-encapsulated herbicides alone and applied to bark mulches. Proc Northeast Weed Sci Soc 58:47
Mathers HM, Lowe SB, Scagel C, Struve DK, Case LT (2007) Abiotic factors influencing root growth of woody nursery plants in containers. Hort Technol 17:151–162
Mizell R, Knox G, Knight P, Gilliam C, Arthers S, Austin R, Baldwin H et al (2012) Woody ornamental and landscape plant production and pest management innovation strategic plan. USDA Southern Region IPM Center, North Carolina State University, Raleigh
Murashige T (1974) Plant propagation through tissue cultures. Ann Rev Plant Physiol 25:135–166
Murashige T, Skoog F (1962) A revised medium for rapid growth and bioassays with tobacco tissue cultures. Physiol Plant 15:472–497
Preece JE, Huetteman CA, Ashby WC, Roth PL (1991) Micro- and cutting propagation of silver maple. I. Results with adult and juvenile propagules. J Am Soc Hort Sci 116:149–155
Preece JE, Read PE (2005) The biology of horticulture: an introductory textbook, 2nd edn. Wiley, Hoboken
Rinehart TA, Edwards NC, Spiers JM (2013) *Vernicia fordii* 'Anna Bella', a new ornamental tung tree. Hort Sci 48:123–125
Read PE (1992) Environmental and hormonal effects in micropropagation. In: Kurata K, Kozai T (eds) Transplant production systems. Kluwer, Netherlands, pp 231–246

Read PE (1990) Environmental effect in micropropagation. In: Ammirato PV, Evans DA, Sharp WR, Bajaj YPS (eds) Handbook of plant cell culture. Ornamental species, vol 5. McGraw-Hill, New York, pp 95–125

Read PE, Bavougian CD (2012) In vitro rejuvenation of woody species. In: Lambardi M, Ozudogru AE, Jain M (eds) Protocols for micropropagation of selected economically-important horticulture plants. Springer-Humana, Louisville, pp 383–396

Read PE, Fellman CD, Economou AS, Yang Q-C (1986) Programming stock plants for propagation success. Proc Int Propagators Soc 35:84–91

Read PE, Preece JE (2013) Cloning: plants—micro-propagation/tissue culture. Encyclopedia of agriculture and food systems. Elsevier Academic Press, San Diego

Resh VH, Cardé RT (2009) The encyclopedia of insects, 2nd edn. Elsevier Academic Press, San Diego

Rideout W (2012) Steam treating for weed control. American Nurseryman Magazine, April, 2012, pp 12–14

Samtani JB, Kling GH, Mathers HM, Case L (2007) Rice hulls, leaf-waste pellets, and pine bark as herbicide carriers for container-grown woody ornamentals. Hort Tech 17:289–295

Skoog F, Miller CO (1957) Chemical regulation of growth and organ formation in plant tissues cultured in vitro. Symp Soc Exp Biol 11:118–131

South DB, Carey WA (2005) Weed control in bareroot hardwood nurseries. In: USDA forest service proceedings RMRS-P-35, pp 34–48

Soyler D, Khawar KM (2006) Effects of prechilling, scarification, incubation temperature, photoperiod, KNO_3 and GA_3 treatments on germination of caper (*Capparis ovata* Desf. Var. *Palaestina* Zoh.) seeds. Propag Ornam Plants 6:159–164

Thammina C, He M, Lu L, Cao K, Yu H (2011) In vitro regeneration of triploid plants of *Euonymus alatus* 'Compactus' (burning bush) from endosperm tissues. Hort Sci 46:1141–1147

Thimann KV, Went FW (1934) On the chemical nature of the root-forming hormone. Proc K Ned Akad van Wet 37:456–459

Thomas P (2000) Trees: their natural history. Cambridge University, Cambridge

Whitcomb CE (1984) Plant production in containers. Lacebark Publ, Stillwater

Wilen CA, Elmore CL (2009) Floriculture and ornamental nurseries. University of California Statewide Integrated Pest Management Program. http://www.ipm.ucdavis.edu/PMG/r280700211.html. Accessed 17 Sept 2012

Yang G, Read PE (1989) Effects of BA, GA, IAA and NAA in the forcing solution on bud break and shoot elongation in forced woody stems. In: Proceedings of the Plant Growth Regulator Society of America, pp 150–155

Yang Q-C, Read PE, Fellman CD, Hosier MA (1986) Effect of cytokinin, IBA and rooting regime on Chinese chestnut cultured in vitro. HortScience 21:133–134

Zampini M (2013) Pete Kruger: a woody plant aficionado shares three of his current favorites. Horticulture, January/February, p 24

Chapter 18
Medicinal and Aromatic Plants—Uses and Functions

Maiko Inoue and Lyle E. Craker

Abstract Medicinal and aromatic plants contain the chemical constituents first used by humans as medicines for healing, as flavoring agents for food and drink, and as mental stimulants for mystic interactions with super natural gods. These plant materials continue to play positive roles in human life, as sources of modern pharmaceuticals to treat medical problems, as herbs and spices to tempt the palate, and in a multitude of other applications. Demand and trade for these plant materials initiated globalization that spread new ideas, new diseases, and new settlements along with native resentment. This chapter presents a detailed and thorough review of plant metabolites used in modern and conventional medicine and those providing spices and condiments. The taxonomy, biochemistry and extraction are described. Truly it may be said that medicinal and aromatic plants have changed the world and undoubtedly play an important role in the future as they have in the past.

Keywords Healing · Flavoring · Pharmaceuticals · Herbs · Spices · Metabolites

Introduction: Special Plants

Medicinal and aromatic plants represent a broad group of plant materials valued primarily for their chemical constituents. Since ancient times, these plants have been used to help restore and maintain health, to preserve and flavor foods, to add color and aroma to daily life, and to serve as pathways to spiritual dreams and mystical adventures. While sometimes referred to as medicinal herbs, culinary herbs, potherbs, spices, seasonings, essential oil plants, or other terms to distinguish plant use, collectively these plants have been used for a variety of applications throughout recorded history, enhancing the quality of human life. Currently, an estimated 70,000 plant species are used in traditional medicine (Farnsworth and Soejarto 1991).

The discovery and initial uses of medicinal and aromatic plants by various groups of people are unknown, but possibly the plants were chosen for use as alternative

M. Inoue (✉) · L. E. Craker
Medical Plant Program, University of Massachusetts, Amherst, MA 01003, USA
e-mail: den8mai@yahoo.co.jp

L. E. Craker
e-mail: craker@umass.edu

G. R. Dixon, D. E. Aldous (eds.), *Horticulture: Plants for People and Places, Volume 2,*
DOI 10.1007/978-94-017-8581-5_18, © Springer Science+Business Media Dordrecht 2014

Table 18.1 Terms used with medicinal and aromatic plants. (Source: American Botanical Council (2013))

Herbal term	Description
Herb or culinary herb	Aromatic plant material from temperate regions used in minor quantities to flavor foods and beverages, but has little or no nutritional value
Spice	Aromatic plant material from tropical regions used in minor quantities to flavor foods and beverages, but has little or no nutritional value
Medicinal plant	Various plants used for treatment of disease or other body afflictions
Essential oil plant	Plants from which a volatile oil that can be extracted by distillation, solvents, or expression
Aromatic plant	Plants with an aroma that comes from the presence of an essential oil in the plant tissue
Poisonous plant	Plants containing alkaloids or other substances that can produce toxic effects when introduced into the body
Endangered plant	Plant species of which few remain, are in danger of becoming extinct, and are generally protected by law and trade restrictions

food sources and then recognized for an attractive aroma or enhanced healing when applied to cuts and scrapes. Undoubtedly, these first trials with plants were tentative and perhaps came from watching the plant materials selected and eaten by animals and earlier humanoids. From this beginning, plant trade among groups would lead to trade among near-by countries and eventually to global trade of medicinal and aromatic plants. The activities of early humans in medicinal and aromatic plants continue today with plant exploration and botanical sourcing for new medicines, new aromas, and new uses. Global trade in medicinal and aromatic plants has become a US$ 30 to US$ 60 billion business (Anon 2008).

Although the populations of many countries remain dependent on traditional uses of medicinal and aromatic plants for health care, these plants are also recognized as major contributors to modern life, pharmaceuticals for human diseases, seasonings to flavor foods and beverages, and beauty through cosmetic goods and herbal gardens. Laboratory chemists work to modify plant constituents to improve bioactivity. In contemporary societies, individualsfrequently seek to develop and protect health through self-treatment with herbal materials as foods, food additives, and dietary supplements meant to improve the normal diet and reduce medical costs. Although known by several names, medicinal and aromatic plants continue to make useful and valuable contributions to world societies (Table 18.1).

Discovery: Historical Progress

To fully understand and appreciate medicinal and aromatic plants and their uses and functions, some history on their interactions with people is necessary (Table 18.2). The anthropological residues provide clues in the search for plant materials to be used as pharmaceuticals. The discovery of medicinal and aromatic plants that could produce delightful aromas, flavor foods, enhance health, and cure illnesses was an

Table 18.2 Significant events in the history of medicinal plants and medicine. (Source: Mamedov and Craker (2012))

Timeline	Significant events	Related developments
~200,000–5,000BC	Humans with ailments	Testing of plants for medicinal activity; development of traditions and myths about medicines
~5,000–3,000 BC	Ötzi, the Iceman, is killed and frozen	Evidence that early humans used plant medicines
~3,000–500 BC	Formal training initiated for useful plant medicines and applications; Ayurvedic medicine founded in India; recognition of willow bark and *Cannabis* as medicines	Initiation of spice trade; in Asia and Middle East; development of written records on medicine; Imhotep and Papyrus Ebers in Egypt, Pen Ts'ao Shen Nung in China; Rigveda in India
~500–300 BC	*Hippocratic corpus* written by Hippocrates; *De Causis Plantarum* and *Historia Plantarum* written by Theophrastis	Formalized medicine practices in diagnosis and treatments; lists of medicinal plants and applications; medical school established in Cyrene
~300–100 BC	Formal medical education is increased; trade in medicinal herbs and spices between east and west	Sharing of medicinal recipes and medicinal plants; medical school formed in Alexandria; study of anatomy
~100 BC–100 AC	Celsus wrote *De Medicina*; Dioscorides wrote *De Materia Medica*; opium recognized as medicinal drug	Development of materia medica and pharmacopoeias
~100–200 AC	Galen made medicinal preparations	Medicinal doses calibrated; initiation of pharmacies
~200–500 AC	Mayans practiced medicine; Greek culture encouraged learning	Continued study of medicine and medicinal practices
~500–1,000 AC	Avicenna did clinical trials on medicines and developed medical books; Albucasis wrote medical encyclopedia; Hildegard of Bingen wrote about medicinal plants	Medicinal plants tested for various health problems; medical reference books available for study; Avicenna writes Book of Healing and Canon of Medicine
~1000–1500 AC	Marco Polo traveled to China; Gutenberg built printing press; medicinal books were written in European languages with descriptions of plants; Ibn al-Baitar wrote books on botany and pharmacy, pioneered veterinary medicine	Travelers and traders told of spices and medicines in Asia; need for medicines and spice trade stimulated exploration with ocean-going ships; ship-building school by Henry the Navigator in Lisbon; voyage of Columbus to America for spices
~1500–1900 AC	Vasco da Gama reached India; Withering discovered heart medicine in foxglove; Seishu Hanaoka used atropine and scopolamine	Medicinal plants and spices start being widely used as medicine throughout the world
~1,900–2,012 AC	Traditional medicine being used by 70–80% of world population; many chemically synthesized medicines based on plant extracts; revived interest in medicinal plants	Development of vaccines and antibiotics discouraged use of medicinal plants in advanced countries; recognized medicinal benefits of plants and plant extracts encourage expanded research in plants for medicine

important development during the early history of humans. Such information would be incorporated into traditional rituals, stories, and myths that could transmit the learned uses of these plants to future generations. Medicine men and women with specialized knowledge and skills in the use of plant materials for healing would become significant and honored members of societies.

With the development of mnemonic symbols and subsequent writing skills, instructions on medicinal plants and their use was recorded in several locations, including Egypt, India, China, the Middle East, and the Americas. The oldest written document on the use of medicinal plants in the preparation of a pharmaceutical is an approximately 5,000 year old Sumerian clay tablet that refers to 250 plants. Early written instructions, which were frequently a mixture of plant material and mysticism, provide historical detail on plants and plant preparations used in the treatment of medical issues. The Ebers papyrus, dated 1,550 BC, but thought to be copied from even older written sources, is the most complete of the medical papyri. A recommended treatment for urinary problems in the Ebers papyrus includes the plants, carob (*Ceratonia siliqua*), celery (*Apium graveolens*), date (*Phoenix dactylifera*), fig (*Ficus carica*), grape (*Vitis vinifera*), gum arabic (*Senegalia senegal* and *Vachellia seyal*), and wheat (*Triticum* spp.) (Carpenter et al. 1998).

In India, the traditional medicine system, Ayurveda, is based on the *Charaka Samhita* (100 BC–100 AC), the *Sushruta Samhita* (300–400 AC), and the Bower manuscript (400–600 AC). Unani, a second traditional medicine system of India, is a Greco-Arab medicinal practice imported to India from the writings of Avicenna, a Persian physician that wrote a series of treatises on medicine. Ayurvedic medicine contains over 600 plant species for use in treatments and Unani medicine approximately 40 plant species, including the Damson tree (*Terminalia* spp.), Malacca tree (*Emblica officinalis*), Ajowan (*Ptychotis ajowan*), and Senna (*Cassia angustifolia*). Both the Ayurvedic and the Unani systems of medicine had spread to other parts of Asia by 1,300 AC.

In China, the traditional medicine system originated primarily from a 2,000 year-old Chinese medical text, *Huangdi Neijing*, containing a series of chapters, which along with earlier writings and later revisions became the foundation for Traditional Chinese Medicine, a health system based on humoral balances. In the Chinese medicine system, good health requires equity among the natural forces associated with diet, lifestyle, emotions, and environment. Thus, good health requires a balance among natural forces, such as Yin-Yang (two opposites form the whole), Qi (life energy), and the Five Elements (interactions and relationships). Medicinal and aromatic plants, along with animal and other natural substances are used to bring the body into balance. The first recognized Chinese herbalist is Shennong, a legendary ruler of China of 5,000 years ago who taught the people the use of medicinal herbs (Anon 2013).

Early European medicine markets made use of spices that were undoubtedly part of the trade goods coming from Asia via routes that followed the Silk Road and via crossings at the Isthmus of Suez located between the Red Sea and the Mediterranean Sea. Although limited western trade with India and China occurred as early as 1,600

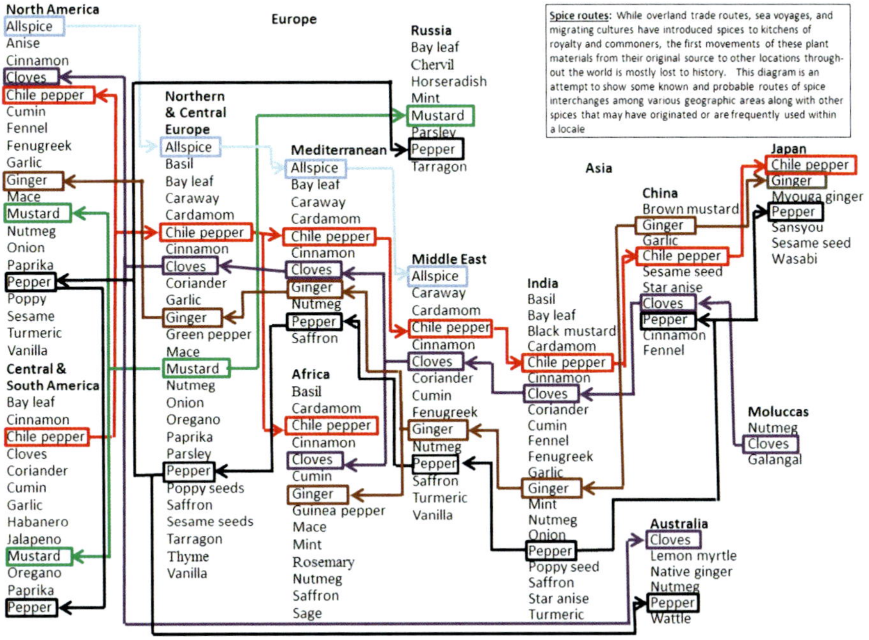

Fig. 18.1 Some early trade routes of spices along with spices frequently used in various regions. (Sources: Czarra (2009); Food University (2013); Keay (2007); Kew Gardens (2013); Spices & Herbs (2013))

BCE, the conquests of Alexander the Great into Northern India by 325 BCE probably introduced other European countries to spices as flavorings and medicines.

A book by Marco Polo, *The Travels of Marco Polo* (Polo and Rossabi 2012), which detailed the riches in Asia he observed during travel (1271–1291) with his father and uncle, became popular reading in Europe, bringing an awareness of other civilizations and spices. Large demands for Asian spices occurred during the fourteenth to seventeenth centuries as possible preventatives or cures for the plagues that swept through Europe during this time period. The limited quantities of spices available via the overland Asian-European trade routes, especially after the Ottoman victory at Constantinople in 1453, promoted the search for a sea route to Asia. An ocean-going ship building and learning facility built in southern Portugal by Prince Henry the Navigator in 1418, trained sailors in ship building, navigation, and other trades necessary for an ocean voyage into unknown waters.

Sailing East for Portugal, as agreed upon in the Treaty of Tordesillas in 1494, Vasco da Gamma reached Calicut, India, opening a new trade route to Asian spices in 1498. Sailing West for Spain, Columbus reached America, a continent unknown to Europe, in 1492, but instead of the pepper (*Piper nigrum*) of Asia, he returned to Europe with samples of allspice (*Pimenta dioica*), and red pepper (*Capsicum* spp.) (Fig. 18.1).

Constituency: Chemical Extracts

Medicinal and aromatic plants are valued for their chemical constituents known as secondary metabolites. These secondary metabolites, which include essential oils, are organic compounds with no known necessity for the normal growth, development, or reproductive activity of an organism, but are bioactive, interacting with tissues and provide the aromas, tastes, and medicinal activity associated with such plants. For example, in sensing the aroma of aromatic plants, the volatile chemical constituents from the plants interact with sensory neurons within the olfactory epithelium system to produce nerve impulses that communicate the aroma signal along the olfactory nerve to the brain. Anticancer activity exhibited by the American Mayapple (*Podophyllum peltatum*), for example, contains constituents that interfere with DNA replication, thus stopping the rapid cell division associated with cancer (Chaurasia et al. 2012).

Positive effects from secondary metabolites are associated with pharmacological action beneficial to the consuming organism, while negative effects are associated with toxicity that harms those consuming the plant tissue. As with other drugs, the bioactivity exerted by secondary metabolites is directly related to absorption, distribution, metabolism, and excretion (ADME) of the metabolite. Beneficial effects from a secondary metabolite may, however, become toxic if too much is consumed.

While secondary metabolites may not be essential for bodily functions, they can enhance life through improvement in health, spirit, and aesthetics. The exact reason why some plants contain essential oils and other secondary metabolites is unknown, but the presumptions are that these chemical constituents could attract or repel insects and other life forms that benefit or stress the plant, respectively, or could act as storage centers for metabolic waste products. Secondary metabolites are used by humans to improve life through use in disease prevention and healing, food flavorings and preservation, beauty products and aesthetics, and recreational drugs and mysticism.

Secondary metabolites are extracted from plant tissue by distillation, expression, or solvents, depending on the constituents and plant material (Table 18.3; Fig. 18.2). Extracts (Table 18.4) are frequently grouped according to the extraction method, application, physical properties, chemical structure, or synthesis pathway of the extract (Figs. 18.3 and 18.4). Grouping of medicinal and aromatic plants by chemical properties would include alkaloids, anthraquinones, flavonoids, glycosides, lipids, phenylpropanoids, resins, saponins, and terpenoids (Figs. 18.5, 18.6, and 18.7). Variations in chemical structures are numerous, differing in chemical constituency, isomerism, and substituents.

Although a number of plants have been tested with some becoming new pharmaceuticals or innovative commercial products, many more plants have not yet been fully evaluated for bioactive properties. Thus, continued exploration, but not exploitation, of plants with health or other benefits is necessary. Over collection and habitat destruction of wild medicinal and aromatic plants used in commercial trade can lead to shortages and possible extinction of species. An estimated 400 species are currently at risk. These at risk species and other species that become endangered

Table 18.3 Methods for extracting essential oil from plant tissues. (Source: Soysal and Öztekin (2007); Esoteric Oils (2013))

Distillation—The conversion of volatile liquids (essential oils) into vapor and then back to liquid by cooling
Water distillation—plant tissue covered in water and brought to boil so that essential oil vaporizes
Steam distillation—steam moved up through plant tissue to vaporize essential oil
Hydro diffusion—steam moved down through plant tissue to vaporize essential oil
Expression—the plant is cold pressed to squeeze essential oil from the tissue; used primarily for citrus oil extractions
Sponge expression—tissue macerated and essential oil squeezed on top of collection sponge
Écuelle à piquer—tissue punched by a machine with spikes to release oil
Machine abrasion—tissue subjected to abrasion to remove oil cells and then centrifuged for oil collection
Solvent—the plant material is placed in a solvent system to extract essential oil from tissue
Solvent—plant material placed in a selected organic solvent to leach essential oil from tissue
Enfleurage—plant tissue pressed into bowl of fat to extract essential oil; used primarily with flower petals
CO_2 extraction—plant material placed in sealed container with hypercritical carbon dioxide

Fig. 18.2 Preparing for distillation of essential oil in Nepal. (Source: Photo courtesy of Han Saran Karki)

must be cultivated to ensure survival of the species and that an adequate supply of the plant is available for commercial use. Some future supplies of medicinal and aromatic plant bioactives may come from production of the chemical constituents in bioreactors (McCoy and O'Connor 2008), and modifications to plant genomes to produce variations in constituents (Runguphan and O'Connor 2009).

The World Health Organization has published quality control methods for plant materials (Anon 2003b) to help ensure public safety and conservation and protection of plants collected in the wild (Table 18.5). In the European Union herbal medicines are regulated under the European Directive on Traditional Herbal Medicinal Products and in the United States dietary supplements and herbal remedies are controlled by the Food and Drug Administration. These agencies, along with the processing and manufacturing companies are responsible for maintaining safe

Table 18.4 Medicinal and aromatic plant extracts. (Source: Leung and Foster (1996))

Extracts	Description
Absolute	Pure essence, the pure odoriferous extract of flowers for use in perfumes
Balsams	Resinous mixtures that contain large proportions of benzoic acid and/or cinnamic acid
Concréte	Water insoluble fraction of a plant remaining in retort after distillation
Decoction	A water based plant extract made by boiling plant material in water
Essential oil	Plant extract obtained from plant tissue by distillation, solvents, or expression
Fixed oil	Plant extract that contains glycerol esters of fatty acids
Infusion	A water based plant extract made by steeping or soaking plants in hot water
Oleoresins	Homogeneous mixtures of resins and volatile oils
Plant extract	Substance or chemical removed from a plant, usually the essential oil
Resins	Solid or semisolid amorphous plant product of a complex chemical nature
Stimulant/tonic	A plant or plant extract that energizes the body and/or mind
Tincture	Pharmacologically, an alcoholic solution of a nonvolatile medicine

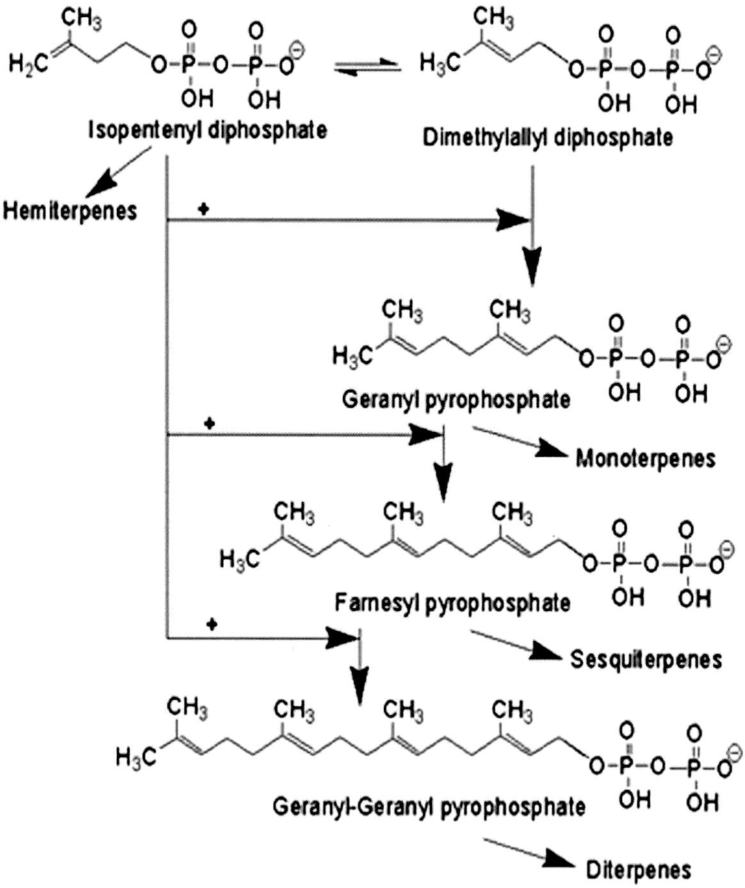

Fig. 18.3 The mevalonic acid synthesis pathway. (Source: Samuelsson (1992))

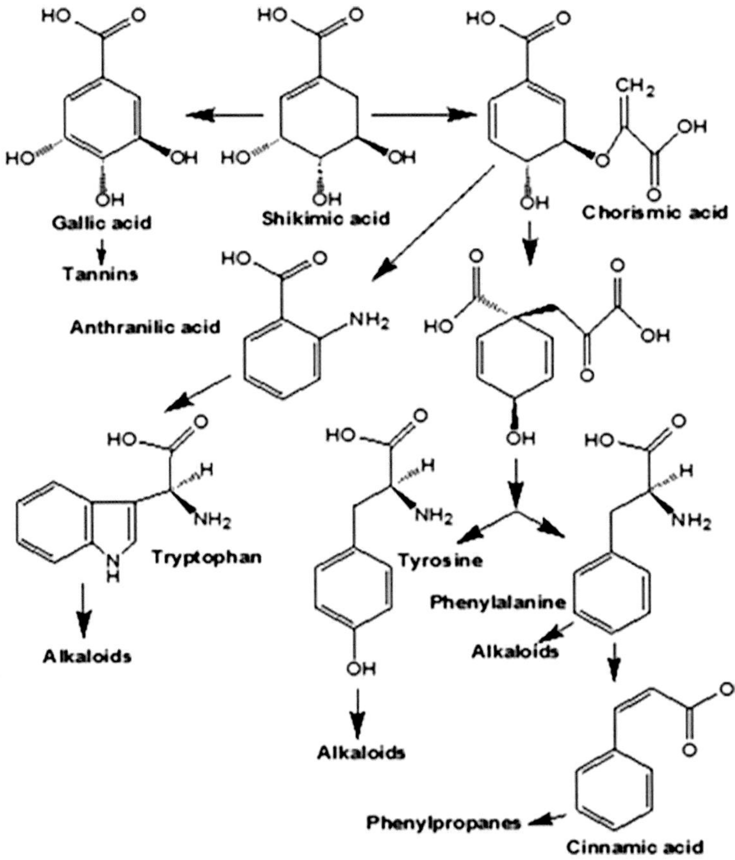

Fig. 18.4 The shikimic acid synthesis pathway. (Source: Samuelsson (1992))

marketable medicinal and aromatic products. Plants and plant extracts go through a series of steps to determine the safety and biological activity of the chemical constituents (Fig. 18.8).

Applications: Pharmaceutical

The use of medicinal and aromatic plants in medicine obviously began when humans first discovered that plants or parts of plants could be used to make ill people feel better. From this beginning of pharmacognosy (the study of medicines derived from natural sources), plant medicines have moved from crude preparations to an important source of pharmaceuticals in the treatment of several diseases. Today, the exploration of plant materials to develop new and effective medicines involves

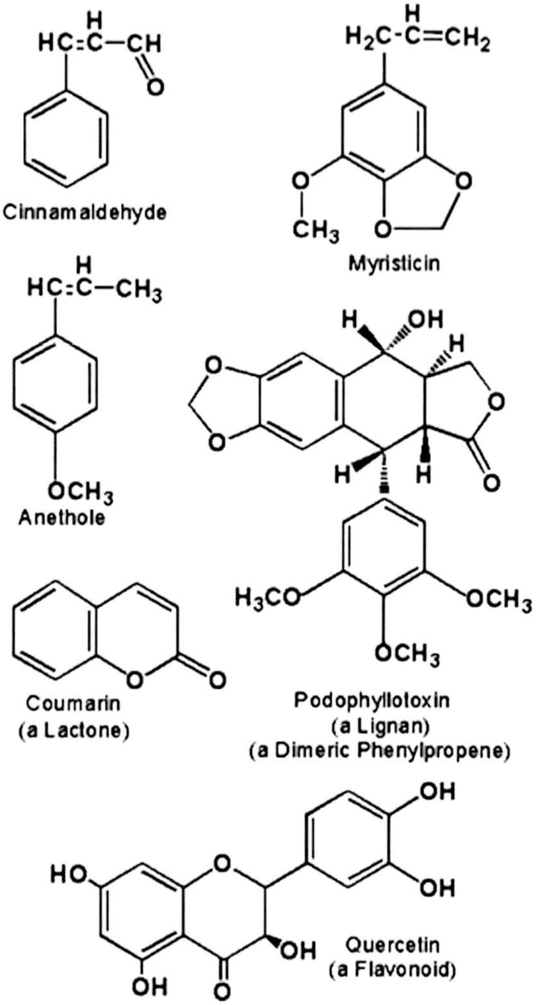

Fig. 18.5 Phenylpropanes. (Source: Jiangsu New Medical College (1978); Samuelsson (1992))

partnerships among plant explorers, ethnobotanists, physicians, medical chemists, and many others interested in developing better medicines.

Phytotherapy, the medical practice of using medicinal and aromatic plants and extracts to promote health, has been criticized by Western medicine due to the unregulated variations within alternative medicines that contrast with the standardized drugs and treatments of conventional medical therapies. The use of extracted constituents that have pharmacological benefits has led to an accepted status for natural product medications in most countries. A number of medications have originated with plants, but may be currently synthesized in chemical laboratories or modified to improve drug standardization, administration, and/or bioactivity (Table 18.6). Use of plants as a source of medicine is approached in different ways, depending on the society (traditional to modern) (Table 18.7).

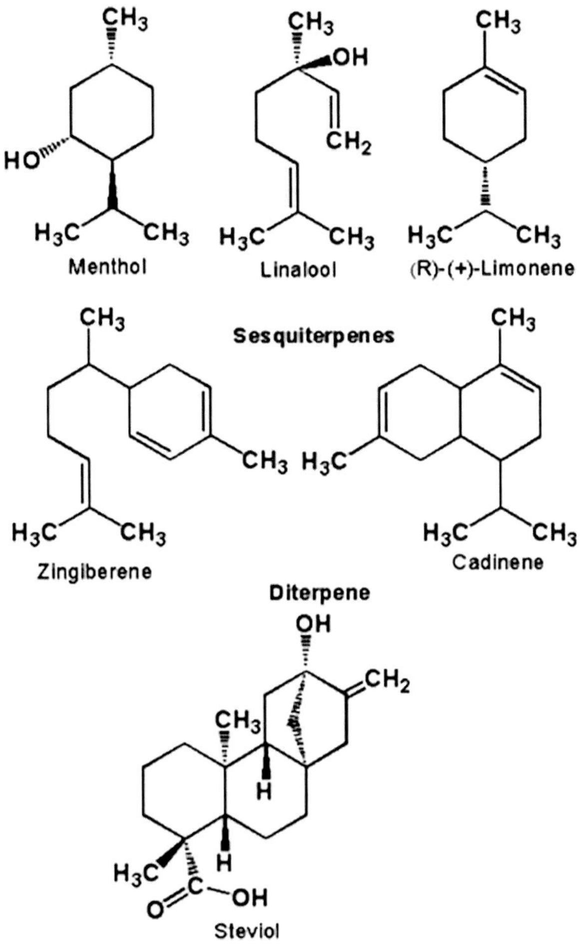

Fig. 18.6 Terpenoids. (Source: Jiangsu New Medical College (1978); Samuelsson (1992))

The World Health Organization indicates plants are used as the primary healthcare for 80% of the world's population (Anon 2008). Most plant based healthcare systems occur in areas lacking modern facilities and pharmaceuticals or in locations that have established a traditional healthcare system based on herbal products (Table 18.8). As both traditional and modern health care become more available, the trend is for people to choose that medicine that will best return them to good health.

The types of plant materials used in traditional medical systems throughout the world reflect the available plant material and the experience of generations of people that have tested various plant materials when ill. Traditional healers generally learn their trade by apprenticeships to established practitioners, by following healing practices used within a family group, or by personal experience from being a patient. Supernatural spirit powers are frequently invoked during the treatment process.

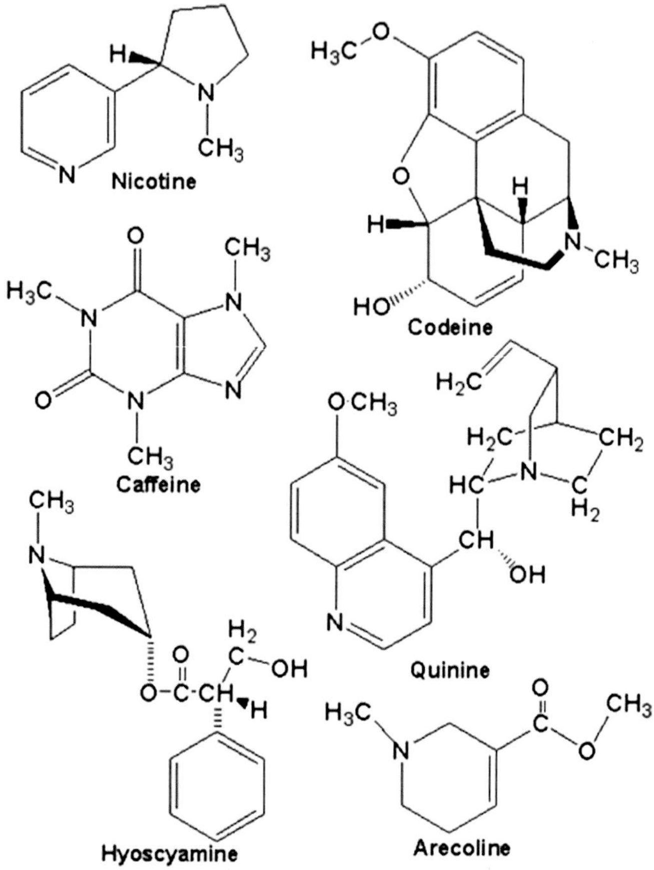

Fig. 18.7 Alkaloids. (Source: Jiangsu New Medical College (1978); Samuelsson (1992))

Table 18.5 World Health Organization guidelines for good practices in production, collection, and manufacturing of medicinal and aromatic plants

Practices	Title	Listing website
Agricultural production and collection	Good Agricultural and Collection Practices (GACP) for medicinal plants	Anon (2003a)
Manufacturing	WHO guidelines on good manufacturing practices (GMP) for herbal medicines	Anon (2007)

Although recognizing that traditional herbal medicines may sometimes be effective, conventional medicine practitioners generally reject the use of traditional medicines, primarily due to the lack of cleanliness and hygiene associated with the practitioner and plant materials, the lack of defined dosages, the lack of practitioner education, the mystification of the medicine and medical treatment, and the reluctance to share information on treatments except to those initiated into a code

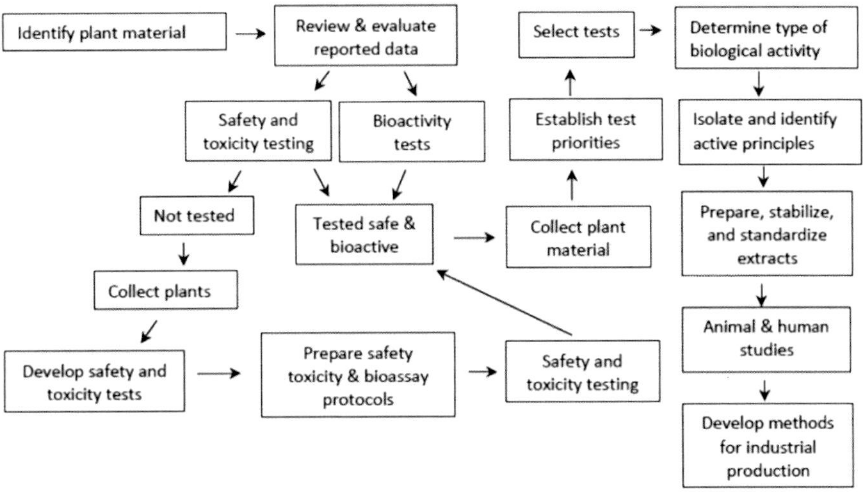

Fig. 18.8 Procedures to determine bioactivity and safety of chemical constituents in plants. (Modified from Fabricant and Farnsworth (2001))

Table 18.6 Some modern medicines originating from plants. (Source: MMS Gardens (2012))

Medicine	Plant material	Medicinal actions
Aspirin	Willow (*Salix alba*)	Antipyretic, analgesic, anti-inflammatory
Cardiac glycosides	Foxglove (*Digitalis purpurea* and *Digitalis lanata*)	Congestive heart failure, edema, atrial fibrillation
Morphine	Poppy (*Papaver somniferum*)	Analgesic
Paclitaxel	Pacific yew tree (*Taxus brevifolia* and endophytic fungi in bark)	Cancer treatment for lung, ovarian, breast, head, neck, Kaposi's sarcoma
Vincristine and Vinblastine	Madagascar periwinkle (*Catharanthus roseus*)	Cancer treatment for lymphomas, leukemia, Wilms' tumor, and other cancers

Table 18.7 Approaches for use of medicinal plants. (Source: Modified from Hirani (2012))

Approach	Application
Magical	Used in traditional societies; the practitioner (shaman) uses herbs to affect the spirit of the patient
Energetic	Used in Chinese, Ayurveda, and Unani medical practices; the practitioner uses herbs to balance energy within patient
Functional	Used by physiomedical practices; the practitioner uses natural products with action thought linked to a physiological function
Chemical	Used by modern phytotherapists: the practitioner prescribes treatments according to known constituents in plant tissues

Table 18.8 Some traditional medicinal plants of Africa and Asia. (Source: Ayensu (1978); Kokwaro (2009); Simon et al. (2007); van Wyk et al. (2011); Wiart (2001, 2012))

Plant material	Applications
Africa	
Aframomum melegueta	Leprosy, aphrodisiac, dysentery, toothache, migraine, rheumatism, fever, diarrhea, antifeedant, control lactation, postpartum hemorrhage
Aspalathus linearis	Colic, antispasmodic, block the nose, emetic, insomnia, headaches, hepatoprotective, suppresses skin tumor formation
Asystasia gangetica	Asthma
Cinnamomum camphora	Coughs, fevers, pneumonia, diarrhea
Cryptolepis sanguinolenta	Blood pressure, fevers
Harpagophytum procumbens	Fever, indigestion, malaria, allergies, rheumatism, arthritis, diabetes, senility, anti-inflammatory, low back pain
Hoodia spp.	Indigestion, blood pressure, diabetes, stomach aches
Mondia whitei	Stomach pain, indigestion, pain, post-partum bleeding
Prunus africana	Fever, improve appetite, chest and stomach pain, gonorrhea, inflammation, kidney disease, urinary tract complaints, prostate hypertrophy
Securidaca longepedunculata	Laxative, nerve disorders, treat wounds and sores, coughs, venereal diseases, snakebites, schistosomiasis, headaches
Voacanga africana	Leprosy, diarrhea, edema, convulsion, madness, carious teeth, cardiac spasms, asthma, anti-ulcer
Asia	
Astragalus propinquus	Treat hepatitis strengthen immune system
Azadirachta indica	Worms, malaria, rheumatism, skin infections
Ephedra sinica	Respiratory ailments, weight loss, low energy and athletic performance
Euphorbia hirta	Bronchial asthma, laryngeal spasms
Panax ginseng	Aphrodisiac, stimulant, type II diabetes
Piper methysticum	Soporific, asthma, urinary tract infection
Saraca indica	Gynecological disorders, edema, swelling
Zingiber officinale	Stimulant, carminative, dyspepsia, gastroparesis, constipation, colic

of secrecy. Plant materials used in treatments are frequently exposed to "unclean" environments (Fig. 18.9). Production, collection, and processing of medicinal and aromatic plants should be done under clean and sanitary conditions (Fig. 18.10).

By contrast, conventional medicines used in "modern" nations requires trained medical doctors and an associated professional staff, diagnostic equipment, and laboratory tests to confirm a patient's health status, and uses manufactured pharmaceuticals in the form of pills, lotions, capsules, liquids, or injections to bring the body and/or mind to a "normal," healthy state. These pharmaceuticals, chemical substances used for treatment, curing, or preventing a disease, are sometimes synthesized in a chemistry laboratory, but due to difficulty in chemical synthesis could also be extracted from medicinal plants. According to Taylor (2004), at least 120 current pharmaceuticals are sourced from plant materials (Table 18.10). Such

Fig. 18.9 Supply of plant material for use in traditional medicine-herbal market in Nigeria. (Source: Herbal market in Nigeria. Photo courtesy of Dr. Ade-Ademilua, University of Lagos)

plant based pharmaceutical drugs are used to treat a variety of medical problems, exemplifying the range of medicines extracted from plant tissues.

A cultural divide separates the traditional herbal healing (in which the medical practitioner brings the body into balance using natural products to modify the factors that caused the imbalance) to conventional healthcare (in which the medical practitioner determines the biological cause and treats the ailment with clinically approved, but expensive medicines). In the United States, an increasing number of healthcare consumers are beginning to seek natural herbal remedies along with a healthy approach to diet and lifestyle as an alternative to the high costs and side effects of chemically-based medicines (Dickinson et al. 2012). In the United Kingdom, 25 % of the population regularly uses herbal medicine (Behrens 2004). Herbal medicine, sometimes referred to as traditional medicine or complementary/alternative medicine (CAM), uses herbal materials, preparations, and finished herbal products that contain bioactive ingredients. Drug terms indicate the plant parts being used in herbal medicinal preparations (Table 18.9).

Pharmaceuticals originating from plants arise from literature reports or with a study of plant materials used in traditional medicines by plant explorers. These medicine hunters visit native villages to observe plant use, identify the plant material, and source plant samples for analyses in chemical laboratories. Plant constituents are subsequently bio-assayed and those with positive indications are tested in clinical trials. The plant constituent to be used as a pharmaceutical may be modified to increase solubility, selectivity, and effectiveness (Lipinski 2004). If the chemical cannot be synthesized in the laboratory, an adequate supply of the plant material must be sourced through collection or cultivation.

Natural products remain an important source for developing medicines in the foreseeable future. Of the 877 new medicines developed during 1981–2002, 52 were from natural products, 237 were derivatives of natural products, and 140 were synthetics modeled after a natural product (Newman et al. 2003). Of the drugs considered essential by the World Health Organisation (Anon 2011), at least 10 % are

Ginseng (*Panax quinquefolius*) production in Wisconsin.

Maca (*Lepidium meyenii*) harvest in Peru.

Chamomile (*Matricaria chamomilla*) processing in Hungary.

Goldenseal (*Hydrastis canadensis*), an endangered species.

Fig. 18.10 Medicinal plant cultivation, harvesting, processing, and wild collection. (Source: All photographs courtesy of the authors, Maca photograph courtesy of Christopher S. Kilham, Medicine Hunter Incorporated)

Table 18.9 Drug terms used with traditional natural product medicines. (Source: Samuelsson (1992))

Drug term	Plant part
Aetheroleum	Essential or volatile oil
Balsamum	Balsam, solution of resin in volatile oil
Bulbus	Underground stem with leaves
Cortex	Bark, all tissue outside of cambium, can be roots, stems, or branches
Flos	Flower, single flowers or entire inflorescences
Folium	Leaf, generally middle leaves of plant
Fructus	Fruit
Herba	Herb, aerial parts of the plant, stem, leaves, flowers, and fruit
Lignum	Wood, from plants with secondary thickening; woody parts of xylem (sometimes thin bark)
Oleum	Oil, fixed oil prepared from plant by expression
Pericarpium	Fruit peel
Pyroleum	Tar, prepared by dry distillation of plant material
Radix	Root, could include rhizome
Resina	Resin obtained from secretory structures or residue after distillation of balsam
Rhizoma	Rhizome
Tuber	Root or rhizome
Semen	Seed or part of seed

Table 18.10 Examples of current pharmaceuticals extracted from plants. (Source: Li and Adair (1994); Rajwar et al. (2011); Taylor (2004))

Pharmaceutical	Plant source
Antitussives	
Bergenin	Marlberry (*Ardisia japonica*)
Glaucine	Yellow hornpoppy (*Glaucium flavum*)
Noscapine	Poppy (*Papaver somniferum*)
Rorifone	Indian Field Cress (*Rorippa indica*)
Antihypertensives and tranquilizers	
Deserpidine	Rauwolscine (*Rauvolfia canescens*)
Reserpine	Snakeroot (*Rauvolfia serpentina*))
Rhomitoxin	Rhododendron (*Rhododendron molle*)
Tetrandrine	Han Fang (*Stephania tetrandra*)
Cancer and tumor treatments	
Betulinic acid	Birch (*Betula alba*)
Camptothecin	Tenuous leaf happytree (*Camptotheca acuminata*)
Colchicine	Autumn crocus (*Colchicum autumnale*)
Demecolcine	Autumn crocus (*Colchicum autumnale*)
Etoposide	Mayapple (*Podophyllum peltatum*)
Irinotecan	Happytree (*Camptotheca acuminata*)
Lapachol	Trumpet tree (*Tabebuia* spp.)
Monocrotaline	Crotalaria (*Crotalaria sessiliflora*)
Podophyllotoxin	Mayapple (*Podophyllum peltatum*)
Taxol	Pacific yew (*Taxus brevifolia*)
Teniposide	Mayapple (*Podophyllum peltatum*)
Topotecan	Happytree (*Camptotheca acuminata*)
Vinblastine	Madagascar periwinkle (*Catharanthus roseus*)
Vincristine	Madagascar periwinkle (*Catharanthus roseus*)
Cardiotonics	
Acetyldigoxin	Grecian foxglove (*Digitalis lanata*)
Adoniside	Pheasant's eye (*Adonis vernalis*)
Deslanoside	Grecian foxglove (*Digitalis lanata*)
Digitalin	Purple foxglove (*Digitalis purpurea*)
Digitoxin	Purple foxglove (*Digitalis purpurea*)
Gitalin	Purple foxglove (*Digitalis purpurea*)
Lanatosides A, B, C	Grecian foxglove (*Digitalis lanata*)
Ouabain	Ouabain tree (*Strophanthus gratus*)
Scillaren A	Jimson weed (*Datura* spp.)

exclusively from plant material. In the United States, interest in natural products decreased with the development of vaccinations and antibiotics and many in the American medical establishment instructed patients that natural products were ineffective for healing (Craker and Gardner 2006). This negative attitude by American medical personnel began to change after the Dietary Supplement Health and Education Act of 1994 was passed, and the medical doctors began to understand natural product supplements could help individuals remain healthy.

In addition to personal use of traditional medicines, farmers and others are using natural products to treat livestock and other domestic animals (Table 18.11).

Table 18.11 Plant materials used in veterinary practice. (Source: Day (2012))

Veterinary medicine	Plant material
Analgesic	Bergamot (*Citrus aurantium*)
Anesthetic	Clove (*Syzygium aromaticum*)
Antibacterial	Oregano (*Origanum vulgare*)
Astringent	Myrrh (*Commiphora myrrha*)
Carminative	Lavender (*Lavandula* spp.)
Digestive	Basil (*Ocimum basilicum*)
Disinfectant	Garlic (*Allium sativum*)
	Tea tree (*Melaleuca bracteata*)
Expectorant	Eucalyptus (*Eucalyptus* spp.)
Galactagogue	Fennel (*Foeniculum vulgare*)
Insect repellent	Cedar (*Cedrus libani*)
	Lemongrass (*Cymbopogon citratus*)
Nervine	Chamomile (*Matricaria recutita*)
Stimulant	Camphor (*Cinnamomum camphora*)
	Rosemary (*Rosmarinus officinalis*)
Tonic	Frankincense (*Boswellia carteri*)

These ethnoveterinary practices frequently include holistic approaches (physical exercise and touch therapy) along with the use of medicinal plants. In Nigeria, 92 plant species are reportedly used in the treatment of animals (Nwude and Ibrahim 1980), and many plants are used for animal care in Asia, Africa, Central Europe and South America. As with using plants for personal healthcare, availability, low cost, effectiveness, cultural appropriateness, and ease preparation make ethnoveterinary practices attractive to farmers and to animal owners interested in a natural approach to animal health.

Applications: Herbs and Spices

Herbs and spices are terms used to distinguish two types of aromatic plants based on their use and preparation. The herbs are generally fresh, but sometimes dry, leafy tissues are used in small portions as a flavoring for foods and as a garnish. The spices are dried flowers, seeds, roots, or bark used in small portions for flavoring, coloring, and preserving food. The separation of aromatic plants into herbs and spices, however, is not clearly defined and some plants, such as dill (*Anethum graveolens*) fit into both categories as dill weed (a herb) and dill seed (a spice).

The herbs and spices used in various cuisines are territorial, a taste and aroma gained as a child when introduced to foods in the home. As individuals and groups migrate from their home country to another, they generally favor the taste and aroma of food from the area in which they came. Thus, these migrations bring new foods and flavors to areas to which they immigrate and introduce these foods and flavors to others, making these new tastes and aromas available as alternatives to the usual

Table 18.12 Some herbs and spices used in Asian dishes. (Sources: Kingsely (2009); Miyano and Miyano (2011))

Spice	Plant source	Cuisines
Asafoetida	*Ferula assa-foetida*	Indian and other Asian
Black cardamom	*Amomum subulatum*	Nepalese, Indian, Chinese, Vietnamese
Chinese cassia	*Cinnamomum cassia*	Southern and Eastern Asian
Cloves	*Syzygium aromaticum*	Asian
Indian bay leaf	Cinnamomum tamala	Indian, Pakistani
Indonesian bay leaf	Syzygium polyanthum	Indonesian, Southeast Asian
Green cardamom	*Elettaria cardamomum*	Indian, Malay, Australian, Asian
Kaffir lime leaf	*Citrus hystrix*	Lao, Thai, Cambodian, Indonesian
Kalonji	*Nigella sativa*	Pakistani
Laksa leaves	*Persicaria odorata*	Vietnamese, Cambodian, Thai, Lao
Lemongrass	*Cymbopogon* spp.	Eastern and Southern Asian
Mace	*Myristica fragrans*	Malay, Indian, Indonesian
Mitsuba	*Cryptotaeniajaponica*	Japanese
Pippali	*Piper longum*	Indian,Malaysian, Indonesian
Rice paddy herb	*Limnophila aromatica*	Vietnamese, Thai
Radhuni	*Trachyspermum roxburghianum*	South Asian
Saffron	*Crocus sativus*	Central Asian and Indian
Sesame leaf	*Perilla frutescens* var. *frutescens*	Korean
Shiso	*Perilla frutescens* var. *crispa*	Japanese
Star anise	*Illicium verum*	Chinese, Asian
Szechuan pepper	*Zanthoxylum simulans* *Zanthoxylum bungeanum*	East Asian, Tibetan, and Chinese
Thai basil	*Ocimum basilicum* cv. *Horapha*	Thai, Taiwanese, Vietnamese
Thai holy basil	*Ocimum tenuiflorum*	Asian
Tia to	*Perilla frutescens*	Vietnamese, Japanese
Turmeric	*Curcuma domestica*	Indian, Vietnamese, Asian

cuisine. This observation is particularly apparent in countries, such as the United States, to which a number of immigrants have arrived from several countries.

Immigrants to the United States from Mexico have brought traditional foods flavored with chili peppers, oregano, cumin, Mexican oregano, and epazote. Immigrants to the United States from China have brought traditional foods flavored with ginger, star anise, fennel seed, Szechuan pepper, cassia, and cloves. Immigrants to the United States from Mediterranean countries have brought traditional foods flavored with basil, bay leaves, caraway seeds, cardamom, chervil, cilantro, parsley, rosemary, saffron, tarragon, and other herbs and spices. The popularity and use of various herbs and spices are associated with familiar foods and culture and generally defined by location (Table 18.12).

Whether used by the pinch or by the bunch, herbs and spices can be used to infuse a meat or vegetable dish with unparalleled aroma and flavor (Fig. 18.11). In heated dishes, the aromatic oils within the herbs and spices diffuse from the plant tissue into the liquid in the cooking dish, flavoring the goods being prepared. In salads and other cool dishes, herbs and spices are usually added using a salad dressing

Fig. 18.11 Some culinary herbs and their uses. (Source: Authors)

or vinegar containing extracts of these plants. Recipes for using herbs and spices exist in several languages in numerous cookbooks for preparation of local dishes.

Herbs and spices are a large global business with a complex market chain utilizing established brokers, importers, exporters, processors, and marketers (Fig. 18.12). As the largest herb and spice consumer in the world (Buzzanell et al. 1995), the United States imports over 40 separate spices from over 50 countries (Table 18.13). The end uses for these herbs and spices vary with about 40 % going to retail outlets and most of the rest being used in industrial food preparation and food service. Most herbs and spices are cultivated, as opposed to being collected in the wild because of the large quantities needed for world markets that could not be supplied by wild collection and by the need to provide authenticated, clean plant material to the market. Consumer demand is moving towards organically produced herbs and spices under fair trade conditions. A number of herb gardens have been established to familiarize the public with medicinal and aromatic plants (Fig. 18.13) (Table 18.14).

Pepper (*Piper nigrum*) plants.

Spice market in Turkey.

Ginger (*Zingiber officinale*) tin.

Processing saffron (*Crocus sativus*) in Iran.

Saffron (*Crocus sativus*) plants.

Fig. 18.12 Some spices of the modern world. (Source: Upper row photographs by the authors, bottom row photos courtesy of Ákos Mathe, West Hungarian University)

Conclusions

The market for medicinal and aromatic plants can be expected to continue into the future, serving as the primary medicine for most people in the world and as an alternative medicine for those individuals accustomed to more conventional medicine. The continued exploration for new medicines among the numerous plant species will most likely discover new chemical constituents that can be used to treat human and animal diseases. The challenge will be to protect and conserve medicinal and aromatic plants and plant habitats from poachers and land developers.

The exact level of the global market for medicinal and aromatic plants is difficult to discern due to the large number of plants, the multitude of countries importing and exporting, and the increasing number of sales outlets, including sales to mass markets via the internet. Over the past 20 years, consumers previously unfamiliar with medicinal and aromatic plants have become more aware of the many contributions these plants make to medicine, food products, cosmetics, and other goods used daily. The acceptance of natural medicines by the conventional medicine system in the United States has encouraged many previously cautious people into trying

Table 18.13 The production and use of some spices. (Sources: Kingsely (2009); Miyano and Miyano (2011))

Common name	Scientific name	Plant part used	World production (metric tons)	Some common uses
Allspice	*Pimenta dioica*	Dried unripe fruit	40,000–50,000	Used in Caribbean cuisine, jerked meats, sausage preparations, and curry powders
Anise	*Pimpinella anisum*	Seed oil	8–10	Used in various foods, drinks, and candies
Cinnamon	*Cinnamomum* spp.	Inner bark	150,000–155,000	Used in baked goods and candies
Cloves	*Syzygium aromaticum*	Flower buds	90,000–105,000	Used in meats, curries, and sweet dishes
Ginger	*Zingiber officinale*	Rhizomes	1,610,000–1,640,000	Used in Asian dishes and sweet foods
Nutmeg	*Myristica fragrans*	Seeds and aril	15,000–20,000	Used in Indian cuisine, potato dishes, and processed meat products
Pepper	*Piper nigrum*	Dried fruit	450,000–460,000	Used in meats, soups, pickling, and curry powders
Turmeric	*Curcuma longa*	Rhizomes	200,000–210,000	Used in Asian and other dishes and used as coloring agent
Vanilla	*Vanilla planifolia*	Beans	9,000–10,000	Used in flavoring ice cream, custard chocolate, coffee and other dishes

Knot garden Formal garden Informal garden

Fig. 18.13 Herb gardens. (Source: Authors)

medicinal plants. The immigration of individuals and families from other countries and the expansion of global travel have introduced several people to new cuisines.

The value of medicinal plant extracts has encouraged scientists and businesses to invest time and money into developing alternative methods for producing these extracts. Chemists are beginning to synthesize constituents in some plant extracts

Table 18.14 Some plant materials used in perfumes. (Sources: Bukisa (2013); Ellena (2011); Kole et al. (2005))

Common name	Scientific name	Common name	Scientific name	Common name	Scientific name
Aloe	*Aloe vera*	Fennel	*Foeniculum vulgare*	Myrrh	*Commiphora myrrha*
Basil	*Ocimum basilicum*	Fir	*Abies balsamea*	Nutmeg	*Myristica fragrans*
Bergamot	*Citrus aurantium*	Frankincense	*Boswellia carteri*	Patchouli	*Pogostemon cablin*
Calendula	*Calendula officinalis*	Rose geranium	*Pelargonium capitatum*	Rose	*Rosa species*
Camphor wood	*Cinnamomum camphora*	Jasmine	*Jasminum sambac*	Rosemary	*Rosmarinus officinalis*
Cardamom	*Elettaria cardamomum*	Jojoba	*Simmondsia chinensis*	Rosewood	*Pterocarpus indicus*
Cedarwood	*Cedrus libani*	Juniper berry	*Juniperus communis*	Sage	*Salvia officinalis*
Chamomile	*Matricaria recutita*	Lavender	*Lavandula species*	Sweet violets	*Viola odorata*
Cinnamon	*Cinnamomum verum*	Lemon grass	*Cymbopogon citratus*	Thyme	*Thymus vulgaris*
Clove	*Syzygium aromaticum*	Mignonette	*Reseda odorata*	Vanilla	*Vanilla planifolia*
Cucumber	*Cucumis sativus*	Mimosa	*Acacia dealbata*	Vetiver	*Vetiveria zizanioides*
Dill	*Anethum graveolens*	Mint	*Mentha species*	Ylang-ylang	*Cananga odorata*

that can perhaps replace the need for limited plant materials or can make changes in a constituent that will enhance bioactivity. Plant scientists and bioengineers are attempting to produce extracts via cell and microorganism cultures. In the future, geneticists may make modifications in a plant genome to enhance extract production or enable other plants to produce bioactive constituents. The success of these alternative production methods may limit the need for cultivation and wild harvesting of some plant materials.

Medicinal and aromatic plants have not cured all the diseases, but the interest in new medicines has led to the development of standardized screening processes for bioactivity and has advanced the use and science of medicinal and aromatic plant materials and their extracts. Scientific investigations on medicinal and aromatic plants are conducted in many countries with results presented at professional conferences and published in professional journals available to the public. Numerous books, websites, and blogs publicize the benefits of natural products. Medicinal and aromatic plant material can be observed in private and public display gardens.

Progress in good practices associated with crop production, wild plant collecting, and processing practices provide businesses and consumers with quality plant materials. Various conservation practices, laws, and trade restrictions help protect endangered species and ecosystems, although the destruction of plant habitats remains a serious issue for survival for some species. Medicinal and aromatic plants have played an important role in human life and will undoubted continue to be important for the foreseeable future.

References

American Botanical Council (2013) Terminology. http://abc.herbalgram.org/site/PageServer?pagename=Terminology/. Accessed 30 Aug 2013

Anon (2003a) Good Agricultural and Collection Practices (GACP) for medicinal plants. http://apps.who.int/medicinedocs/en/d/Js4928e/. Accessed 15 Sept 2013

Anon (2003b) World Health Organization WHO guidelines on good agricultural and collection practices (GACP) for medicinal plants. whqlibdoc.who.int/publications/2003/9241546271.pdf. Accessed 29 Mar 2013

Anon (2007) WHO guidelines on good manufacturing practices (GMP) for herbal medicines. apps.who.int/medicinedocs/documents/s14215e/s14215e.pdf. Accessed 15 Sept 2013

Anon (2008) World Health Organization Traditional medicine. Fact sheet N°134. http://www.who.int/mediacentre/factsheets/fs134/en/. Accessed 25 Oct 2012

Anon (2011) World Health Organization WHO model list of essential medicines, 17th list. http://www.who.int/medicines/publications/essentialmedicines/en/index.html. Accessed 12 Dec 2012

Anon (2013) Shennong. http://en.wikipedia.org/wiki/Shennong. Accessed 15 Sept 2013

Ayensu ES (1978) Medicinal plants of West Africa. Reference Publications, Algonac

Behrens J (2004) In: Vines G (ed) Herbal harvests with a future—towards sustainable sources for medicinal plants. Plantlife International, Wiltshire

Bukisa (2013) 15 most important plant sources of expensive perfumes. http://www.bukisa.com/articles/26629_15-most-important-plant-sources-of-expensive-perfumes. Accessed 10 Aug 2013

Buzzanell PJ, Dull R, Gray F (1995) The spice market in the United States: recent developments and prospects. United States Department of Agriculture, Agriculture Information Bulletin No. 709

Carpenter S, Rigaud M, Barile M, Priest TJ, Perez L, Ferguson JB (1998) An interlinear transliteration and english translation of portions of the Ebers Papyrus—possibly having to do with diabetes mellitus. http://biology.bard.edu/ferguson/course/bio407/Carpenter_et_al_(1998).pdf. Accessed 15 Mar 2013

Chaurasia OP, Ballabh B, Tayade A, Kumar R, Phani Kumar G, Singa SB (2012) Podophyllum: an endergered and anticancerous medical plant—an overview. Indian J Tradit Knowl 11(2):234–241

Craker LE, Gardner Z (2006) Medicinal plants and tomorrow's pharmacy—an American perspective. In: Bogers RJ, Craker LE, Lange D (eds) Medicinal and aromatic plants—agricultural, commercial, ecological, legal, pharmacological, and social aspects. Springer, Dordrecht

Czarra F (2009) Spices: a global history. Reakton Books-Edible, University of Chicago Press, Chicago

Day C (2012) Alternative veterinary medicine: a brief introduction. Lulu Press, Inc. http://www.lulu.com/. Accessed 13 Sept 2013

Dickinson A, Bonci L, Boyon N, Franco JC (2012) Dietitians use and recommend dietary supplements: report of a survey. Nutr J 11(1):14

Ellena J-C (2011) Perfume-the alchemy of scent. Arcade Publishing, New York

Esoteric Oils (2013) Extraction of essential oils. http://www.essentialoils.co.za/extraction-methods.htm. Accessed 26 Aug 2013

Fabricant DS, Farnsworth NR (2001) The value of plants used in traditional medicine for drug discovery. Environ Health Perspect 109:69–75

Farnsworth NR, Soejarto DD (1991) Global importance of medicinal plants. In: Akereb O, Heywood V, Synge H (eds) Conservation of medicinal plants. Cambridge University Press, Cambridge

Food University (2013) Clove history. http://www.fooduniversity.com/foodu/produce_c/producereference/Resources/Spices/Cloves/HistoryN.htm. Accessed 23 July 2013

Hirani S, (2012) Herbal therapy (nd). Dr Hirani online. http://www.drhiranionline.com/herbal-therapy.htm. Accessed 10 July 2012

Jiangsu New Medical College (1978) Traditional Chinese medicine dictionary, 1st edn. Shanghai Scientific & Technical Publishers, Shanghai

Keay J (2007) The spice route: a history. California studies in food and culture. University of California Press, Berkeley

Kew Gardens (2013) Plant cultures—exploring plants & people. http://www.kew.org/plant-cultures/plants/. Accessed 15 July 2013

Kingsely F (2009) Williams-Sonoma essentials of Asian cooking: recipes from China, Japan, India, Thailand, Vietnam, Singapore, and more. Oxmoor House, Birmingham, AL

Kokwaro J (2009) Medicinal plants of East Africa, 3rd edn. University of Nairobi Press, Nairobi

Kole PL, Jadhav HR, Thakurdesai P, Nagappa AN (2005) Cosmetics potential of herbal extracts. Nat Prod Radiance 4(4):315–321

Leung AY, Foster S (1996) Encyclopedia of common natural ingredients used in food, drugs, and cosmetics. Wiley, New York

Li S, Adair KT (1994) Camptotheca acuminata Decaisne, Xi Shu, a promising anti-tumor and anti-viral tree for the twenty-first century. Stephen F. Austin State University, Nacogdoches

Lipinski CA (2004) Lead and drug-like compounds: the rule-of-five revolution. Drug Discov Today Technol 1(4):337–341

Mamedov NA, Craker LE (2012) Man and medicinal plants: a short review. Acta Hortic (ISHS) 964:181–190

McCoy E, O'Connor SE (2008) Natural products from plant cell culture. Prog Drug Res 65:331–370

Miyano H, Miyano C (2011) Herb cultivation for mind and body, Somjili. Seobodo

MMS Gardens (2012) Hortus Medicus. http://www.piam.com/mms_garden/plants.html. Accessed 10 July 2013

Newman DJ, Cragg GM, Snader KM (2003) Natural products as sources of new drugs over the period 1981–2002. J Nat Prod 66(7):1022–1037

Nwude N, Ibrahim MA (1980) Plants used in traditional veterinary medical practice in Nigeria. J Vet Pharmacol Ther 3(4):261–273

Polo M, Rossabi M (2012) The travels of Marco Polo: the illustrated edition (the illustrated editions). Sterling Publishing, Asheville

Rajwar S, Khatari P, Patel R, Dwivedi S, Dwived A (2011) An overiew on potent herbal anticancer drugs. Int J Res Pharm Chem 1(2):202–210

Runguphan W, O'Connor SE (2009) Metabolic reprogramming of periwinkle plant culture. Nat Chem Biol 5(3):151–153

Samuelsson G (1992) Drugs of natural origin. Swedish Pharmaceutical, Stockholm

Simon JE, Koroch AR, Acquaye D, Jefthas E, Juliani R, Govindasamy R (2007) Medicinal crops of Africa. In: Janick J, Whipkey A (eds) New crops and new uses. ASHS Press, Alexandria

Soysal Y, Öztekin S (2007) Extraction. In: Öztekin S, Martinov M (eds), Medicinal and aromatic crops. The Haworth Press, Inc., Binghamton

Spices & Herbs (2013) History of allspice. http://www.spicemedicinalherbs.com/. Accessed 15 June 2013

Taylor L (2004) The healing power of rainforest herbs. Square One Publishers, Inc., New York

van Wyk B-E, van Oudtshoorn B, Gericke N (2011) Medicinal plants of South Africa, 2nd edn. Briza publications, Pretoria

Wiart C (2001) Medicinal plants of Southeast Asia. Pelanduk Publications, Kuala Lumpur

Wiart C (2012) Medicinal plants of China, Korea, and Japan: bioresources of tomorrow's drugs and cosmetics. CRC Press, Boca Raton

Chapter 19
Urban Greening—Macro-Scale Landscaping

Gert Groening and Stefanie Hennecke

Abstract With late eighteenth and early nineteenth centuries macro-scale landscaping was introduced to several European countries. In eighteenth century Europe, some ideas were drawn from Chinese examples. In the course of the nineteenth and twentieth centuries open space policies were developed as part of increasing urbanization both in Europe and in North America. Land embellishment examples from Europe such as Ermenonville, France; Woerlitz and Muskau in Germany; and Aranjuez in Spain are given. The development of open space policies, especially park planning, on national, regional and local levels in Europe, North America, Australia, and Asia are provided with examples from Adelaide, Australia; Berlin, Germany; Boston, Massachusetts; Chicago, Illinois, Los Angeles, California and Washington, D.C. in the US; Kyoto, Japan; Padova, Italy; and St. Petersburg, Russia. Rural examples of macro-landscaping such as those in Westchester County, New York, and the Green Acres Program in New Jersey are briefly touched. Green belt ideas in London and Vienna, as well as the idea of garden cities in England, Scandinavia, and India are mentioned. Also considered are issues of fruit landscapes, landscapes of exemplification and allotment and community gardens worldwide.

Keywords Land embellishment · Garden city · City park · Landscape park · Open space system

G. Groening (✉)
Forschungsstelle Gartenkultur und Freiraumentwicklung,
Institut für Geschichte und Theorie der Gestaltung, Universität der Künste,
Postfach 12 05 44, 10595 Berlin, Germany
e-mail: groening@udk-berlin.de

S. Hennecke
Fachgebiet Freiraumplanung, Universität Kassel, Fachbereich Architektur Stadtplanung
Landschaftsplanung (ASL), Gottschalkstrasse 26 (Raum 2111), 34127 Kassel, Germany
e-mail: hennecke@uni-kassel.de

G. R. Dixon, D. E. Aldous (eds.), *Horticulture: Plants for People and Places, Volume 2*,
DOI 10.1007/978-94-017-8581-5_19,

Introduction: Macro-Scale Landscaping and Urban Greening

For thousands of years settlements, agriculture, mining, forest clearing, and river regulation are clear signs of human terrestrial intervention. Here we focus on human changes and their design aspects. When and where has nature been transformed into a landscape by a designer? Masterpieces in literature, fine arts, film and music have helped to shape visions of cultural landscapes and have expanded a social conception of the treatment and transformation of nature. Especially telling is the development of the ratio between buildings and the surrounding open space. Why had a special place been chosen to set up a new settlement or a sanctuary?

When settlements turned into cities the era of urban open space provision for recreation and leisure started. Early on the embellishment of urban settlements with green structures reflected notions of luxury. So a look at the ratio between open space and built structures and between cities and open spaces promises a fruitful and stimulating dialogue.

Macro-scale landscaping and urban greening demand the participation of professionals from many disciplines. Landscape architects came in fairly late in macro-scale landscaping and landscape planners even later. Macro-scale landscaping examples range from the early Mesopotamian cities to the implementation of eighteenth century land embellishment which marked the beginning of an ongoing professionalization in landscaping. This includes urban greening from the Gardens of Babylon to the nineteenth century urban park programmes. The twentieth century vision of a “citylandscape” encompasses totalitarian and democratic aspects of the idea in connecting landscaping and urban greening. Recent projects of macro-scale landscaping and urban greening serve to open the discussion on a twenty-first century perspective. An inverse vision of macro-scale landscaping is micro-scale landscaping as displayed in miniaturized landscapes. Examples are a miniature Fujiyama (Mt Fuji) in Suizenji, a famous park in the city of Kumamoto, Kyushu, Japan, which replicates the 53 stations of the Tokaido trail from Kyoto to Tokyo, a miniature Chinese garden in the gardens of Biddulph Grange, United Kingdom, and a miniature Mount Vesuvius volcano at Woerlitz Park, Saxony-Anhalt, Germany, and last but not least the infinite number of miniaturized landscapes known as bonsai in Japan, p’en ching in China, and bunjae in Korea (Stein 1990). However, these will not be addressed here.

Landscaping—A Multidisciplinary Activity

By 3300 BCE one third of the territory of the city of Uruk in Mesopotamia was covered by gardens (Van Ess 2001). In some ancient cities, such as Alexandria in Egypt, there were in all probability residential quarters which may have looked like a park landscape (Carroll-Spillecke 1989). As Gilbert Highet explains, Vergil’s "poems of escape" and his "irresistible pleasure of fantastic dreaming" (Highet 1959, p. 63), in "The Bucolics" written about 40 BCE, are about "a beautiful land which may be northern Italy or a decorative vision of the primitive Greek region of Arcadia"; also, Vergil's "idyllic landscapes are sunny, but they have long sloping shadows and areas of cold blackness" (Highet 1959, p. 60). The "eroticized meadows" as well as "the orchards and gardens of Aphrodite" and some of these Vergilian shadows came from Greece (Calame 2008; Motte 1973; Gothein 1909). From a twenty-first century global perspective this may be considered a typical Eurocentric approach (Groening 2012a) that ignores conceptions held elsewhere in the world. Aware of the culturally coloured nature of this perspective we try to refer to a global perspective. Nonetheless an overview of all cultural contributions to macro-scale landscaping such as e.g. the peach blossom fountain in China (Xu 2012), the mythical qualities associated with gardens in sacred places of ancient Egypt (Hugonot 1989; Hornung 2011), the celestial properties of Ravana's vimana, a magical palace within gardens described in the Indian Ramayana epic (Gothein 1926) and the ideas associated with Islamic gardens (MacDougall and Ettinghausen 1976; Gadebusch 1998; Wescoat and Wolschke-Bulmahn 1996; Zangheri et al. 2006) are beyond the scope of this article.

All these examples associate our macro-landscapes with beauty, dreams and visions as well as with primitivism, shadows and coldness. In addition to the built landscape the written and the painted landscapes influence the view upon Chinese gardens described in the eighteenth century "Dream of the Red Chamber" novel (Cáo 1759/1791; Xiao 2001), the picturesque parks from "Elective Affinities" by Johann Wolfgang von Goethe (1749–1832), the landscapes in the paintings of Nicolas Poussin (1596–1665) and the descriptions of foreign landscapes by Alexander von Humboldt (1769–1859).

Literature (Bann 2012; Fischer et al. 2010), painting (Woermann 1871), mathematics (Remmert 2004), and the various kinds of construction technologies involved in garden wall making, walks, terraces, terrain modulation, and water provision show many connections to macro-scale landscaping. The knowledge of plants is also articulated in the visual (e.g. Tongiorgi and Hirschauer 2002) and literary arts (O'Malley and Meyers 2010). Landscaping is an expression of civilization and urbanization. For many centuries such landscaping may have been restricted to small pieces of land; a garden surrounding a home (Jashemski 1979/1993) and, in rare cases, a palace in a park (Nielsen 2001; Sonne 1996). In Europe some of the older open spaces to offer the joys of nature were the Horti Lucullani, the gardens of Lucullus on the Pincian Hill in Rome, Italy, created around 60 BCE (Kaster 1974), and gardens of the Campanian villas owned by affluent Romans (Häuber 1994; Jashemski 1981).The ancient Romans already employed four different kinds of gardeners: the topiarius (the art gardener); the arborator (the tree gardener); the olitor (the vegetable gardener); and the vinitor (the vine gardener) (Azzi Visentini 2004).

Clearly, urban greening was an early sign of civilization, and it engaged many disciplines. Macro-scale landscaping issues included the expansion of towns and cities and suburban villas and gardens were re-shaped for particular purposes. The area may have been a swamp, a forest, a river bank or a seashore, a valley or a

Fig. 19.1 Chongsheng temple site; eastern slopes of Cangshan near Dali overlooking and reaching down to Erhai Lake in Yunnan province, China, 2011. (Photo Gert Gröning)

mountain, a desert or a virgin forest. In developing cities rulers frequently felt they should have an additional residential area outside the city. Once selected it would often trigger further development in a part of the territory that previously had not been substantially developed. In sixteenth-century Aranjuez in Spain, some 40 kms south of Madrid, and far beyond the palace and the village some 10,000 trees were planted to form navigation lines; the design of these projects illustrated a model of the universe (Luengo Añón 2008; Nard 1851). Versailles in France may serve as an example where André Le Nôtre (1613–1700) (Orsenna 2000) created for the Sun King Louis XIV (1638–1715) an outstanding "Jardin à la Française" park setting (Berger 1985), the design of which radiated almost all over Europe in the decades to follow. In Asia macro-landscaping came as a result of the decision to establish large temple sites which then stimulated urbanization and interpretations by monks, poets, painters, and contemporary urbanites (Gröning and Hennecke 2009). The Chongsheng temple site on the eastern slopes of Cangshan near Dali overlooking Erhai Lake in Yunnan province, China (Fig. 19.1), exemplifies such a case for macro-scale landscaping. In imperial residences such as the Shugakuin Rikyu in Kyoto, Japan, a country villa for the Emperor Go-Mizunoo (1596–1680) (Kuck 1968) the unique shakkei, the borrowed landscape (jiejing in Chinese), suggests an imaginary aspect of macro-scale landscaping.

As allowed by their political and economic power, the ruling classes would attempt to follow their ruler's direction. In the European context this became addressed in a study about the Italian "villeggiatura" where many sixteenth-century Venetians migrated to the countryside (Bentmann and Müller 1970). This macro-scale landscaping project needed to be bolstered through the glorification of agriculture as economic and ideological basis. In the course of history an ever larger proportion of the population became able to put into practice the old dream of life in the countryside, of "Arcadia as 'regressive utopia', as perversion of the utopia notion" (Bentmann and Müller 1970, p. 136). In some countries, such as Sweden (Qviström 2012; Lantmäteriet 1977) and large parts of Russia (Struyk and Angelici

1996), this idealized conception of country living is fairly developed in the early part of the twenty-first century, in others, such as Brazil and China, it is on the threshold. In some areas of the world, these processes have been relatively slow for many centuries; whilst in other countries, macro-landscaping has gained momentum with increasing urbanization.

Land Embellishment—A Combination of Usefulness and Beauty in the Eighteenth and Nineteenth Centuries

The interest in combining joy and use of land embellishment relates to the ancient Roman "dulce et utile", with Horace (65–8 BCE) explaining this principle in his letter to the Pisones in the "Ars Poetica". During the eighteenth century this principle spread all over Europe. The combination of useful economy, advanced technology and beauty challenged numerous theoretical approaches in different countries.

In a number of cases, the ideology of agriculture included education for the rural population as a precondition for improved productivity and this was reflected through higher taxation. An early example of this are the late seventeenth-century "svenska matrikel kartorna över Vorpommern", where Swedish register maps for West Pomerania on the southern coast of the Baltic Sea precisely described rural land use in order to assess possible taxation (Asmus 2005; Historische Kommission 2000; Helmfrid 1995).

In late eighteenth-century France, René-Louis Girardin (1735–1808), the Vicomte d'Ermenonville, implemented his ideas about the macro-scale composition of a landscape by combining the useful and enjoyable aspects at his estate some 50 kms northeast of Paris (Girardin 1777). Girardin supported the ideas of the French philosopher Jean Jacques Rousseau (1712–1778) whom he admired for envisioning a rural society of peasant owners. In those years French was the lingua franca in Europe. Interest in the ferme ornée in France (Watelet 1774) and the ornamented farm in England (Goodchild 2010) spread easily in the late eighteenth century. At the same time the German Prince Franz von Anhalt Dessau (1740–1817) developed a macro-scale landscaping concept that turned the overall appearance of his estate into an irenopolis, or a city of peace (Hirsch 1988). This impulse related to physiocracy, a doctrine proposed by François Quesnay (1694–1774) the "fondateur du système physiocratique" (Quesnay 1766). Quesnay admired Confucianism. He presented the emperors of China as examples for other enlightened rulers who supposedly exercised their power as a result of natural law (Quesnay 1767; Maverick 1946). For Quesnay "la classe des propriétaires" and their economy of agriculture provided the base for levying taxes only (Kellner 1847). Franz von Anhalt's showpiece of macro-landscaping was Woerlitz Park, which is part of the Garden Kingdom of Dessau-Woerlitz in Germany. To a large extent he modelled it on parks he had seen on his trips to England in 1763 and 1766 (Losfeld and Losfeld 2012). Woerlitz Park also included a stretch of "Japan" and a part of "Italy" in the so-called "Japanbreite" landscape that comprised the growing of grain within the park, and the planting of Italian poplars on the Po plains. A contemplative look at it from various points such as the rock seat and the iron bridge which had been modelled after the world-wide first cast iron bridge opened in Coalbrookdale, England, in 1781 clearly reminded the viewer of the Horatian idea of "dulce et utile" where the useful agricultural pursuits and enjoyable landscape were combined. Franz von Anhalt-Dessau who had an interest in far eastern culture may have known about the Koishikawa Korakuen built in early seventeenth century Tokyo. The Japanese scholar Seiko Goto compared the park at Woerlitz to the Korakuen at Koishikawa and found a number of related issues, concluding that: "The rice paddies in Korakuen, like the farmland of Woerlitz Park, were made for a purpose. These garden concepts were also compatible with the philosophy of practical beauty held by Prince Franz, for in Korakuen, Mitsukuni also created a bamboo forest for the making of spears and a plum tree forest for harvesting plums in war time" (Goto 1996, p. XXXV–XXXVI).

The nineteenth century proved especially fruitful for macro-scale landscaping. Technological and social developments, the emergence of journals as well as their distribution, and an ever increasing system of railroads and better public schooling allowed a larger percentage of the population to read on these topics and were essential for spreading knowledge. Although potatoes were already known in Spain in the second half of the sixteenth century, their cultivation in the German states spread only in the eighteenth century. In Bavaria Gustav Vorherr (1778–1847) promoted landscape embellishment, *Landesverschönerung*, in his newly created journal "Monatsblatt für Verbesserung des Landbauwesens und für zweckmäßige Verschönerung des baierischen Landes". This was a monthly journal devoted to the improvement of rural architecture and appropriate embellishment within the state of Bavaria (Gröning 1992). The journal appeared from 1821 to 1829 as a supplement to the "Weekly of the Agricultural Association in Bavaria"(Wochenblatt des landwirthschaftlichen Vereins in Baiern). This association started the still popular annual "Oktoberfest" on the Theresienwiese in Munich, Germany. The first reference to macro-scale landscaping appeared in the "Monatsblatt" in 1821 (Felix 1821). It related to "beautiful garden art" and how it stimulated an interest in new gardens, including plantations of new and unknown species of trees and shrubs meant to have a benevolent effect on the inhabitants of the city of Regensburg in Bavaria. The 1825 land embellishment program of Carl Gottlieb Bethe (1778–1840) with a plan by Peter Joseph Lenné (1789–1866), who is considered Germany's first modern landscape architect (Günther 1985; Von Buttlar 1989), for Reichenbach in Pomerania, Prussia (Fig. 19.2), was equally a bourgeois endeavour (Gröning 2010). Far exceeding the design for the park of some 30 ha that lay next to the manor, this land embellishment project included macro-scale landscaping for a rural estate of almost 1,000 ha. A similar programme planned with the assistance of Lenné was the aristocratic programme for the surroundings of Potsdam, just west of Berlin, Germany. This programme was implemented from 1833 onwards by the kings Friedrich Wilhelm III (1770–1840) and Friedrich Wilhelm IV (1795–1861) of Prussia. A somewhat comparable macro-scale landscaping project also developed around St. Petersburg, Russia, with the parks and gardens of the Peterhof perspective (which included Peterhof) and the parks and gardens at Zarskoje Selo, Pavlovsk and Gattchina (Ananieva 2011; Vergunov and Gorokhov 1988).

Fig. 19.2 Planting plan by Peter Joseph Lenné for a land embellishment program for the estate of Carl Gottlieb Bethe at Reichenbach in Pomerania, Prussia, 1826; Plansammlung Staatliche Schlösser und Gärten Potsdam, Inventarnummer 3548

Urban Greening—Nineteenth Century Representation and Education

The hanging gardens of Babylon built by king Nebuchadnezzar II (around 600 BCE) are a famous early symbol for greening urban space. However, it took many more centuries to establish the green as an important component of cities. With the exception of small private gardens hidden behind walls or located in the inner courtyards of urban residences cities had little green space before the seventeenth century (Moldi-Ravenna and Sammartini 1996; Hunt 2009). In spatial terms most cities were small and it was easy to reach the surrounding open space.

Early examples for greening public urban space are the squares of London, England, of the early eighteenth century (Hennebo and Schmidt 1977). These areas were first used as public spaces for market places and parade grounds and later were turned into lawns and planted out with trees and shrubs. After becoming inner city green spaces, these squares were fenced in and turned into exclusive parks for "keyholders only", i.e. for those who were fortunate to own a house and a plot adjacent to the park. The idea of the green square became popular in St. Petersburg, Russia, and Paris, France over the early part of the nineteenth century. In contrast to London the Paris green system, laid out by Jean-Charles Alphand (1817–1891) and Georges-Eugène Haussmann (1809–1891) (Alphand 1867–1873/1984), was open to the public from the start (Henard 1909). Green and decorative public squares combined with broad urban boulevards constituted important elements of late eighteenth and nineteenth centuries city planning in many places in Europe and North America. These early green plans were developed by city engineers such as Ildefons Cerdà (1815–1876) for Barcelona, James Hobrecht (1825–1902) for Berlin, and Pierre Charles L'Enfant (1754–1825) for Washington, D.C.

In numerous European cities the first step towards macro-scale landscaping was the deconstruction of the surrounding fortification walls during the eighteenth and nineteenth centuries. It allowed to connect intra and extra mural open spaces. Some of these fortifications consisted of several rings and fractions of rings which gradually turned into green open spaces. In the United States of America in 1869 the "Illinois Health Board member Dr. John H. Rauch (1868), along with real estate developers and civic boosters, persuaded voters and the Illinois General Assembly to create three tax districts to fund an ambitious ring of parks several miles beyond the built-up neighbourhoods" (Duis 1993, p. 9) of Chicago. In 1870 Frederick Law Olmsted (1822–1903), the "father" of American landscape architects, influenced early green space theory by reading his influential paper on "Public Parks and the Enlargement of Towns" before the American Social Science Association, at the Lowell Institute in Boston, Massachusetts (Olmsted 1871). In Europe the increasingly popular bourgeois fashion to stroll through nature either as a mode of reflection and inner dialogue or as a way of displaying increasing social prominence fostered the layout of promenades (König 1996; Bernatzky 1960). Early green belt ideas rose in the inner city in Vienna, the "Ringstraße", where an almost circular boulevard replaced the former city fortification (von Eitelberger 1859) in 1862, the same year as the "Stadtpark", the first public park was developed by the Viennese City Council (Hajós 2007) which included a separate "Children's Park".

Early on the idea of a green belt attracted the imagination of royal gardeners of king Friedrich Wilhelm IV (1795–1861) who then acted as urban planners. In his 1840 "Projectirte Schmuck- u. Grenzzüge v. Berlin mit nächster Umgegend", Lenné, the then Director-General of the Royal Prussian Garden Administration, proposed that green open spaces for the population of the rapidly growing city of Berlin, Germany, should be in the shape of a semicircular green structure. Already in 1580 Queen Elisabeth I (1533–1603) proclaimed a green belt for the City of London, England (Hennebo and Schmidt 1977) to ensure healthy and controlled city growth. However due to increasing land competition from industrialization (Reinisch 1992) such ideas became only partially implemented.

Gradually city growth enclosed these aristocratic and feudal gardens originally designed as rural retreats and as hunting grounds that were beyond city limits. In the course of the Age of Enlightenment, which was a cultural movement of intellectuals in the seventeenth and eighteenth centuries, many of these formerly private gardens and retreats became accessible to the public. For example, in 1637, Hyde Park in London, England, was opened to the public by King Charles I (1630–1685). In 1740, Frederick the Great (1712–1786) opened to the public the Tiergarten, a large open space in Berlin, Germany, which had been used as royal hunting ground.

At the end of the eighteenth century Christian Cay Lorenz Hirschfeld (1742–1792) described special requirements for a public park in his final of five volumes "Theorie der Gartenkunst", theory of garden art (Hirschfeld 1779–1785), which became fairly popular in Europe. In Hirschfeld's opinion, shared by many of his contemporaries and followers like Friedrich Ludwig von Sckell (1750–1823), the Royal Court Garden Director in Munich, Bavaria, parks should help to educate the public. The placement of educational elements such as monuments and botanical instructions in public parks should support notions of civic decency and social interaction among the people and keep the public in contact with nature (Gröning 2012b). Hirschfeld believed that different social classes would meet in public parks, raise awareness for manners and needs of others and through them develop a sense of common purpose and social cohesion.

Projects for people's parks were realized all over Europe and North America (Appelshäuser 1994). In Paris, France, the Park des Buttes Chaumont designed by Alphand may serve as an example (Komara 2009). In Munich, Germany, the Englischer Garten carried out by Sckell between 1804 and 1823 is another example (Lauterbach 2002). Parks developed into more and more multifunctional places for playing, sports, public education and recreation (Hennecke 2008). In 1858 the "Greensward" plan for Central Park of New York City (Rosenzweig and Blackmar 1992) was developed by Frederick Law Olmsted (Capen McLaughlin 1977) with Calvert Vaux (1824–1895) envisioning rapid urban growth impacting on Central Park in the second half of the nineteenth century. Olmsted's vision for macro-scale landscaping continued with his "Emerald Necklace" plan for the park system in Boston, Massachusetts, which the Boston park commissioners had promoted since the early 1870s (Zaitzevsky 1982). Similar systems for macro-scale landscaping

Fig. 19.3 A Greater West Park System, West Chicago Park Commissioners, booklet title with "urbs in horto" motto, Chicago 1920

were developed for Chicago, Illinois (Cleveland 1871) with its "urbs in horto" motto (Fig. 19.3), then the "Empire City of the West", St. Louis, Missouri and San Francisco, California (Young 2004) and many more cities in the United States of America (Heckscher 1977). Another influential macro-scale landscaping project by Olmsted and Vaux was the 1869 plan for the Chicago suburb Riverside. It was the antecedent to late nineteenth and early twentieth centuries proposals for garden cities (Howard 1898/2010; Will and Lindner 2012) and green-belt cities (Osborn 1946). Howard refered to the City of Adelaide, South Australia, and included a special map "showing park lands all round city, and its mode of growth" (Howard 1902, between p. 128–129).

The European model of urban greening was also exported to the colonies of expanding imperialism. This model needed the next generation of garden architects to transform the European scheme and develop an individual design and use of public open space in the independent young nations. For example Roberto Burle Marx (1909–1994) criticized the Portuguese way of greening Rio de Janeiro with decorative plantings and colonial monuments. He looked for a Brazilian way to create public parks (Sá Carneiro et al. 2013). His proposals for the Aterra de Flamengo Park (1961) and the beach promenade Calçadão de Copacabana (1970) in Rio de Janeiro, Brazil, displayed a twentieth century style of design (Cavalcanti and el-Dahdah 2009).

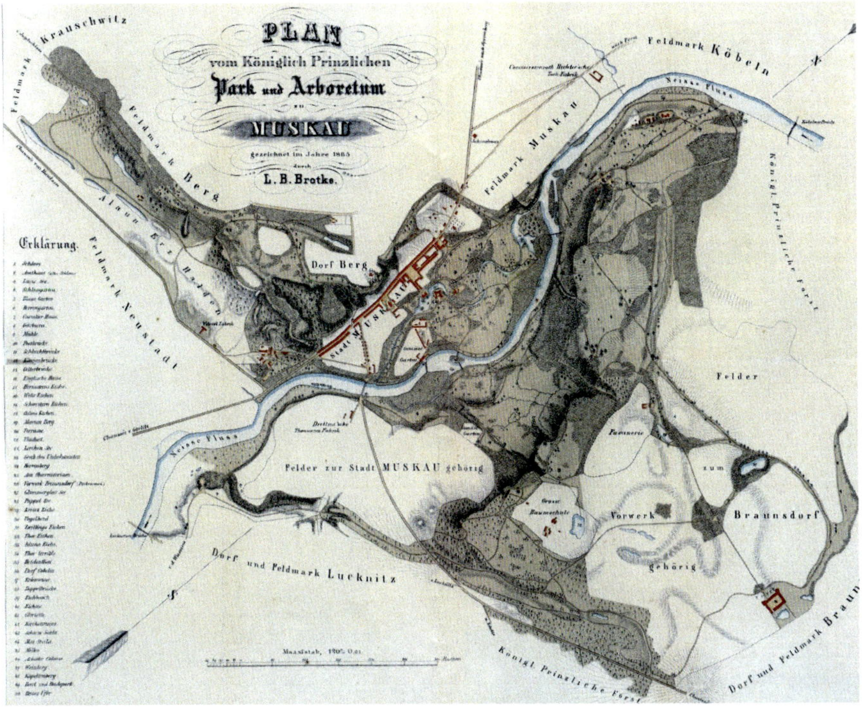

Fig. 19.4 Plan of park and arboretum in Muskau, Germany. (By L. B. Brotke 1865)

Twentieth-Century Merging of Urban Greening and Macro-Scale Landscaping

Prince Hermann von Pückler-Muskau (1875–1871) deliberately included the little town of Muskau into his macro-scale landscaping idea (Gröning 2010, 2008, 2006, 2003) (Fig. 19.4). To some extent his vision influenced twentieth century modern architecture, landscape architecture and city planning (Brantz and Dümpelmann 2011; Tishler 2000). Rapidly growing cities make it increasingly difficult to draw clear lines between city borders and "free" nature.

Founded in 1771 by the Franciscan monk Junipero Serra (1713–1784) (Oltra Perales 1988) from Petra, Mallorca, the Spaniards of San Gabriel Mission in Southern California established a new settlement named El Pueblo de la Reyna de Los Angeles, The Pueblo of the Queen of the Angels, with eleven families. In early twenty-first century the City of Los Angeles and Los Angeles County in the United States of America are home to about 14 million people. For late eighteenth century Los Angeles macro-scale landscaping would appear unfitting. Given the increase in population, and the impact that green open space has on the environment, macro-scale landscaping for early twenty-first-century Los Angeles appears appropriate.

Fig. 19.5 Berlin and surroundings, Germany, open space scheme; Denkschrift II des Amtes für Stadtplanung, Berlin, plan 1, Koeppen 1929

Indeed, in 1930 the Olmsted Brothers and Bartholomew and Associates submitted to the Citizens' Committee on Parks, Playgrounds, and Beaches their report "Parks, Playgrounds and Beaches for the Los Angeles Region" (Hise and Deverell 2000) which outlined the green space requirements for the City of Los Angeles.

Martin Wagner (1885–1957) who was Chief Planner for the City of Berlin in the 1920s envisioned a city plan where every inhabitant could participate in an open space structure (Wagner 1915). Pointing to the lack of green open spaces in industrialized cities, Wagner espoused the building of public parks and the introduction of allotment gardens within or near modern residential areas (Fig. 19.5). Garden architect Leberecht Migge (1881–1935) developed this idea further in his socially engaged concept of a self-sufficiency garden for everyone, "*Jedermann Selbstversorger*" (Haney 2010; Migge 1919). Frank Lloyd Wright (1867–1959) exaggerated this principle to one of an overall utopia of the decentralized "Broadacre City" where every family in the United States should garden on one acre, about 4,000 m^2, of land (Pimlott 2007; Wright 1932). Wright's radical intention to deny and restructure the existing cityscape was shared by modernist architects and city planners of the day. In Germany after World War I the idea of "Gärtnersiedlungen", gardeners' settlements, and "Fruchtlandschaften", fruit landscapes, was proposed (Reinhold 1932). After World War II Georg Béla Pniower (1896–1960) suggested a fairly differentiated scheme for a social program to provide a city of one million inhabitants with gardens (Pniower 1948). Pniower also commenced a macro-scale landscaping research program on an area of 130 km^2 in the Huy-Hakel area near Magdeburg, Germany, for a "Beispiellandschaft", a landscape of exemplification, in the 1950s (Heinrichsdorff 1959). In early twentieth century Westchester County which borders New York City to the north developed and implemented a famous macro-scale landscaping project, its parks, parkways and reservations system (Panetta 2006) with the Bronx-River-Parkway the first in the US to open in 1925 and the Bronx River Parkway Reservation adjacent to it (Fig. 19.6). In 1961 the US Congress appointed the Outdoor Recreation Resources Review Commission (ORRRC) which studied outdoor recreation in America, its history, place and future in current American life. It meant a big boost to macro-scale landscaping. The Green Acres Program of New Jersey serves as another example of macro-scale landscaping by interconnecting open spaces protected and preserved New Jersey's natural environment (Foresta 1979). Established in 1920 the "Siedlungsverband Ruhrkohlenbezirk" under its long time director Robert Schmidt (1869–1934, director 1920–1932) developed a macro-scale landscape idea of "Regionale Grünzüge", regional green spaces, which still today shape the spatial pattern in this part of Germany (Regionalverband 2010).

The destruction of cities and landscapes after World War II stimulated some planners to develop utopias for a restart of modern city planning without the "freight" of existing structures. Sculptural compositions of high-rise buildings, traffic infrastructure and green spaces formed this modern "city-landscape". New cities envisioned by Le Corbusier (1887–1965), like the "Ville Contemporaine"and the "Ville Radieuse", reflected still stronger impulses of macro-scale landscaping (Le Corbusier 1935). Newly founded capitals such as Canberra, Australia (1927), Chandigarh, Punjab, India (1947) (Randhawa 1961; Högner 2010) and Brasilia, Brazil (1960) followed this vision. However, the idea of multifunctional open space programmes lost momentum in this combination of macro-scale landscaping and city planning. The huge open spaces provided in some of these newly founded cities were rarely used and vandalism is frequent and costly.

The democratic idea that enabled every citizen to participate in modern living conditions like green structures had a totalitarian counterpart, however with technical and scientific progress supporting ideologies of exclusion and control. For example the National Socialist "Generalplan Ost" (General Plan East) planned to transform a number of East European countries into a macro-scale landscaped "German" homeland. For this the plan provided a rationale for genocide. The "Basics of Planning for Building Up of the Eastern Areas" clarified that the elimination of the inhabitants was a fundamental planning factor: "In the following material, it is taken for granted that the entire Jewish population of this area, roughly 560,000, have already been evacuated or will leave the area in the course of this winter" (Wolschke-Bulmahn and Gröning 2002, p. 87–88).

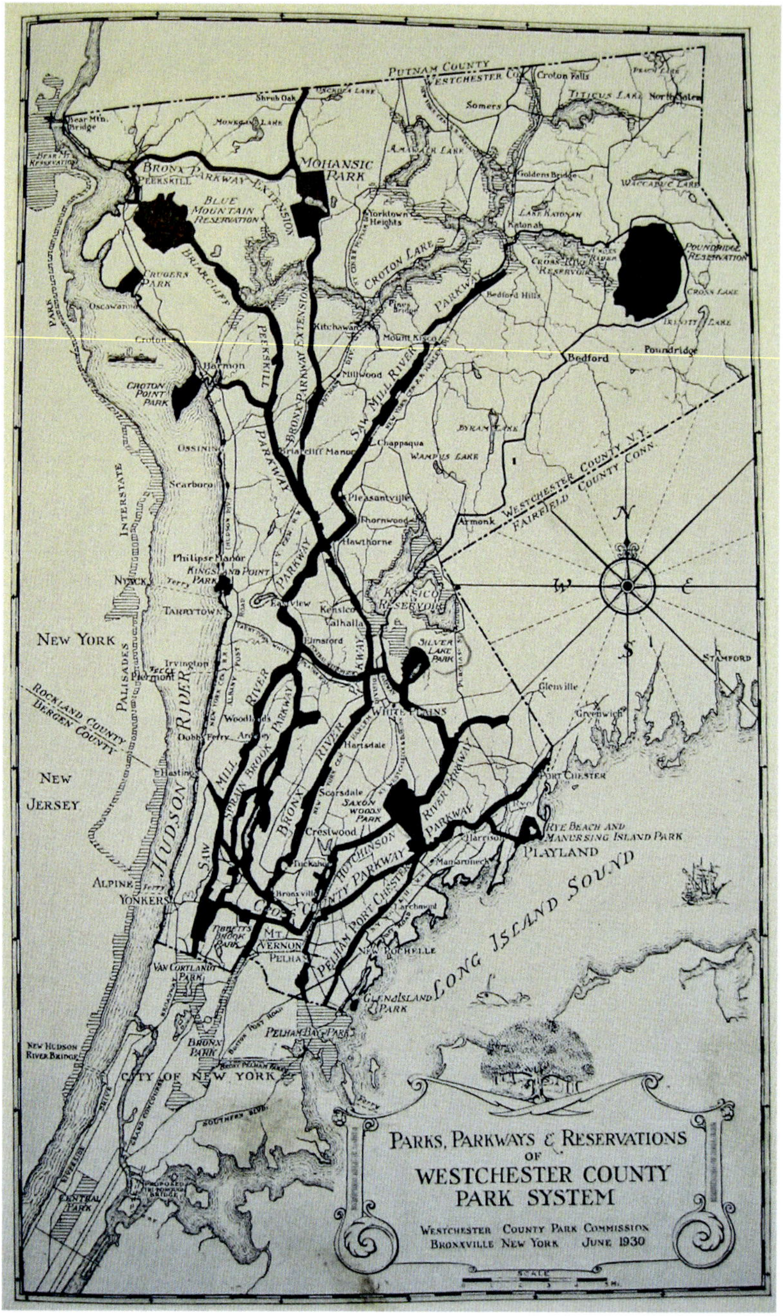

Fig. 19.6 Westchester County, New York; map of parks, parkways and reservations system; Report of the Westchester County Park Commission 1930

The Perspective: Macro-Scale Landscaping and Green Cities Between Branding and Participation

In 2050 more than 70 % of the world population will live in cities, that is, in urbanized landscapes. To identify the border of a city will then prove difficult. Urbanization, intensive agriculture and the exploitation of resources lead to an overall transformation of the surface of the earth. What impact can macro-scale landscaping and urban greening have in this scenario? In 1987 the Netherlands publicly presented four spatial development models, "careful", "dynamic", "critical", and "relaxed", for the development of their entire territory in 2050 (Van der Cammen 1987). Surveying projects undertaken at the turn of the twenty-first century suggest professional development may take one of two directions. On one hand, projects seem to conquer new dimensions of massive macro-scale landscaping visible even from outer space. On the other hand, increasing numbers of citizens proclaim the design of urban spaces as a concern of grass roots activism. Is professionalized urban greening about to vanish at this level? Not so if one takes the outstanding example of Padova, Italy, and its "Project for the restoration and development of the river network and interconnecting public open spaces" under its dynamic director of open spaces, parks, gardens and street fittings, Gianpaolo Barbariol (Debiasio Caimani 1996).

For several decades macro-scale landscaping has been applied as a means of representing national strength. Recently new islands were created in Dubai in the landscape of a palm tree and of the continents of this earth. Their design refers to baroque park projects visualizing the spatial power of a sovereign. In the era of *Google Earth* their visibility from a satellite seems an important branding factor for a nation, a city or a company. Recently the building of the new Kansai airport in Osaka Bay, Japan, needed a gigantic transportation of stones from one location to another. The stones came from nearby Awaji Island where they left a huge hole. As compensation for this material loss Awaji Island was given the "Hanahaku Japan Flora 2000", an international garden exhibition, and a conference center designed by Tadao Ando (1941-). Awaji Yumebutai is a topographical landscape project with stairways, bridges, including the Hyakuda-en, a garden of 100 flower bed squares (Fig. 19.7), built in memory of the 1995 Kobe earthquake victims (Gröning 2001). New design challenges are provided when remodelling landscapes for energy production. Where and how to locate wind, water and solar power stations in a landscape? Is this a question of engineering only or can nineteenth century approaches to land embellishment serve as examples? More recently the issue of "land grabbing", the large-scale acquisition of land by domestic or transnational companies, governments and individuals points to the careful need for macro-scale landscaping.

The dynamics of rapidly growing cities provoke enormous efforts to maintain and expand inner city open space and green structures competing with projects for infrastructure, office and housing projects (Wascher 2004) with such growth often neglecting the provision of publicly accessible green space. Cities like Tokyo (Schmidt 2012) and São Paulo have developed plans to reduce urban green deficits

Fig. 19.7 Awaji Yumebutai, Japan, a topographical garden show project with stairways, bridges, and the Hyakuda-en, a garden of 100 squared flower beds, 2000. (Photo Gert Gröning)

(Smaniotto Costa 2012; Cavalheiro 1981). They start from a low average level of 1 to 5 m^2 of green space per inhabitant. Cities such as London and Rio de Janeiro use big events like the World Cup (2014) and Olympic Games (2012 and 2016) to redevelop urban areas and install green open spaces. In Germany "Bundesgartenschauen", federal garden shows, help to pursue macro-scale landscaping projects (Theokas 2004). A recent example is the "Internationale Gartenschau", the international garden show, in Hamburg 2013. In macro-scale landscaping terms it connected the open spaces of the Wilhelmsburg district south of the Elbe River to those of the city of Hamburg districts north of the Elbe River.

More recently the decentralized disposition of small and self-organized community gardens and rooftop gardens (MacLean 2012; Wolschke-Bulmahn 2001) complement urban open space planning. At times the worldwide community garden movement (Gröning 2012c, 2005) seems more successful than urban parks and recreation departments in the provision of green open space for their citizens. The "Green Thumb" programme of New York City has been successfully managed by small groups of neighbours (Mees and Stone 2012). The management of private and public open space in the cities by grass roots movements could be read as a sign of democratic empowerment. It also alerts city governments and administrations to a strong interest in the provision of more green open spaces and their maintenance in order to keep growing cities liveable and sustainable.

References

Alphand J-C (1867–1873/1984) Les Promenades de Paris. Two volumes. J. Rothschild, Paris; reprint 1984, Princeton Architectural Press, New York

Ananieva A (2011) Russisch Grün. Eine Kulturpoetik des Gartens im Russland des langen 18. Jahrhunderts. Histoire, vol 17. Transcript, Bielefeld

Appelshäuser K (1994) Die öffentliche Grünanlage im Städtebau Napoleons in Italien als politische Aussage. Frankfurter Fundamente der Kunstgeschichte vol 11, Frankfurt a M

Asmus I (2005) Erfarenheter kring en nyutgåva av den svenska matrikelkartorna över Vorpommern. In: Roeck Hansen B (ed) Nationalutgåva av de äldre geometriska kartorna. Kungliga Vitterhets Historie och Antikvitets Akademien, konferenser, 57. Kungliga Vitterhetsakademien, Stockholm 81–94

Azzi Visentini M (2004) Topiaria: architetture e sculture vegetali nel giardino occidentale dall'antichità a oggi. Ed. Fondazione Benetton Studi Ricerche, Treviso

Bann S (ed) (2012) Interlacing words and things. Bridging the nature-culture opposition in gardens and landscapes. Dumbarton Oaks colloquium on the history of landscape architecture, vol XXXIII. Trustees for Harvard University, Washington, DC

Bentmann R, Müller M (1970) Die Villa als Herrschaftsarchitektur. Edition Suhrkamp 396. Suhrkamp, Frankfurt a M

Berger RW (1985) In the garden of the sun king. Studies on the park of Versailles under Louis XIV. Dumbarton Oaks Research Library and Collection, Trustees for Harvard University, Washington, DC

Bernatzky A (1960) Von der mittelalterlichen Stadtbefestigung zu den Wallgrünflächen von Heute. Ein Beitrag zum Grünflächenproblem deutscher Städte. Patzer, Hannover

Brantz D, Dümpelmann S (eds) (2011) Greening the city. Urban landscapes in the twentieth century. University of Virginia Press, Charlottesville

Calame C (2008) The poetics of eros in ancient Greece. Princeton University Press, Princeton

Cáo Xuěquin (1759/1791) Hong Lou Meng. Beijing

Capen McLaughlin C (ed) (1977) The papers of Frederick Law Olmsted. vol 1. The formative years, 1822–1852. The Johns Hopkins University, Baltimore

Carroll-Spillecke M (1989) ΚΕΠΟΣ. Der antike griechische Garten. Deutsches Archäologisches Institut, Architekturreferat. Wohnen in der klassischen Polis, vol III. Deutscher Kunstverlag, München

Cavalcanti L, el-Dahdah F (eds) (2009) Roberto Burle Marx. Art and Landscapes. Embassy of Brazil, Berlin

Cavalheiro F (1981) Die kommunale Freiraumverwaltung in São Paulo/Brasilien, gegenwärtige Situation und Chancen zukünftiger Entwicklung. Doctoral Dissertation. Fakultät für Gartenbau und Landeskultur der Universität Hannover, Hannover

Cleveland HWS (1871) A few hints on landscape gardening in the West. Jansen, McClurg & Co, Chicago

Debiasio Caimani L (ed) (1996) Padova. Il verde urbano: riconversione ecologica della città. Piccin, Padova

Duis PR (1993) The shaping of Chicago. In: Sinkevitch A (ed) AIA guide to Chicago, Harcourt Brace & Company, San Diego, p 3–24

Felix G (1821) Die nächsten Umgebungen von Regensburg. Monatsblatt für Verbesserung des Landbauwesens und für zweckmäßige Verschönerung des baierischen Landes 1(4):13–15

Fischer H, Matveev J, Wolschke-Bulmahn J (eds) (2010) Natur- und Landschaftswahrnehmung in deutschsprachiger jüdischer und christlicher Literatur der ersten Hälfte des 20. Jahrhunderts. CGL-Studies, vol 7. Martin Meidenbauer Verlagsbuchhandlung, München

Foresta RA (1979) Aquiring public open space in an urban state: the politics and dynamics of New Jersey's Green Acres Program. Doctoral Dissertation, Rutgers, The State University of New Jersey, New Brunswick

Gadebusch RD (1998) Das Paradies liegt in Hindustan … Die Darstellung des islamischen Gartens in der Moghulmalerei. Indo-Asiatische Zeitschrift 2:64–80

Girardin R-L (1777) De la composition des paysages, ou des moyens d'embellir la Nature autour des Habitations, en joignant l'agréable à l'utile. P. M. Delaguette, Paris; reprint 1979, Éditions du Champ Urbain, Paris

Goodchild P (2010) The 'ornamented farm' and its history in England: from Switzer to Repton. In: Stiftung "Fürst, Pückler-Park Bad Muskau" (ed) Die 'ornamental farm'. Gartenkunst und Landwirtschaft. Muskauer Schriften, vol 7, Verlag Graphische Werkstätten, Zittau p 13–28

Gothein ML (1909) Der griechische Garten. Mitteilungen des kaiserlich deutschen archäologischen Instituts, Athenische Abteilung XXXIV:100–144

Gothein ML (1926) Indische Gärten. Drei Masken Verlag, München

Goto S (1996) The philosophy of Koishikawa Korakuen in its design and usage. Doctoral dissertation no. 104, Chiba University, Japan

Gröning G (1992) The idea of land embellishment. As exemplified in the Monatsblatt für Verbesserung des Landbauwesens und für zweckmäßige Verschönerung des baierischen Landes (Monthly for improvement of rural architecture and appropriate embellishment of the state of Bavaria), from 1821 to 1829. Journal of Garden History 12(3):164–182

Gröning G (2001) Die Insel Awaji—Ein Zentrum gartenkultureller Aktivitäten in Japan. Stadt und Grün 50(8):572–577

Gröning G (2003) Nulla regula sine exceptione, Von der Idealisierung des Eigentums über die Kultur des Bodens zum wahren Kunstgenuss, Ein Kommentar zu Pücklers Andeutungen über Landschaftsgärtnerei. In: Mattenklott G, Weltzien F (eds) Entwerfen und Entwurf, Praxis und Theorie des künstlerischen Schaffensprozesses. Dietrich Reimer Verlag, Berlin, pp 149–171

Groening G (2005) The world of small urban gardens. Chronica Horticulturae 45(2):22–25

Gröning G (2006) Die goldene Axt. Wachstum und Kontrolle in Pücklers Schriften—autopoietische Kräfte in der Konzeption des Landschaftsgartens. In: Weltzien F (ed) Von selbst, Autopoietische Verfahren in der Ästhetik des 19. Jahrhunderts. Dietrich Reimer Verlag, Berlin, pp 155–170

Gröning G (2008) Hermann Fürst von Pückler-Muskau und Humphry Repton: a map of influence. In: Bosbach F, Gröning G (eds) Landschaftsgärten des 18. und 19. Jahrhunderts: Beispiele deutsch-britischen Kulturtransfers. Prinz-Albert-Studien, vol 26. K. G. Saur Verlag, München p 49–78

Gröning G (2010) Das Gut Reichenbach (Radaczewo), Pommern—eine musterhaft verschönerte Feldflur? In: Stiftung "Fürst-Pückler-Park, Bad Muskau" (ed) Die 'ornamental farm'. Gartenkunst und Landwirtschaft. Muskauer Schriften, vol 7. Verlag Graphische Werkstätten, Zittau p 73–90

Groening G (2012a) A European View of Asian-Pacific Perspectives in Garden Culture and Open Space Development. In: Groening G (ed) Proceedings of the International Symposium on advances in ornamentals, landscape and urban horticulture, IHC 2010. Acta Horticulturae 937, vol 2, p 1287–1292

Gröning G (2012b) Zur 'Öffentlichen Didaktik' und Kulturvermittlung in der Gartenkultur und Freiraumentwicklung. In: Dannecker W, Thielking S (eds) Öffentliche Didaktik und Kulturvermittlung. Hannoversche Beiträge zu Kulturvermittlung und Didaktik, vol 2. Aisthesis Verlag, Bielefeld, p 21–51

Gröning G (2012c) Guerilla Gardening, Community Gardening—neue Formen urbaner Gartenkultur? In: Fachhochschule Osnabrück, Fakultät Agrarwissenschaft und Landschaftsarchitektur

(ed) Guerilla Gardening—von unsichtbaren Akteuren und sichtbaren Aktionen, Schriftenreihe Freiraumplanung, p 21–38. http://www.al.hs-osnabrueck.de/22482.html
Gröning G, Hennecke S (eds) (2009) Hwa Gye und Da Guan Yuan. Beiträge zur koreanischen und chinesischen Gartenkultur. Universität der Künste, Berlin
Günther H (1985) Peter Joseph Lenné. Gärten/Parke/Landschaften. Deutsche Verlags-Anstalt, Stuttgart
Häuber C (1994) … endlich lebe ich wie ein Mensch. Zu domus, horti und villae in Rom. In: Hellenkämper SG (ed) Ausstellungskatalog Das Wrack. Der antike Schiffsfund von Mahdia, Rheinisches Landesmuseum, Bonn p 911–926
Hajós G (ed) (2007) Stadtparks in der österreichischen Monarchie 1765–1918. Studien zur bürgerlichen Entwicklung des urbanen Grüns in Österreich, Ungarn, Kroatien, Slowenien und Krakau aus europäischer Perspektive. Böhlau, Wien
Haney D (2010) When modern was green: life and work of landscape architect Leberecht Migge. Routledge, London
Heckscher A (1977) Open spaces. The life of American cities. Harper & Row Publishers, New York
Heinrichsdorff G (1959) Erkenntnisse und Erfahrungen aus den Forschungs- und Entwicklungsarbeiten im Huy-Hakel-Gebiet. Naturschutz und Landschaftsgestaltung im Bezirk Magdeburg. Rat des Bezirkes Magdeburg, Magdeburg
Helmfrid S (1995) Det svenska lantmäteriet i Pommern 1692-1709. In: Baudou E, Moen J (eds) Rannsakningar efter antikviteter—ett symposium om 1600-talets Sverige. Kungliga Vitterhets Historie och Antikvitets Akademien, Konferenser 30, Kungliga Vitterhetsakademien, Stockholm p 197–200
Henard E (1909) Les espaces libres à Paris. Les fortifications remplacées par une ceinture de Paris. Le Musée Social, no 4. Rousseau édition, Paris
Hennebo D, Schmidt E (1977) Entwicklung des Stadtgrüns in England von den frühen Volkswiesen bis zu den öffentlichen Parks im 19. Jahrhundert. Geschichte des Stadtgrüns, vol 3. Patzer, Hannover
Hennecke S (2008) Der Volkspark für die Gesundung von Geist und Körper—Das ideologische Spannungsfeld einer bürgerlichen Reformbewegung zwischen Emanzipation und Disziplinierung des Volkes. In: Schweizer S (ed) Gärten und Parks als Lebens- und Erlebnisraum: Funktions- und nutzungsgeschichtliche Aspekte der Gartenkunst in Früher Neuzeit und Moderne, Wernersche Verlagsgesellschaft, Worms p 151–165
Highet G (1959) Poets in a landscape. Pelican books A 445. Penguin Books Ltd., Harmondsworth
Hirsch E (1988) Dessau-Wörlitz. Zierde und Inbegriff des XVIII. Jahrhunderts. C. H. Beck, München
Hirschfeld CCL (1779–1785) Theorie der Gartenkunst, 5 vol. M. G. Weidmanns Erben und Reich, Leipzig
Hise G, Deverell W (2000) Eden by design. The 1930 Olmsted-Bartholomew plan for the Los Angeles region. University of California Press, Berkeley
Historische Kommission für Pommern in Verbindung mit dem Landesarchiv Greifswald und der Gesellschaft für pommersche Geschichte, Altertumskunde und Kunst e.V. (ed) (2000) Die schwedische Landesaufnahme von Vorpommern 1692–1709, Karten und Texte. Steinbecker Verlag Ulrich Rose, Greifswald
Högner B (2010) Chandigarh—living with Le Corbusier. Jovis, Berlin
Hornung E (2011) Schöpfungsmythen. In: Tietze C (ed) Ägyptische Gärten, Arcus-Verlag, Weimar, p 15–26
Howard, E (1898) To-morrow: A Peaceful Path to Real Reform. Swan Sonnenschein & Co., Ltd. Reprint 2010, Cambridge University Press, Cambridge, UK
Howard E (1902) Garden cities of tomorrow. Swan Sonnenschein & Co. Ltd., London
Hugonot J-C (1989) Le jardin dans l'Egypte ancienne. Publications Universitaires Européennes, Série XXXVII, Archéologie, vol 27. Peter Lang, Paris
Hunt JD (2009) The Venetian city garden. Birkhäuser, Basel

Jashemski WF (1981) The Campanian peristyle garden. In: Macdougall EB, Jashemski WF (eds) Ancient Roman gardens. Dumbarton Oaks colloquium on the history of landscape architecture, vol VII, Trustees for Harvard University, Washington, DC p 29–48

Jashemski WF (1979/1993) The gardens of Pompeii, Herculaneum and the villas destroyed by Vesuvius vol I (1979), vol II (1993) Caratzas Brothers (1979) and Aristide Caratzas (1993), New Rochelle, NY

Kaster G (1974) Die Gärten des Lukullus. Entwicklung und Bedeutung des Pincio-Hügels in Rom. Doctoral Dissertation. TU München

Kellner G (1847) Studien zur Geschichte des Physiokratismus. I. Quesnay. Dieterichsche Universitäts-Buchdruckerei, Göttingen

König G (1996) Eine Kulturgeschichte des Spaziergangs. Spuren einer bürgerlichen Praktik 1780–1850. Böhlau, Wien

Komara AE (2009) Measure and map. Alphand's contours of construction at the Parc des Buttes Chaumont, Paris 1867. Landscape Journal 28(1):22–39

Kuck L (1968) The world of the Japanese garden. From Chinese origins to modern landscape art. Weatherhill, New York

Lantmäteriet (1977) Fritidsbebyggelsen i Sverige 1975/1976. Meddelande 7. Departementens Offsetcentral, Stockholm

Lauterbach I (ed) (2002) Friedrich Ludwig von Sckell (1750–1823). Gartenkünstler und Stadtplaner. Wernersche Verlagsgesellschaft, Worms

Le C (1935) La Ville radieuse. Editions de l'Architecture d'Aujourd'hui, Bologne-sur-Seine

Losfeld A, Losfeld C (eds) (2012) Die Grand Tour des Fürsten Franz von Anhalt-Dessau und des Prinzen Johann Georg durch Europa. Aufgezeichnet im Reisejournal des Georg Heinrich von Berenhorst 1765 bis 1768. Kataloge und Schriften der Kulturstiftung Dessau Wörlitz, vol 33. Mitteldeutscher Verlag, Halle

Luengo Añón A (2008) Aranjuez. Utopía y realidad. La construcción de un paisaje. Ediciones Doce Calles, S.L., Madrid

Macdougall EB, Ettinghausen R (eds) (1976) The Islamic garden. Dumbarton Oaks colloquium on the history of landscape architecture vol IV. Trustees for Harvard University, Washington, DC

MacLean A (2012) Up on the roof, New York's hidden skyline spaces. Princeton Architectural Press, New York

Maverick LA (1946) China a Model for Europe. vol I: China's Economy and Government Admired by Seventeenth and Eighteenth Century Europeans. vol II: Despotism in China. A translation of Francois Quesnay's Le Despotisme de la Chine (Paris 1767). Paul Anderson Company, San Antonio

Mees C, Stone E (2012) Food, homes and gardens: public community gardens potential for contributing to a more sustainable city. In: Viljoen A, Wiskerke JSC (eds) Sustainable food planning: evolving theory and practice. Wageningen Academic Publishers, Wageningen, p 431–452

Migge L (1919) Jedermann Selbstversorger. Eine Lösung der Siedlungsfrage durch neuen Gartenbau. Diederichs, Jena

Moldi-Ravenna C, Sammartini T (1996) Secret gardens in Venice. Arsenale Editrice, San Giovanni Lupatoto

Motte A (1973) Prairies et Jardins de la Grèce Antique. De la Religion à la Philosophie. In: Académie Royale de Belgique (ed) Mémoires de la Classe des Lettres, Collection in-8°- 2e série, T. LXI—Fascicule 5 et dernier, Bruxelles

Nard F (1851) Guia de Aranjuez. Imprentada de la V. de D. R. J. Dominguez, Madrid

Nielsen I (2001) The gardens of the Hellenistic palaces. In: Nielsen I (ed.) The Royal palace institution in the first millennium BC. Monographs of the Danish institute at Athens, vol 4, Aarhus Universitetsforlag, Aarhus, p 165–185

O'Malley T, Meyers ARW (eds) (2010) The art of natural history: illustrated treatises and botanical paintings, 1400–1850. Studies in the history of art series. Yale University, New Haven

Olmsted FL (1871) Public parks and the enlargement of towns. Journal of Social Science 3:1–36

Oltra Perales E (1988) Vida de Fray Junípero Serra. Narrada para el hombre de hoy. Ed. Asis, Valencia

Orsenna É (2000) Portrait d'un homme heureux. André Le Nôtre 1613-1700. Fayard, Paris
Osborn FJ (1946) Green-belt cities. Faber and Faber Limited, London
Panetta R (2006) Westchester: the American suburb. Fordham University Press, New York
Pimlott M (2007) Without and within: essays on territory and the interior. Episode Publishers, Rotterdam
Pniower GB (1948) Bodenreform und Gartenbau. Siebeneicher Verlag, Berlin
Quesnay F (1766) Analyse du Tableau Économique. Reprinted in: Oncken A. (1888) Oeuvres économiques et philosophiques du F. Quesnay, Fondateur du système physiocratique, Joseph Baer & Cie, Francfort S/M p 563–660
Quesnay F (1767) Le Despotisme de la Chine. In: Éphémérides du citoyen ou bibliothèque raisonnée des sciences morales et politiques, Mars, Avril, Mai, Juin, Paris. Reprinted in: Oncken A (1888) Oeuvres économiques et philosophiques du F. Quesnay, Fondateur de système physiocratique, Joseph Baer & Cie, Francfort S/M p 305–328
Qviström M (2012) Contested landscapes of urban sprawl: landscape protection and regional planning in Scania, Sweden, 1932–1947. Landscape Research 37(4):399–415
Randhawa MS (1961) Beautiful trees and gardens. Sree Saraswaty Press Ltd., Calcutta
Rauch JH (1868) Public parks: their effects upon the moral, physical, and sanitary conditions of the inhabitants of large cities: with special reference to the city of Chicago. S.C. Griggs & Company, Chicago
Regionalverband Ruhr (ed) (2010) Unter freiem Himmel. Birkhäuser, Basel
Reinhold J (1932) Die gärtnerische Siedlung in Deutschland. Berichte über Landwirtschaft, Sonderheft 72. Paul Parey, Berlin
Reinisch U (1992) Stadtplanung im Konflikt zwischen absolutistischem Ordnungsanspruch und bürgerlich-kapitalistischen Interessen. Peter Joseph Lennés Wirken als Stadtplaner von Berlin. In: Brandenburgisches Landesamt für Denkmalpflege (ed) Peter Joseph Lenné. Gartenkunst im 19. Jahrhundert. Beiträge zur Lenné-Forschung, Berlin, pp 34–85
Remmert V (2004) Hortus mathematicus. Über Querverbindungen zwischen Gartentheorie und -praxis und den mathematischen Wissenschaften in der frühen Neuzeit. Wolfenbütteler Barock-Nachrichten 31(1):3–23
Rosenzweig R, Blackmar E (1992) The park and the people. A history of central park. Cornell University, Ithaca
Sá Carneiro AR, de Figueiroa Silva A, Marques da Silva J (2013) Jardins de Burle Marx no Nordeste do Brasil. Editora Universitária UFPE, Recife
Schmidt H (2012) Grün in Tokyo. Zwischen historischen Parks und hoher Verdichtung. Stadt und Grün 61(10):15–19
Smaniotto Costa C (2012) Die Wiederentdeckung der Stadträume. Ein Blick auf die brasilianische Metropole São Paulo. Stadt und Grün 61(10):7–14
Sonne W (1996) Hellenistische Herrschaftsgärten. In: Hoepfner W, Brands G (eds) Die Paläste der hellenistischen Könige. Philipp von Zabern, Mainz p 136–143
Stein RA (1990) The world in miniature. Container gardens and dwellings in far eastern religious thought. Stanford University, Stanford
Struyk R, Angelici K (1996) The Russian dacha phenomenon. Housing Studies 11(2):233–250
Theokas AC (2004) Grounds for review. The garden festival in urban planning and design. Liverpool University Press, Liverpool
Tishler WH (ed) (2000) Midwestern landscape architecture. University of Illinois, Chicago
Tongiorgio LT, Hirschauer GA, National Gallery of Art (2002) The flowering of Florence: botanical art for the medici. National Gallery of Art, Washington, DC
Van der Cammen H (ed) (1987) Nieuw Nederland 2050, onderwerp van ontwerp, 2 vols. Staatsuitgeverij, S'Gravenhage
Van Ess M (2001) Uruk. Architektur II, Ausgrabungen in Uruk-Warka. Endberichte vol 15/1. Philipp von Zabern, Mainz
Vergunov AP, Gorokhov VA (1988) (Russian) gardens and parks. Nauka, Moskwa
Von Buttlar F (ed) (1989) Peter Joseph Lenné. Volkspark und Arkadien. Nicolaische Verlagsbuchhandlung, Berlin

Von Eitelberger R (1859) Die preisgekrönten Entwürfe zur Erweiterung der inneren Stadt Wien. Kaiserlich-Königliche Hof- und Staatsdruckerei, Wien

Wagner M (1915) Das sanitäre Grün der Städte. Ein Beitrag zur Freiflächentheorie. Heymann, Berlin

Wascher DM (2004) Landscape-indicator development: steps towards a European approach. In: Jongman RHG (ed) The new dimensions of the European landscape. Wageningen UR frontis series, vol 4, Springer, Dordrecht, p 237–252

Watelet C-H (1774) Essai sur les jardins. Reprint, Gérard Monfort Éditeur, Saint Pierre de Salerne

Wescoat JL, Wolschke-Bulmahn J (eds) (1996) Mughal gardens, sources, places, representations, and prospects. Dumbarton Oaks colloquium on the history of landscape architecture, vol XVI. Dumbarton Oaks Research Library and Collection, Washington, DC

Will T, Lindner R (eds) (2012) Gartenstadt, Geschichte und Zukunftsfähigkeit einer Idee. Thelem, Dresden

Woermann K (1871) Ueber den landschaftlichen Natursinn der Griechen und Römer. Vorstudien zu einer Archäologie der Landschaftsmalerei. Theodor Ackermann, München

Wolschke-Bulmahn J (2001) Anmerkungen zur historischen Entwicklung von Dachgärten im frühen 20. Jahrhundert. In Lösken G, Wolschke-Bulmahn J (eds) Festschrift anlässlich des 70. Geburtstages von Prof. Dr. Hans-Joachim Liesecke, Schriftenreihe des Fachbereichs Landschaftsarchitektur und Umweltentwicklung der Universität Hannover, vol 57, Hannover p 270–283

Wolschke-Bulmahn J, Gröning G (2002) The national socialist garden and landscape ideal. In: Etlin RA (ed) Art, culture, and media under the third Reich. The University of Chicago, Chicago, pp 73–97

Wright FL (1932) The disappearing city. W. F. Payson, New York

Xiao C (2001) The chinese garden as lyric enclave: a generic study of the story of the stone. Center for Chinese Studies Publications, Ann Arbor

Xu H (2012) Die Idee der Pfirsichblütenquelle und die Freiraumentwicklung in chinesischen Städten. Dargestellt an Wohnanlagen in Kunming im späten 20. und frühen 21. Jahrhundert. Dr. Kovač, Hamburg

Young T (2004) Building San Francisco's parks, 1850–1930. The Johns Hopkins Press, Baltimore

Zaitzevsky C (1982) Frederick Law Olmsted and the Boston park system. The Belknap Press of Harvard University Press, Cambridge, MA

Zangheri L, Lorenzi B, Rahmati NM (2006) Il giardino islamico. Giardini e paesaggio, vol 15. Leo S. Olschki, Firenze

Chapter 20
Urban Trees

Mark Johnston and Andrew Hirons

Abstract Urban trees and woodlands provide a wide range of environmental, economic and social benefits to those that live and work in towns and cities. Urban trees need careful selection, planting and establishment if they are to develop into a healthy and vibrant urban forest that will deliver these ecosystem services. Professional tree care is required to maintain mature trees in urban landscapes so that their contributions to society are sustained and potential conflicts with urban infrastructure are minimised. To be efficient and effective, urban forest management must be undertaken in a planned, systematic and integrated manner. While knowledge of urban trees and their management has progressed substantially over the past few decades, the challenge facing professionals is to gain public and political support for an effective urban forest programme.

Keywords Arboriculture · Urban forestry · Tree management · Tree selection · Tree establishment · Green infrastructure · Ecosystem services

Introduction

Trees have been planted and cultivated in towns and cities throughout the world since the dawn of civilisation (Campana 1999). The motivation behind this has embraced a wide range of economic, environmental, social and even spiritual objectives depending on the different circumstances and culture of these urban societies. The people engaged in this work have always required a degree of specialist

M. Johnston (✉)
Arboriculture and Urban Forestry, Myerscough College, Preston,
Lancashire PR3 0RY, UK
e-mail: mjohnston@myerscough.ac.uk

A. Hirons
Lecturer in Arboriculture, Myerscough College, Preston,
Lancashire PR3 0RY, UK
e-mail: ahirons@myerscough.ac.uk

G. R. Dixon, D. E. Aldous (eds.), *Horticulture: Plants for People and Places, Volume 2,*
DOI 10.1007/978-94-017-8581-5_20,

knowledge, not least because of the risks involved in maintaining mature trees in close proximity to residents and built development.

Over the centuries, the planting and care of urban trees has been embraced by a number of different types of professionals. This has included horticulturists, landscapers, foresters, civil engineers and many others. Since the early twentieth century, the planting and care of trees for amenity purposes has generally come to be regarded as the province of arboriculture, although the original use of that term was almost synonymous with forestry or silviculture (Campana 1999). Modern arboriculture as a science and profession can trace much of its origins to both horticulture and forestry. Its approach to planting and cultivation, together with its emphasis on the aesthetic benefits of trees, were largely derived from horticulture. However, much of the equipment and many of the techniques involved in mature tree maintenance have their origins in forestry.

The recognised scope of arboriculture embraces all woody plants and not just trees. Its emphasis has traditionally been on the so-called amenity benefits of trees and shrubs, with much attention given to the aesthetic. Arboriculture generally focuses on individual or small groups of trees and shrubs, in both urban and rural situations. Since the 1960s, the planning and management of tree populations throughout an urban area has become known as 'urban forestry' and the totality of trees and woodland in and around a town or city is now referred to as the 'urban forest' (Johnston 1996). The emergence of urban forestry as a distinct discipline embracing urban trees has been promoted by a number of factors. Not least of these is the extensive research over the past few decades that has led to rapid expansion in our knowledge of the many environmental, economic and social benefits of urban trees (Konijnendijk et al. 2005; Anon 2010a). Our urban forests have a vital role to play in creating healthy and sustainable communities. Maximising these benefits for urban communities requires a holistic management view. Among their many benefits, urban trees and woodlands can help mitigate the effect of climate change on towns and cities through urban cooling (Ennos 2010) and their role in sustainable urban drainage systems (Denman et al. 2012). They can also help reduce air pollution, promote economic regeneration and improve the physical and mental health of urban residents (Hiemstra et al. 2008; Anon 2010a). Collectively, these benefits are now often described as urban ecosystem services.

The recycling of urban tree debris has significant environmental and economic benefits by reducing transport and disposal costs while also providing a range of useable products

Urban trees in towns and cities are a crucial element of green infrastructure and provide a wide range of environmental, economic and social benefits

As well as providing numerous benefits, there are also some risks and constraints associated with the presence of trees in urban areas. These can include potential damage to property, increased construction expenditure and risks to people from falling trees or branches. Urban tree management is often about achieving a balance between securing the maximum benefits with the least possible conflicts.

Planting and Care of Urban Trees

Urban trees need careful management if they are to deliver abundant benefits without conflicting with urban infrastructure or society. Appropriate species selection, careful planting and aftercare assist in the establishment of new trees in our urban environments. Correctly specified, professional care to the tree canopy and rooting environment help sustain mature trees in a challenging urban environment. The care of young and mature trees is vital to the security and development of green infrastructure within our towns and cities.

Selection, Planting and Establishment

The urban environment may present a number of challenges to the growth and performance of trees. Changes in land-use, reduced evaporative cooling from vegetation and heat absorbing structures combine to elevate temperatures by up to 12 °C in urban areas, when compared to adjoining rural areas (Oke 1994). In cold climates, this 'urban heat island effect' may have tangible benefits, both for the population and for tree growth, but this same effect in warmer regions can lead to increased energy use by buildings (via air conditioning), reduce outdoor comfort for people and an increased likelihood of physiological stress in trees (Sieghart et al. 2005).

Impervious surface characteristics lead to increased precipitation run-off and, frequently, localized flooding. These same surfaces may alter the gas exchange properties of soils and act to limit biological processes, such as nutrient recycling.

Anthropogenic activity can substantially alter the soil resource within an urban environment. This may lead to contamination of soils (e.g. salt and heavy metals), changes in the physical properties of soils (e.g. soil bulk density and surface sealing), reduced soil microbiota and diminished nutritional status (Meuser 2010). However, some soils within urban areas, particularly in parks and gardens have been amended, modified and improved to offer excellent conditions for plant growth (Pouyat et al. 2010). Consequently, soils within urban environments are highly heterogeneous, and can differ substantially within narrow spatial limits.

While some functions of the urban environment may improve tree performance, in general, tree growth is reduced (Quigley 2004). High quality urban sites can deliver tree life expectances in excess of 60 years (Skiera and Moll 1992). However, Roman and Scatena (2011) estimate the mean life expectancy of street trees to be 19 to 28 years, substantially below that expected of established forest trees.

In central Europe, approximately two hundred and fifty woody species are used within urban green infrastructure (Roloff et al. 2009). In spite of this, typically <5 genera account for 50–70 % of all street trees planted in central and north-western European countries (Pauleit et al. 2002; Sæbø et al. 2005; Bühler et al. 2007). A slightly higher genotypic diversity can be found in southern European cities (Sæbø et al. 2005). However, in Spain five genera accounted for 56 % of trees planted in

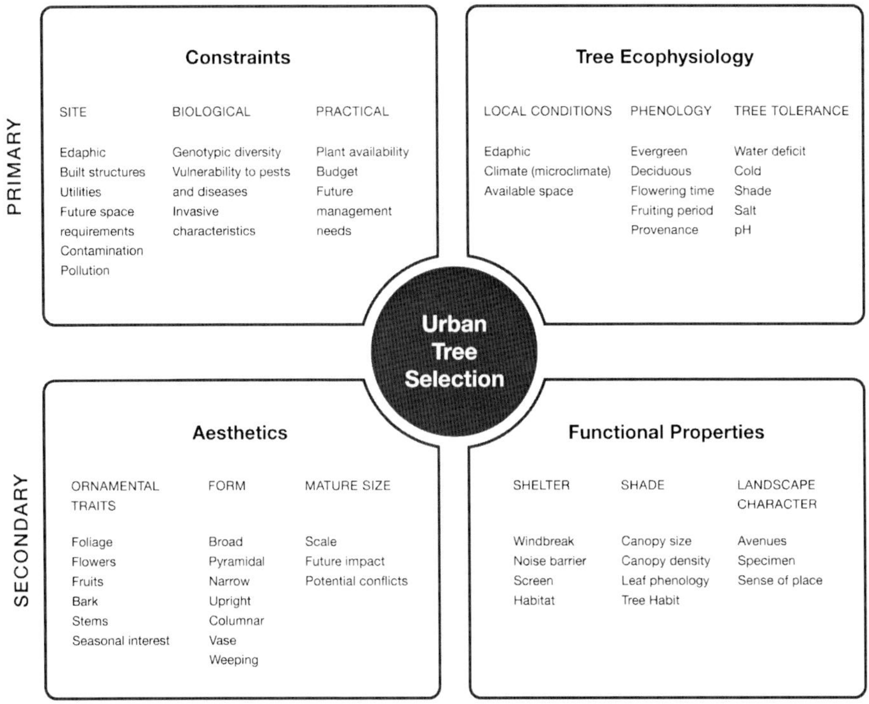

Fig. 20.1 Urban tree selection criteria. Primary factors will include *constraints* and *tree ecophysiology*, secondary factors will include desired *aesthetics* and *functional properties*

paved areas (Garcia-Martin and Garcia-Valdecantos 2001). In other regions similar trends can be observed: just five species made up over half the urban trees in Hong Kong (Jim 1987); in Argentina (Mendoza), an urban street tree assessment found that just five species accounted for 86 % of the trees (Breuste 2012) and in eastern North America two genus, *Acer* and *Fraxinus*, were highly dominant in the urban forest (Raupp et al. 2006). Limited genotypic diversity in urban trees is, therefore, a prevailing international issue in urban forests and one that arboriculturists and urban foresters should seek to address in strategic planning. Failure to increase species diversity, where appropriate, is likely to leave the urban forest vulnerable to invasive pathogens and the impacts of climate change.

The evaluation of tree performance in natural habitats, comparable to urban sites, can assist in the selection of species currently under-represented in urban environments. Natural habitats with steep, south facing slopes, low air humidity and rapidly drying shallow soils are analogous to paved urban sites. Dendroecological studies of these habitats are helping to identify stress tolerant species and genotypes that may be successfully utilized within urban areas (Sjöman et al. 2012).

Selecting the appropriate species for an urban site requires careful consideration of four principle factors: *constraints, tree ecophysiology, aesthetics* and *functional properties* (Fig. 20.1). Potential *constraints* must be considered as part of the tree selection process; these will vary but are likely to include site features, biological considerations and practicalities. *Tree ecophysiology* links a tree's performance and capacity for growth in its environment with particular plant traits. The understanding of these can be highly instructive when matching species to local edaphic and climatic conditions. Of particular importance is the inherent stress tolerance that species' exhibit to water deficits, cold and shade. Where uncertainty exists regarding the capacity of a species to establish; knowledge of the natural species distribution (and ecology), observations from arboreta and established literature should contribute to decision making.

When the compatibility between tree ecophysiology and potential constraints has been reconciled, *aesthetics* and *functional attributes* may be considered. Frequently, the drivers for tree recruitment to the urban environment may be closely allied to the aesthetics and/or functional properties but if success in a planting scheme is to be fully realized these factors should be secondary to considerations of constraints and tree ecophysiology. It is clear that expanding genotypic diversity will be necessary to maintain urban forest stability and resilience under future climate scenarios.

The successful establishment of trees relies on a number of factors beyond tree selection. High quality landscape trees are sourced from high quality nursery stock. The procurement of trees for the urban environment should be specified using measurable criteria of nursery stock. These should include: above and below ground parameters; and precise guidance on the handling of the tree stock between the nursery and the planting site (Clark 2003; Hirons and Percival 2012).

A high quality rooting environment must be provided if trees are to thrive. Soil provides an essential medium for tree growth through the provision of water and mineral elements and acting as a substrate for tree anchorage (Kozlowski et al. 1991).

The degree of soil compaction has particular importance for the process of tree establishment because it acts on a range of soil characteristics which are deleterious to root development (Fig. 20.2). As soil is compacted and bulk density increases, physical resistance to roots is increased; soil aggregates breakdown and pore space is greatly reduced. Soil aeration is diminished which compromises biological respiration of roots and soil biota and, as a result, nutrients become limited and root function is lost. Modification of soil structure also changes hydraulic properties and significantly slows water movement through the soil presenting both water deficits and water-logging as potential problems (Kozlowski 1999).

Typically, roots are unable to penetrate moist soils of a bulk density greater than 1.4–1.6 g cm^3 in fine textured soils and 1.75 g cm^3 in more coarsely textured soils. As soil dries, these thresholds are reduced, variation across species also exists (Kozlowski 1999; Brady and Weil 2008). Soil compaction in excess of these values frequently exists in urban situations and may act to reduce the available rooting volume of trees.

The importance of enhancing soil volumes has led to the development of artificial substrates known as 'structural soils' (e.g. Amsterdam tree soil; Cornell University structural soil; Stalite) and structural cells (e.g. SilvaCell® and StrataCell™). Structural soils have been designed to take limited engineering loads while maintaining a structure which still facilitates root development (Couenberg 1994; Grabosky and

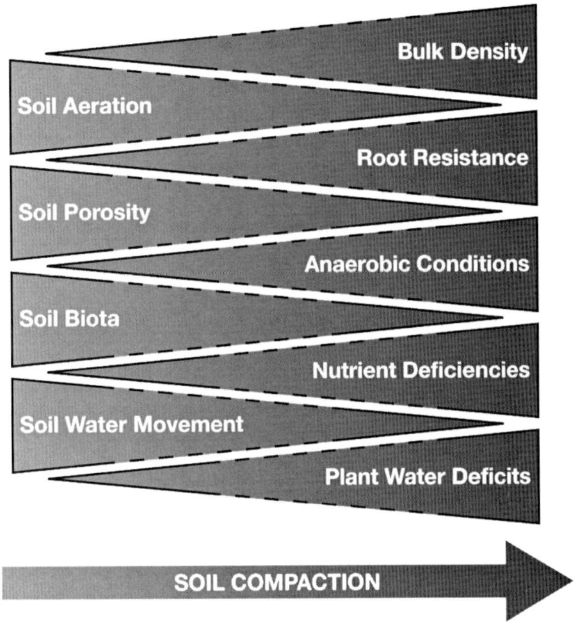

Fig. 20.2 Soil characteristics modified by soil compaction. Increasing or decreasing band width indicates the impact of soil compaction on the named soil characteristic. *Dashed lines* indicate that trends are likely to be non-linear. (redrawn from Hirons and Percival 2012)

Bassuk 1995; Kristoffersen 1998). This approach undoubtedly enhances available rooting volumes but as a result of the high sand and stone fraction in these soils, persistent retention of water and nutrients has been cited as a potential problem (Trowbridge and Bassuk 2004). Smiley et al. (2006) compared growth parameters on trees established in structural and non-compacted soil and surrounded by pavement. Trees in the non-compacted soil treatment out-performed structural and compacted soils in almost every parameter measured. This emphasizes the value of managing soil compaction and extending soil volumes in urban environments to promote tree health. Structural cells combine a rigid framework capable of supporting the physical loads encountered in urban environments with extensive voids designed to host high quality soil. As a result, compaction within the rooting environment is prevented and soil conditions which promote tree vitality can be maintained (Urban 2008). However, long term studies which assess the value of these systems are needed to provide robust evidence of their value. Structural soils and structural cells may also be used to provide 'corridors' through highly compacted regions to areas which offer a more favourable rooting environment. Where the bulk density of the soil is likely to be limiting root development in established urban trees the use of high pressure pneumatic tools to cultivate the rootzone can substantially expand the available area for root development and enhance tree health (Fite et al. 2009).

Often, the right tree has been selected for the right place; a high quality plant has been secured from the nursery; the root environment is capable of providing resources for tree development but deficient planting practices and inadequate post-planting aftercare lead to tree mortality. Detailed planting specifications which

Table 20.1 Key elements and criteria which may be used to generate robust planting specifications for work contracts and method statements

Specification elements	Specification criteria
Tree characteristics before planting	Specimen true to species or variety type
	Graft compatibility (if appropriate)
	Healthy with good vitality
	Free from pests, disease or abiotic stress
	Free from injury
	Self-supporting with good stem taper
	Stem-branch transition height
	Sound branch attachment and structure
	Good pruning wound occlusion
	Canopy symmetry
	High rootball occupancy
	Diversity in rooting direction
	Good root division
	Extensive fibrous root system
	Free from root defects (e.g. circling roots)
	Free pests, disease or abiotic stress
Planting pit and rootzone	Planting pit 2–3 times the diameter of the rootball
	Imported soil is of defined standard (e.g. BS 3882)
	Low soil bulk density (1.2 g cm^3) maintained in planting pit and rootzone
	Potential rooting (soil) volume adequate for mature tree of species planted
Planting practice	Hessian, wire baskets and other containers removed from rootball and correctly disposed of
	Tree planted at stem-root transition
	Tree upright and supported (where necessary) using above or below ground techniques
Formative pruning	Damaged branches removed using natural target pruning methodology
	Rubbing and crossing branches removed
	Sub-ordination of competing stems
Tree aftercare	Mulch to depth of between 5 and 10 cm and to defined width. Stem to remain exposed and not buried by mulch
	Mulch replenishment schedule defined
	Irrigation schedule based on local soil variables (preferably soil matric potential)
	Tree protection and support to have defined timescale for evaluation and/or removal

make planting practitioners accountable to measureable planting attributes should be used much more extensively in planting contracts to avoid these trends. Audits may then be used to monitor work standards. Table 20.1 provides key elements and criteria that may be used to develop tree planting specifications for use in work contracts and method statements.

Table 20.2 Examples of national standards for arboricultural practice

Country	Standard
Great Britain	BS3998—Tree work recommendations
	BS5837—Trees in relation to design, demolition and construction—recommendations
Germany	DIN 18320—German construction contract procedures (VOB)—Part C: General technical specifications in construction contracts (ATV)—Landscape works
	DIN 18919—Vegetation engineering/management in landscaping; development and maintenance of green spaces
	DIN 18920—Vegetation engineering/management in landscaping; protection of trees, plant populations and vegetation areas during construction
	ZTV—Baumpflege—Additional technical contractual terms and guidelines for tree care
US	ANSI A300—for tree care operations—Tree, shrub and other woody plant management
	ANSI Z133—Safety requirements for arboricultural operations
Australia	AS 4373—Pruning of amenity trees
	AS 4970—Protection of trees on development sites
New Zealand	Approved code of practice for safety and health in arboriculture

Tree Care in the Urban Environment

Established, mature trees need managing in urban environments to promote tree health and performance, prevent conflicts with urban infrastructure and ameliorate risk. In many countries, national standards have been developed to present a framework for arboricultural operations (Table 20.2). These can be useful documents for benchmarking practices but should always be supported by current best practice guides (e.g. International Society of Arboriculture—Best Management Practice series).

The reasons for pruning urban trees are varied. Formative pruning is often valuable at an early stage in tree development. The fundamental objective will be to provide good structure for future crown development by the removal of damaged and crossing branches; sub-ordination or removal of competing leaders; and the maintenance of tree form (Gilman 2012). Pruning mature trees may be for reasons of form, tree health, aesthetics, safety, and clearance from infrastructure (Dujesiefken et al. 2005; Johnson 2007). Regardless of objectives, there is now international agreement that the natural target pruning technique proposed by Shigo (1989) should be followed. Further guidance can be found in relevant national standards (Table 20.2) or in established arboricultural texts (e.g. Brown and Kirkham 2004; Harris et al. 2004; Gilman 2012).

Where visual analysis of the tree crown indicates potential branch failure that may lead to substantial damage to persons or property and, it is undesirable to remove the branch by pruning, tree crown stabilization may be used. Non-invasive approaches using high strength, UV resistant, hollow ropes, have been used successfully to increase the stability by bracing vulnerable branches to stable scaffold branches. However, it is vital that tree inspection schedules are established to monitor the tree crowns with an installed stabilization system so that any further deterioration in crown structure can be managed.

Management of Urban Trees

Unless urban trees were part of a grand design or formal landscape, their planting and care in cities throughout the world has traditionally been undertaken in a rather haphazard manner (Lawrence 2006). In many countries, it was only in the nineteenth century with the emergence of local government with civic powers embracing urban trees that there was much coordinated effort in urban tree management. Even now, this often continues to be undertaken in an ad-hoc manner (Johnston et al. 1999; Saretok 2006; Britt and Johnston 2008; Stobbart and Johnston 2012).

Principles of Sustainable Urban Forest Management

With the emergence of the concept of urban forestry in the 1960s came a far more integrated approach to the management of urban trees and woodlands. This integrated approach was evident at the level of the trees themselves and at the level of the different organisations that have some involvement in urban tree management. Urban forestry takes a holistic overview of the urban forest and ensures that its individual elements, such as street trees, park trees or trees in residential housing areas, are not treated in isolation from each other.

Urban forestry is also a multidiscipline approach to the management of the urban tree resource. It draws on a number of established disciplines such as arboriculture, urban and amenity horticulture, traditional forestry, ecology and conservation, urban planning, landscape architecture and the social sciences. While it could be argued that arboriculture is probably the most significant discipline, a wide range of different professions have an important contribution to make to the sustainable management of the urban forest.

While there are various definitions of urban forestry, in essence it can be described as a planned, systematic and integrated approach to the planting, maintenance and management of trees and woodlands in and around urban areas (Johnston and Rushton 1998). Each of these three elements of the urban forestry approach is now examined in more detail.

Planned Management

In common with other forms of resource management, the principle of planned management is central to urban forestry. Urban trees can be a major feature of urban infrastructure but to realise their enormous potential to improve the quality of urban life, they must be considered at an early stage in the planning process. They should not be regarded as an 'add-on extra' to be located in any space left over after most of the other elements of urban development have been established. The management of the urban forest itself needs to be directed by a long-term strategy and management plan that will lead to the sustained yield of benefits (Britt and Johnston 2008; van Wessenaer et al. 2012).

The gathering of a wide range of data relevant to the urban forest is the first stage in the planning process and the development of a long-term strategy. To develop any meaningful strategy urban forest managers need to know the nature and extent of the existing urban forest and the conditions in which it grows. This stage can be described quite simply as 'What do we have?' (Britt and Johnston 2008). With the rapid development of computer and digital technology in recent years, considerable advances have been made in surveying, mapping and evaluating the urban forest. Aerial photography and satellite imagery is now available in many urban areas throughout the world and can be used to map urban forest cover (Myeong et al. 2001). On the ground, information about the species, size, health and condition, location, etc. of individual trees can be gathered using a range of handheld data collection devices and then downloaded into a central computerise tree management system (CTMS). The capabilities and cost of these CTMSs can vary considerably but in Europe, North America, Australasia and many parts of Asia there are now commercial companies offering suitable software and hardware to meet the needs of most local authorities and other potential users. The use of a CTMS can dramatically increase the efficiency and effectiveness of an urban forest programme (Wagner and Smiley 1990). They are capable of producing long and short-term work programmes, management reports, contract specifications and other management documents.

The second stage in the planning process is the selection of management objectives, or simply known as 'What do we want?' (Britt and Johnston 2008). While the relevant local authority usually takes the lead, it should consult widely with the local community and relevant stakeholders to establish their needs and priorities in terms of urban trees and woodlands. Part of that consultation should involve producing a draft urban forest/tree strategy document that can be issued for public consultation and then revised in the light of feedback.

Having decided on strategic policies and objectives, the third stage of the planning process involves developing management plans to achieve them. This stage can be described as 'What do we do?' (Britt and Johnston 2008). Specific management plans need to be formulated to achieve each objective. For example, neighbourhoods that have few trees and private gardens can be targeted as priority areas for local authority tree planting. Other areas where there is a dramatic imbalance of age-classes or species distribution can also be targeted with appropriate tree removal and replacement schemes. It is important that management plans include SMART targets that are Specific, Measurable, Achievable, Relevant and Timed. Whatever practical tree work is specified it should be stated that this will be undertaken to the highest industry standards (Anon 2010b) see also Table 20.2.

The fourth and final stage of the planning process involves the monitoring and revision of the urban forest/tree strategy. This is known simply as 'Are we getting what we want?' (Britt and Johnston 2008). The urban forest is a growing resource that changes over time. Managing it also changes the nature and extent of that resource. These changes need to be regularly monitored and the strategy revised accordingly. If specific objectives are not being met then the reason for this must be investigated and appropriate action taken. The timeframe for many strategies is 5 years with a major revision of the strategy undertaken at the end of that period. However, if a longer timeframe can be established, for example 20 years, this is preferable, assuming that the local authority is prepared to commit sufficient resources over that period (van Wessenaer et al. 2012).

Many urban forest programmes around the world have to function within severe financial constraints (Britt and Johnston 2008; Stobbart and Johnston 2012). Consequently, many operate an essentially reactive programme with very little systematic or pre-planned work. However, in recent years a range of urban forest evaluation systems have been developed that enables local authorities to show just how cost-effective money spent on these programmes actually is in terms of the ecosystem services they provide to communities. The results of these evaluation systems can be used not only to protect existing levels of funding but also to generate additional funding. One of the most successful is called i-Tree, a peer-reviewed software suite from the USDA Forest Service that provides some of the best urban forestry analysis and benefits assessment tools (Nowak et al. 2010; Rogers et al. 2012). The i-Tree Tools help communities of all sizes to strengthen their urban forest management and advocacy efforts by quantifying the structure of community trees and the ecosystem services that trees provide. An example is New York City where the use of i-Tree has enabled substantial additional funding to be attracted to its urban forest programme (Wells 2012).

Systematic Management

The practical management of the urban forest to realise any planned objectives needs be undertaken in a systematic manner (Grey 1996). This involves establishing and operating a wide range of management systems that will ensure the successful implementation of the overall urban forest/tree strategy (Britt and Johnston 2008). The various operations involved in the planting, maintenance and management of the trees should, as far as possible, be conducted in an organised and systematic manner, at the appropriate time. Without this systematic approach, management becomes inefficient and ineffective.

There are many advantages to systematic management in contrast with 'on demand' working (Miller 1997). These include less time spent in transporting crews to and from different work sites, less set-up time for the same reason, a considerable reduction in costs for the same amount of work done, and more environmentally sound ways of working, for example, through reduced fuel and vehicle use. Because the trees are regularly maintained there are fewer complaints and requests from the public, and savings in time and effort dealing with these. Trees are also more likely to retain a high value, rather than letting them deteriorate to the point where a request or complaint is generated. However, it is recognised that any urban forest programme will inevitably involve some element of unplanned work that cannot be accommodated into a systematic work programme. This is because there will always been expected or unpredictable events that impact on the trees, such as severe weather or a sudden outbreak of a pest or disease. Nevertheless, the amount of purely reactive work should be kept to a minimum, particularly if the trees are generally maintained in good condition.

A wide range of tree maintenance and management operations can be undertaken systematically. For example, all newly-planted trees should receive systematic post-planting maintenance, such as watering, mulching, adjustment of stakes and ties, until they are established. Pruning operations, such as dead wooding and crown thinning, can be undertaken systematically, tree by tree, street by street. At a management level, tree inspections need to be undertaken systematically to ensure that all the relevant trees in the neighbourhood are included.

Most urban forest programmes generate a considerable amount of 'green waste' through tree maintenance operations that requires some form of disposal. The utilisation of this green waste not only makes sound environmental sense but the products from this can also be a source of income for the local authority (Mayhead and Blandford 2004; Britt and Johnston 2008)

Modern urban forest management is continually developing new and more sophisticated management systems for monitoring and controlling the growth of the urban forest. The use of CTMSs by an increasing number of local authorities has enabled them to implement systematic maintenance programmes that significantly increase the potential for the cost-effective management of the urban forest.

Integrated Management

The principle of integrated management is central to the concept of urban forestry. Effective and efficient management requires an overview of the entire forest and its individual elements cannot be considered in isolation from each other or from the rest of the urban environment. Integrated management can only be applied practically when the different organisations and groups that have some ownership, responsibility or concern for the urban forest work in partnership. While all this activity in individual urban areas should be coordinated centrally through the relevant local authority, the existing and potential contribution of the local private and voluntary sectors should not be underestimated (Grey 1996) (Fig. 20.3). In many respects urban forestry is as much about people as it is about trees (Johnston 1989).

An extensive programme of community involvement is a vital component of any urban forestry initiative. It encourages local residents to make their own contribution to the planting and care of their urban forest and helps promote positive attitudes and behaviour towards urban trees (Johnston 1985). This involvement is often

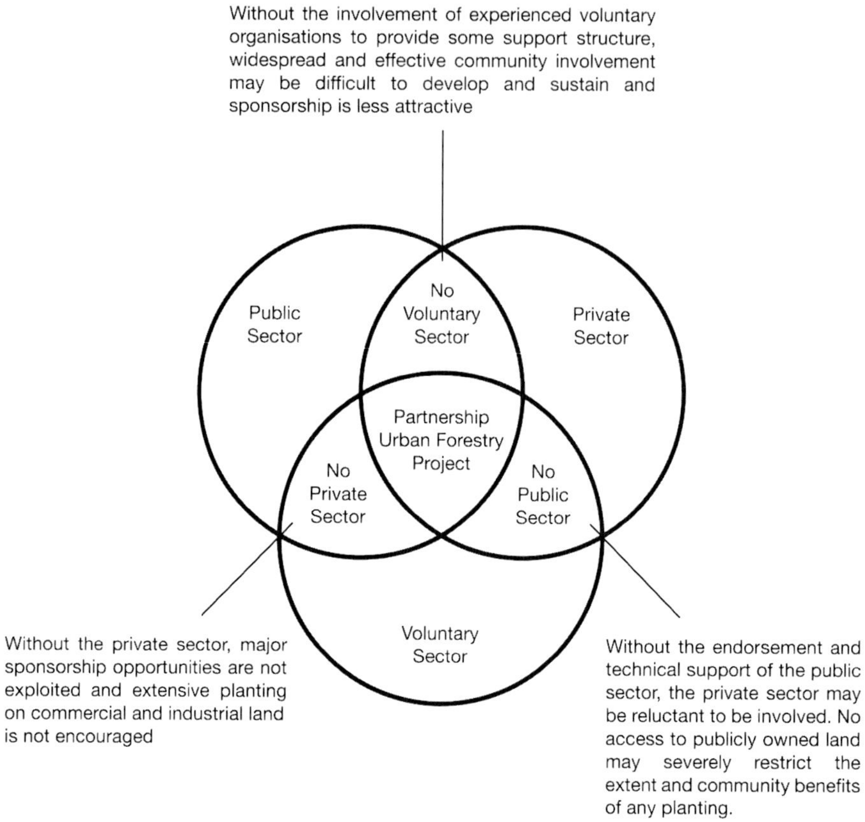

Fig. 20.3 A city or district-wide urban forestry project should involve a partnership between the public, private and voluntary sectors. If any sector is not represented, the overall impact of the project is likely to be limited. (After Johnston 1989)

expressed in practical planting and tree care but it should also include allowing residents to have an influence of wider policy issues. All community involvement programmes should involve a balance of education, consultation and practical participation strategies. If any of these elements is not represented, the impact of the programme will be limited (Britt and Johnston 2008) (Fig. 20.4). A wide range of techniques can be employed to achieve this, including different types of community events and activities (Van Herzele et al. 2005; Anon 2012a).

Many cities throughout the world are multicultural in character and urban forestry programmes must ensure that they develop policies that promote social inclusion (Johnston and Shimada 2004). Any programme of extensive community involvement must also have some type of organisational structure to support this (Britt and Johnston 2008).

The principle of integrated management should also be applied to the organisation of the local authority's own urban forest programme. Too often, responsibilities for trees are split across different departments and sections. The problem of 'departmentalism' within many local authorities is widely recognised as being responsible for a fragmented and uncoordinated approach to their tree management efforts.

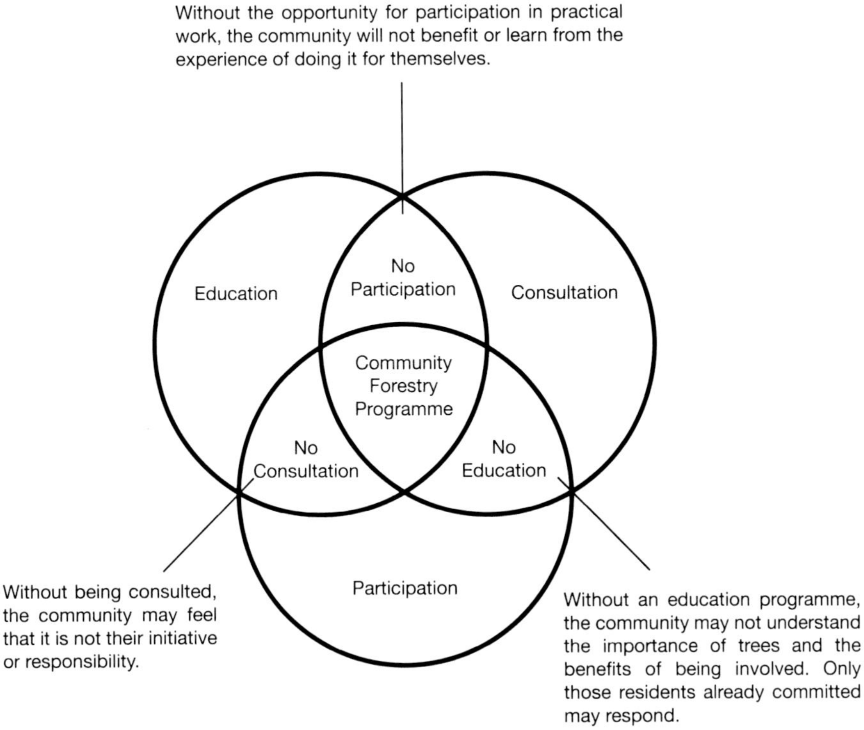

Fig. 20.4 A community involvement or community forestry programme should include a balance of education, consultation and participation strategies. If any element is not represented, the impact of the programme will be limited. (After Johnston 1989)

Trees in Urban Design

Throughout history many people have considered that the design function of urban trees and open spaces was to imitate pastoral landscapes that offered an escape from city life. However, this tired romantic style has been challenged and superseded over the past few decades by a more dynamic approach that shows how trees can be used to enhance urban elements rather than hide them (Arnold 1993; Robinson 2011). Urban foresters, landscape architects and city planners are now encouraged to use trees, not as mere decoration, but as living architecture to create and reinforce urban spaces.

In recent years, some landscape architects and conservationists have rejected that approach and advocated the creation of naturalistic urban landscapes using native species of trees and other plants. This 'natives only' or 'natives are best' approach has sought to exclude or severely limit the use of non-native species in towns and cities. However, recent research, particularly in Europe and North America, has shown we need a more balanced and sustainable approach to urban tree selection that is based firmly on science (Johnston et al. 2012). The city is not a natural habitat but a human one that displays a unique heritage of landscapes involving a mix of native and non-native species. Tree selection needs to favour those species that provide the appropriate environmental, economic and social benefits for urban residents. The use of exclusively native species for urban areas can severely limit that choice, particularly in meeting the challenges of climate change (Knox et al. 2008) and increasing the resilience of the urban forest against tree pathogens (Santamour 1990).

In general, it is the large-growing species of urban trees that provide the most benefits; particularly those that help mitigate the effects of urban climate change (Anon 2010c). However, in increasingly risk averse societies and with restricted tree maintenance budgets there has been a trend away from planting large-growing trees in favour of smaller-growing ornamentals (Anon 2011). While these small trees are seen as posing a reduced threat to people and property and usually require less maintenance work, their benefits are proportionally much less than large forest-type trees (Anon 2010c). From a design viewpoint this trend has also led to fears of a proliferation of 'lollipop landscapes' (Britt and Johnston 2008). In Britain, the Trees and Design Action Group (TDAG), a multidiscipline group of professionals, has attempted to reverse this trend with detailed guidance on the benefits of large-species trees and how these can be successfully integrated into the urban landscape (Anon 2012b). The work of TDAG has attracted much attention and efforts are underway to establish TDAG groups in a number of countries.

The Future for Urban Trees

Knowledge of urban trees and urban tree management has progressed enormously over the past few decades. Furthermore, with the development of the internet and e-learning, access to arboricultural and urban forestry information has never been greater (Johnston and Hirons 2012). It is also now widely recognised that urban forests are a vital element of green infrastructure, delivering a vast range of ecosystem services. Despite this, it is clear that even in so-called developed countries, urban forest cover is often very patchy and standards of maintenance and management are less than satisfactory. Meanwhile, in many developing countries in Africa, Asia, South America and elsewhere, urban forest managers are faced with even greater challenges (Shikur 2012).

Across the world, the real challenge for the professionals is to gain public support for an extensive and healthy urban forest and an effective programme of management that will sustain this. Then, this public support must be translated into political will and action to ensure it actually happens. If all the various professionals who care for urban trees are to have any real impact they must learn how to influence the decision makers who ultimately decide what happens to our urban forests (Anon 2012b).

References

Anon (2010a) Benefits of green infrastructure. Report by forest research. Forest Research, Farnham
Anon (2010b) Tree work-recommendations, 3rd edn. British Standards Institution, London
Anon (2010c) No trees, no future: trees in the urban realm. Revised edition Trees and Design Action Group, London
Anon (2011) Common sense risk management of trees: guidance on trees and public safety in the UK for owners, managers and advisers. Forestry Commission, Edinburgh
Anon (2012a) Public engagement in forestry: a toolbox for public engagement in forest and woodland planning. Forestry Commission, Edinburgh
Anon (2012b) Trees in the townscape: a guide for decision makers. Trees and Design Action Group, London
Arnold H (1993) Trees in urban design, 2nd edn. Van Nostrand Reinhold, Workingham
Brady NC, Weil RR. (2008) The nature and properties of soils, 14th edn. Pearson Education Prentice Hall, New Jersey
Britt C, Johnston M (2008) Trees in towns II: a new survey of urban trees in England and their condition and management. Department for Communities and Local Government, London
Breuste JH (2012) Investigations of the urban street tree forest of Mendoza, Argentina. Urban Ecosyst 15:1–18
Brown GE, Kirkham T (2004) The pruning of trees, shrubs and conifers, 2nd edn. Timber Press, Portland
Bühler O, Kristoffersen P, Larsen SU (2007) Growth of street trees in copenhagen with emphasis on the effect of different establishment concepts. Arboric Urban For 33(5):330–337
Campana RJ (1999) Arboriculture: history and development in North America. Michigan State University Press, Michigan
Clark R (2003) Specifying trees: a guide to the assessment of tree quality. NATSPEC, Sydney, NSW, Australia
Couenberg E (1994) Amsterdam tree soil. In: Watson GW, Neely D (eds) The landscape below ground, 1993. The Morton Arboretum. International Society of Arboriculture, pp 24–33
Denman E, May P, Moore G (2012) The use of trees in urban stormwater management. In: Johnston M, Percival G (eds) Trees, people and the built environment. Forestry commission research report. Forestry Commission, Edinburgh
Dujesiefken D, Drenou C, Oven P, Stobbe H (2005) Arboricultural practices. In: Konijnendijk CC, Nilssoon K, Randrup TB, Schipperijn J (eds) Urban forests and trees. Springer, Heidelberg
Ennos R (2010) Urban cool. Physics World. August issue, 2–5
Fite K, Wells CE, Smiley ET (2009) Impacts of root invigorationTM and its individual components on red maple (Acer rubrum) at four urban sites. In: Watson GW, Costello LR, Scharenbroch BC, Gilman EF (eds) The landscape below ground III. International Society of Arboriculture, Lisle
García-Martín G, García-Valdecantos JL (2001) El arboladourbano en lasciudadesespañolas (The urban tree in Spanish cities). In: Actasdel III CongresoForestalEspañol, pp 467–474
Gilman EF (2012) An illustrated guide to pruning, 3rd edn. Delmar, New York
Grabosky J, Bassuk NL (1995) A new urban tree soil to safely increase rooting volumes under sidewalks. J Arboric 21:187–201
Grey GW (1996) The urban forest: comprehensive management. John Wiley and Sons, New York
Harris RW, Clark JR, Matheny NP (2004) Arboriculture: integrated management of trees shrubs and vines. Pearson Education Prentice Hall, New Jersey
Hirons AD, Percival GC (2012) Fundamentals of tree establishment: a review. In: Johnston M, Percival G (eds) Trees, people and the built environment. Forestry Commission, Edinburgh
Hiemstra J, Schoenmaker E, Tonneijck A (2008) Trees: relief for the city. Plant Publicity, Holland
Jim CY (1987) The status and prospects of urban trees in Hong Kong. Landscape and Urban Planning 14:1–20

Johnson DL (2007) Pruning. In: Kuser JE (ed) Urban and community forestry in the Northeast, 2nd edn. Springer, Heidelberg
Johnston M (1985) Community forestry: a sociological approach to urban forestry. Arboric J 9:121–126
Johnston M (1989) Involving the public. In: Hibberd BG (ed) Urban forestry practice. Forestry Commission Handbook 5, 26-34. HMSO
Johnston M (1996) A brief history of urban forestry in the United States. Arboric J 20:257–278
Johnston M, Hirons A (2012) Going online with arboricultural education. Arboric Urban For 38(3):105–111
Johnston M, Rushton BS (1998) A survey of urban forestry in Britain: part I, aims and method of research. Arboric J 22:129–146
Johnston M, Shimada L (2004) Urban forestry in a multicultural society. J Arboric 30(3):185–192
Johnston M, Collins K, Rushton B (1999) A survey of urban forestry in the Republic of Ireland. In: Collins K (ed) Proceedings of Ireland's third national conference on urban forestry. The Tree Council of Ireland, Dublin
Johnston M, Nail S, James S (2012) Natives versus Aliens? The relevance of the debate to urban forest management. In: Johnston M, Percival G (eds) Trees, people and the built environment. Forestry commission research report. Forestry Commission, Edinburgh
Knox J, Weatherhead K, Kay M (2008) Climate change impacts on water for horticulture. Environment Agency and Cranfield University, Cranfield
Konijnendijk CC, Nilssoon K, Randrup TB, Schipperijn J (eds) (2005) Urban forests and trees. Springer, Heidelberg
Kozlowski TT (1999) Soil compaction and growth of woody plants. Scand J For Res 14:596–619
Kozlowski TT, Kramer PJ, Pallardy SG (1991) The physiological ecology of woody plants. Academic Press, London
Kristoffersen P (1998) Designing urban pavement sub-bases to support trees. J Arboric 24:121–126
Lawrence HW (2006) City trees: a historical geography from the renaissance through the nineteenth Century. University of Virginia Press, Charlottesville
Mayhead G, Blandford G (2004) The mersey and red rose forests timber stations project, September 2003-March 2004. Report to the countryside agency and forestry commission, North West England Conservancy
Meuser H (2010) Contaminated urban soils. Springer, Heildelberg
Miller RW (1997) Urban forestry: planning and managing urban greenspaces. 2nd edn. Prentice Hall, New Jersey
Myeong S, Nowak DJ, Hopkins PF, Brock RH (2001) Urban cover mapping using digital, high-spatial resolution aerial imagery. Urban Ecosyst 5:243–256
Nowak D, Hoehn R, Crane D, Stevens J, Leblanc F (2010) Assessing urban forest effects and values, Chicago's urban forest. Resource Bulletin NRS-37. USDA Forest Service, Radnor, PA
Oke TR (1994) Global change and urban climate. Proceedings of the 13th international congress of biometrics. Calgary, Canada Part III, Vol I
Pauleit S, Jones N, Garcia-Martin G, Garcia-Valdecantos J, LRivière LM, Vidal-Beaudet L, Bodson M, Randrup TB (2002) Tree establishment practice in towns and cities-Results from a European survey. Urban For Urban Gree 1:83–96
Pouyat RV, Szlavecz K, Yesilonis ID, Groffman PM, Schwarz K (2010) Characteristics of urban soils. In: Aitkenhead-Peterson J, Volder A (eds) Urban ecosystem ecology. American Society of Agronomy, Crop Science Society of America, Soil Science Society of America
Quigley MF (2004) Street trees and rural conspecifics: will long-lived trees reach full size in urban conditions? Urban Ecosyst 7:29–39
Raupp MJ, Buckelew Cumming A, Raupp EC (2006) Street tree diversity in Eastern North America and its potential for tree loss to exotic borers. Arboric Urban For 32(6):297–304
Robinson N (2011) The planting design handbook, 2nd edn. Gower Publishing, London
Rogers K, Jarratt T, Hansford D (2012) Torbay's urban forest: assessing urban forest effects and values. A report on the findings from the UK i-Tree Eco Project. Treeconomics, Exeter

Roloff A, Korn S, Gillner S (2009) The climate-species-matrix to select tree species for urban habitats considering climate change. Urban For Urban Gree 8:295–308

Roman LA, Scatena FN (2011) Street tree survival rates: meta-analysis of previous studies and application to a field survey in Philadelphia, PA, USA. Urban For Urban Gree 10:269–274

Sæbø A, Borzan Z, Ducatillion C, Hatzistathis A, Lagerström T, Supuka J, García-Valdecantos JL, Rego F, Van Slycken J (2005) The selection of plant materials for street trees, park trees and urban woodland. In: Konijnendijk CC, Nilssoon K, Randrup TB, Schipperijn J (eds) Urban forests and trees. Springer, Heidelberg

Santamour FS (1990) Trees for urban planting: diversity, uniformity, and common sense. Proc. 7th Conf. Metropolitan Tree Improvement Alliance METRIA 7:57–65

Saretok L (2006) A survey of urban forestry in Sweden. Myerscough College, Lancashire

Shigo AL (1989) Tree pruning: a worldwide photo guide. Shigo and Trees Associates, Durham

Shikur ET (2012) Challenges and problems of urban forest development in Addis Ababa, Ethiopia. In: Johnston M, Percival G (eds) Trees, people and the built environment. Forestry commission research report. Forestry Commission, Edinburgh

Sieghardt M, Mursch-Radlgruber E, Paoletti E, Couenberg E, Dimitrakopoulus A, Rego F, Hatzistathis A, Randrup TB (2005) The abiotic urban environment: impact of urban growing conditions on urban vegetation. In: Konijnendijk CC, Nilssoon K, Randrup TB, Schipperijn J (eds) Urban forests and trees. Springer, Heidelberg

Sjöman H, Gunnarsson A, Pauleit S, Bothmer R (2012) Selection approach of urban trees for inner-city environments: learning from nature. Arboric Urban For 38(5):194–204

Skiera B, Moll G, (1992) The sad state of city trees. Am For (March/April):61–64

Smiley ET, Calfee L, Fraedrich BR (2006) Comparison of structural and noncompacted soils for trees surrounded by pavement. Arboric Urban For 32:164–169

Stobbart M, Johnston M (2012) A survey of urban tree management in New Zealand. Arboric Urban For 38(6):247–254

Trowbridge PJ, Bassuk NL (2004) Trees in the urban landscape: site assessment, design and installation. John Wiley and Sons, New Jersey

Urban J (2008) Up by roots: healthy soils and trees in the built environment. International Society of Arboriculture, Illinois

Van Herzele A, Collins K, Heyens V (2005) Interacting with greenspace-public participation with professionals in the planning and management of parks and woodlands. Ministerie van de Vlaamse Gemeenschap, Brussels

Van Wassenaer P, Satel A, Kenny A, Ursic M (2012) A framework for strategic urban forest management planning and monitoring. In: Johnston M, Percival G (eds) Trees, people and the built environment. Forestry commission research report. Forestry Commission, Edinburgh

Wagner JA, Smiley ET (1990) Computer assisted management of urban trees. J Arboric 16(8):209–215

Wells MP (2012) Using urban forest research in New York City. In: Johnston M, Percival G (eds) Trees, people and the built environment. Forestry commission research report. Forestry Commission, Edinburgh

Chapter 21
Trees in the Rural Landscape

Glynn Percival, Emma Schaffert and Luke Hailey

Abstract Rural trees are trees not used for timber purposes or located within urban landscapes such as towns and cities. Rural tree use and function varies between developed, developing and under-developed countries. Conservation and ecology are the main factors driving rural tree planting in developed countries whereas food production, animal protection and use of rural trees for medicine, fire wood and fuel predominate in developing and under-developed countries. The importance and uses of rural trees with respect to food production, animal feed, shelter, medicine, firewood, fuel, timber, service wood, biological diversity, water purification systems, climate, tree protection, conservation and dead wooding are discussed in a world-wide context.

Keywords Food · Animal feed · Shelter · Pharmaceutical products · Firewood · Fuel · Timber · Service wood · Biological diversity · Water purification systems · Climate · Tree protection · Conservation dead wooding

Introduction

Rural trees relate to those that are located outside of towns and cities to include park land trees, farm land trees, woodlands, road side trees and hedgerow trees. Rural trees are located on agricultural land, built-up areas such as settlements and infrastructure, and bare land (e.g. dunes, former mining areas, abandoned industrial sites). Rural trees may be productive; such as orchards, isolated or stands of field trees, agroforestry systems, or protective; such as trees with an ecological or landscaping function (Fig. 21.1). They may be predominantly natural and thus

G. Percival (✉) · E. Schaffert · L. Hailey
Bartlett Tree Research Laboratory, Shinfield Field Unit, Cutbush Lane,
University of Reading RG2 9AF
e-mail: gpercival@bartlettuk.com

E. Schaffert
e-mail: Scienceadmin@bartlettuk.com

L. Hailey
e-mail: lhailey@bartlettuk.com

G. R. Dixon, D. E. Aldous (eds.), *Horticulture: Plants for People and Places, Volume 2*,
DOI 10.1007/978-94-017-8581-5_21, © Springer Science+Business Media Dordrecht 2014

Rural Trees

Built-on areas (roads/canals/train lines)	**Orchards/ Permanent woody crops supplementing agricultural land**	**Barren land** (rocky outcroppings, former mines)	**Pasture and meadows**
Hedgerows	Annual crops under permanent tree cover:	Isolated and scattered trees	Isolated and scattered trees
Plantings along roads and highways	Coffee	Shelterbelts/	Shelterbelts, windbreaks
Plantings along lakes and rivers	Cocoa	windbreaks	Woodlots
Botanical gardens	Olive trees	Agroforestry systems	Parklands Living fences
Rural parks	Fruit trees		Riparian buffers
Riparian buffers	Oil, date and coconut palms		
	Agroforests		
	Other agroforestry systems		

Fig. 21.1 Classification of rural trees. (adapted from Anon (1999))

not maintained, such as woodland, and riparian buffers. In spatial terms, they may be scattered discontinuously on farmland and pasture, or growing continuously in line-plantings along roads, canals and watercourses, around lakes, or in small aggregates with a spatial continuum such as tree stands or woodlands (Alexandre et al. 1999).

The rural landscape is by nature the result of several cultural and natural factors and processes. In many instances rural landscapes associated with land attributes and human activities express a country's social and economic status and function. Rural treescapes in the form of species composition, tree dimension and performance reflect ecological and environmental functions of that particular country. Rural trees may be naturally formed from seed but in general are usually cultivated. Rural trees may be exotics, domesticated species or bred by local populations for a specific purpose, as Scharma (1996) describes in regard to eucalypts planted on the Palestine coast for many reasons including as a windbreak and for dune stability.

Functions of Rural Trees

Rural trees are multi-purpose. The wealth and diversity of their uses creates a permanent, daily interaction with the human population that can be seen in production systems, leisure pursuits and economic necessity. Shade, shelter, environmental protection, and enhancement of rural and scenic surroundings are the main reasons cited for leaving trees standing in fields (Auclair et al. 2000). In developing countries rural trees are planted which offer a range of functions that in turn influence land management, labour and capital investment (Arnold 1996). In northern Brazil for example babaçu palm trees (*Attalea speciosa*) form an important and integral part of the countries cultivation system (Arnold 1996) providing many practical and commercial benefits from natural medicine, food products and roof coverings to cosmetics and cattle fodder, while oak trees (*Quercus spp.*) play a vital role in the Spanish dehesa agroforestry system (Parsons 1962).

Food

Rural trees can represent a major food source for human populations especially in under-developed and developing countries. Parklands of shea-nut (*Vitellaria paradoxa* syn. *Butyrospermum parkii*, *Butyrospermum paradoxum*), are maintained because they offer pods and nuts used as condiments or butter in Senegal, Ethiopia and Uganda (Anon 2002). In Sub-Saharan Africa and the Sahel, the pericarp of the fruits of the Doum palm (*Hyphaene thebaica*) located along the Nile River in Egypt and Sudan, in the Inner Niger Delta, and in riverine areas of northwestern Kenya are crushed to provide a meal that can substitute for cereals, being used in cakes and sweetmeats (Bernus 1980). Rural trees not only provide enhanced food security but promote dietary balance and diversity. *Tamarindus indica* is indigenous to tropical Africa, particularly in Sudan, where it grows wild; it is also cultivated in Cameroon, Nigeria and Tanzania as its fruits are high in vitamin C. The Baobab (*Adansonia digitata*) are extensively planted through-out regions of Africa as the fruit pulp provides a source of vitamin C while fresh leaves are rich in protein, vitamin A, and C, calcium, phosphorous and iron (Yazzie et al. 1994).

Animal Feed

Rural trees such as *Khaya senegalensis, Daniellia oliveri, Isoberlinia doka, Pterocarpus erinaceus, Mitragyna inermis, Dichrostachys glomerata and Ziziphus mauritiana* are a major source of livestock feed covering 16 countries in the Sahel and Sudano-sahelian regions to include Cameroon, Ghana, Mali, Senegal, and Togo. These trees can make the difference between life and death as grassland resources do not provide sufficient quality feed through-out the year, meaning livestock rely on their survival for tree fruits or leaves to enhance or balance daily fodder rations. In other regions of the world rural trees are the only source of green forage during times before the monsoon season. In India for example, mulberry trees, as well as supplying leaves for silkworms, supply feed for cattle populations that depend solely on their leaves. Trees have the advantage that they can be planted on hills, on wasteland and at the edges of ponds, and canals, etc. Trees can also be grown on boundaries where regular crops cannot be grown. Rural trees are now integrated within some farming systems of southern Africa as livestock feed banks with varying degrees of success. Within the Philippines the White Leadtree, *Leucaena leucocephala*, has been the most popular of fodders for cattle and goats while *Sesbania grandiflora* is commonly found in rice growing areas of Central Luzon on roadsides and garden perimeters.

Shelter

Animals exposed to cold winds and rain use extra feed in keeping warm. Heat loss due to these factors can lower the perceived temperature by tens of degrees making even short windbreaks valuable to livestock and farmers, especially in such climates as those of western and northern Scotland (Quine and Sharpe 1997). Exposed livestock can also develop irregular feeding patterns, becoming susceptible to pest and disease attack. Rural tree shelter can help reduce pneumonia in cattle in winter and can increase lamb survival rates. Windbreaks may also be used to protect crops, for example in the Scilly Isles evergreen trees alleviated high winds allowing flower and bulb production throughout the winter in its relatively warm climate. The crop sold at a premium on the mainland United Kingdom, being weeks earlier than its own production and provided the majority of employment on the islands (Dallimore 1913; Leuze 1966). Trees grown as windbreaks have allowed gardens such as Abbotsbury to exist in areas of high wind exposure. The south-facing slopes of Tresco Abbey Gardens are terraced to minimize exposure, but also windbreaks of Monterey Pine (*Pinus radiata*) and Monterey Cypress (*Cupressus macrocarpa*) have been established (Taylor 2008). Castle of Mey in Scotland is also very exposed and planting is only possible close to the castle behind small shelterbelts of Sycamore trees. The Arduaine Garden situated on the West coast of Scotland has predominantly Larch species planted in shelterbelts for protection (Taylor 2008). *Pinus radiata* is used extensively as windbreaks in New Zealand while River Red Gum (*Eucalyptus camaldulensis*) and wattle (*Acacia spp*) are used for similar purposes in Australia. The rain tree (*Samanea saman* (Jacq.)) is native to tropical America, and has now become widespread throughout the humid and sub-humid tropics. The tree provides protective shade and produces highly palatable pods that are suitable as a dry season feed supplement. Additionally, there are reports of enhanced grass production beneath its canopy indicating a potential role in maintaining or improving the productivity of tropical grasslands. English oak, poplar, birch, alder and hazel are used primarily as shelter belt plantings in the United Kingdom and parts of

Europe. *Euphorbia tirucalli, E. balsamifera*, and *Gliricidia sepium* are planted very closely in Indonesia to form living fences, and their clippings used as fodder while the leaves of *Hibiscus tiliaceus* feed cattle in Vanuatu. Within Europe hawthorn (*Crataegus oxyacantha*) are regularly planted to form animal proof hedgerows.

Pharmaceutical Products from Trees

Leaves, roots, bark, flowers and sap of rural trees are used for medicinal purposes world-wide. In developing countries where access to Western pharmaceuticals is not possible, medicines derived from trees can have a significant impact on human health. Bark from the African locust bean (*Parkia biglobosa*) is used in the Sahara region of Africa as a mouthwash and a vapour inhalant for toothache as well as to treat infectious diseases and digestive problems. Leaves are used for sore eyes and burns while leaf pulp is used as a diuretic (Boffa 2000b; Arbonnier 2000). The bark of *Prunus africanum*, an evergreen tree native to the montane regions of Sub-Saharan Africa and the Islands of Madagascar yields an extract of pygeum, traditionally used for fevers, malaria and wound dressing (Spore 2000). Cupuacu (*Theobroma grandiflorum*) is an important and widely cultivated fruit from the tropical regions of Brazil whose pulp is used for the preparation of a variety of cosmetic products reducing skin degeneration and restoring skin flexibility and softness (Anon 2002).

Essential oils such as Juniper Berry oil have been proven to have antimicrobial properties, and are traditionally used for medicinal purposes by the Similpal tribes in India, treating ailments such as tumours, asthma and epilepsy (Mohanta et al. 2007). *Osmanthus fragans* flowers produce a strong apricot fragrance that is used in teas and planted around temples. The oil can be extracted using solvents and is in demand in China for use in perfumes (Prance and Nesbitt 2005). The Neem Tree is a very valuable resource in developing countries such as India for its oil, which can be used to treat malaria and as a natural pesticide. In India alone there were 18 million trees that are mostly grown rurally (Ilamurugu 2005).

Medicinal uses of flowers are widely recognized by local communities in many countries. In Bangladesh the two main tree species that dominate are *Mangifera indica* and *Zizyphus mauratiana* (Mustafa et al. 2002). In a study of cultural groups in the West Kameng district in India, 3% of medicinal plant use is from flowers (Namsa et al. 2011). Flowers of the False Acacia, (*Robinia pseudoacacia*) and Judas Tree, (*Cercis siliquastrum*), are presently used as a natural laxative infused in hot water, across the European continent (Roberts 2000).

Medicinal properties of various tree saps are mainly discovered by local communities, such as the Canhane village in Mozambique, who use the sap of the Jumping Bean Tree, (*Spirostachys africana*), to treat ear and eye ailments, and Sneezewood, (*Ptaeroxylon obliquum*) to treat stomach ache (Ribeiro et al. 2010). Sap from leaves is harvested from the Marula, (*Sclerocarya birrea*), in areas of southern Africa, for application onto eyes for soreness (Gouwakinnou et al. 2011) (Table 21.1).

Table 21.1 Some medicinal woody species found in Addis Ababa, Ethiopia to treat different health problems. (Tenkir 2011)

Species	Parts used	Disease treated
Ficus thonningii	Bark and root bark	Wounds, cold and influenza
Carissa spinarum	Root	With goat milk used to treat gastric ulcer and chest complaints
	Leaf and the seed	Tooth ache
Vernonia amygadalina	Roots and stem	Intestinal upset
	Bark of young twigs	Appetizer
Cordia Africana	Roots and fruits	Ascaris
Croton macrostachus	Fruits and root	Venereal diseases
	Roots	Purgative and malaria
Bersama abyssinica	Roots	Ascaries and rabies
Olea europea subsp cuspidate	Roots	Hemorrhoids and intestinal complaints
Hagenia abyssinica	Female flowers	Remove tape worm
Maesa lanceolata	Root	Jaundice
Clausena aniseta	Root	Ascaris and influenza
Dodonaea angustifolia	Root and leaves	Hemorrhoids and wound dressing

Symbolic, Cultural, and Religious Value

Rural trees are imbued with symbolic, cultural, and religious value occurring in many aspects of culture, language, history, art, religion, medicine and politics. Trees featured in mythology include the Banyan and the Peepal (*Ficus religiosa*) trees in Hinduism, and the modern tradition of the Christmas tree in Germanic mythology, the Tree of Knowledge of Judaism and Christianity, and the Bodhi tree in Buddhism. In folk religion and folklore, trees are often said to be the homes of tree spirits. Historical Druidism as well as Germanic paganism appears to have involved cultic practice in sacred oak groves while the term *druid* itself possibly derives from the Celtic word for oak. Veteran trees when due to their age and size are valued in the United Kingdom for their beauty, solemnity and mystery (Anon 2002, Watkins 1998). It is an old traditional practice to conserve and protect an entire forest or a small wood lot in the name of the village Deity (*Kul Daveta*). This is prevalent in every village of the mountainous region of India, particularly so in Himachal Pradesh, Garhwal and Kumaon hills and also the Kashmir hills where individuals are not allowed to fell trees.

Firewood and Fuel

Wood is an extensive and widely used source of fuel in many developing countries. Wood-based fuel usage for Africa is 91 %, Asia 82 %, and Latin America 70 % (Anon 1999). In eastern Java, fuelwood meets 63 % of their energy requirements; and in central Java, 49–81 % (Salem and van Nao 1981). Trees such as laurel (*Cordia al-*

liodora), grown for shade in the coffee and cocoa plantations of Central America, supply both fuelwood and timber (Somarriba 1990). Firewood is used by 1.5 million households as the main form of domestic heating. In 1995, approximately 1.85 million m^3 of firewood was used in Victoria annually, with half being consumed in Melbourne's capital city. Species used as sources of firewood include the River Red Gum, from forests along the watercourses over much of mainland Australia, such as the Murray River, Box (*Eucalyptus angophoroides*) and Messmate Stringybark (*Eucalyptus obliqua*), in southern Australia, and Sugar Gum (*Eucalyptus cladocalyx*), a wood with high thermal efficiency, as well as Jarrah (*Eucalyptus marginata*) that usually comes from small plantations in the southwest of Western Australia.

Ethiopia has a large demand for fuel in rural areas, however problems in terms of land degradation and deforestation mean that the demand is not met (Gebreegziabher 2007). This deficit has led to household tree plantings to increase fuel availability (Gebreegziabher 2007). Common Bushweed (*Securinega virosa*) is a tree with close grained wood which is frequently used for firewood by local communities across East Africa (Weiss 1979). In South Africa, three villages were studied to observe their use of woodland resources. Nearly 92.0 % of households used fuelwood from local resources and on average this constituted 30.7 % of household value (Shackleton et al. 2002). The value of woodland resources used by local households was calculated to be between $US 315.71 and 810.61, which is a significant amount of income, as most rural households live below the poverty line. Household gardens in areas of South East Asia (Cambodia, Thialand, Vietnam) can be a valuable source of fuel and the fuel requirement of six coconut palms can support a family of five over a 5 year period (Salem and Van Nao 1981). Coconut palms can also be grown in the gardens as a source of food, and coconut shells and leaves can be used as a biomass fuel. In Southern Iran, *Tamarix* spp. are planted on the peripheral edges of crop fields to protect vegetable crops, but also to provide a source of fuelwood with 200 trees per hectare producing between 4.7 and 6.3 m^3 per tree (Salem and Van Nao 1981).

A broader range of species is used for charcoal and firewood than for other uses in many countries. For example 40 species of tree in Las Pavas, Spain are used by local people for fuelwood (Aguilar and Condit 2001). There are 53 species of tree known for their use as fuelwood throughout Cameroon and Central African Republic, with local people's preferences for tree species in these regions reflecting those most appropriate for fuelwood (Vabi 1996).

Studies of farmland in Nepal have highlighted that of the three most common trees across all the observed sites, two species, *Prunus ceradosa* and *Schima wallichii*, were used primarily as fuelwood (Gilmour 1997). *Manilkara achras* and *Pimenta dioica*, two tree species found in the Petén region of Guatemala are in demand as fuelwood, having slow burning characteristics, and are widely used by local populations (Mutchnick and McCarthy 1997). In Brazil and South Africa, eucalypts are planted along railways to provide fuel for wood-burning locomotives (Turnbull 1999). In numerous countries, e.g. India, fuelwood is the main source of energy for many rural populations (Turnbull 1999). The scale of wood shortages in many countries is such that governments alone cannot meet the needs and often including in Ethiopia, people are dependent on *Eucalyptus* as a source of fuel, and also house building materials (Turnbull 1999). Eucalypts have been grown on an increasing scale in India since 1982 by rural farmers (Turnbull 1999).

Timber and Service Wood

Rural trees provide service wood for planks, poles, beams, roofing materials, furniture and fencing materials and transport items such as wheelbarrows and carts (Anon 2002). In British Colombia, the Paper Birch (*Betula papyrifera*) has many timber uses including furniture, wood for musical instruments and decorative items in less commercially-based production (Byrne and Dow 2006). Spalted wood is particularly in demand; this is timber that has been deliberately infected with the white rot fungus (*Phanerochaete chyrsosporium*) producing attractive patterns (Anon USDA FS 2004). Indigenous trees in the Nangaritza valley, South Ecuador, are of huge importance to the Shuar, where the second largest use (23 %) of the local forest products is use in construction and housing (Pohle and Reinhardt 2004). Spiritweed trees (*Aegiphila*sp.) are commonly used in house construction for their strong fibres, as are *Inga nobilis* Willd. and the Walte Palm (*Wettinia maynensis*) (Pohle and Reinhardt 2004).

Timber from trees such as *Faidherbia albida* and *Sclerocarya birrea* in Mali is also used in local craft industries. Wood from the Pehibaye palm (*Bactris Gasipaes*) is marketed to make parquet flooring in the American tropics (Clément 1989) with 51 % derived from scattered trees in crop fields (Gakou et al. 1994). There are several species of tree grown in rural areas of East Africa that are commonly used by local populations such *Terminalia pruniodes* and *T. spinosa* which are widely sought after for main supports poles in housing (Weiss 1979). *Lannea stuhlmanni* wood is used in households to make pestles to pound maize and rice; staple food substances of this region (Weiss 1979). *Thespesia populnea* is a hard wood that is durable in salt water, often used for minor planking and repairs to small boats by local fishermen in Bagamoyo, Tanzania (Weiss 1979). Building poles are a hugely important use for tree timbers and in western Kenya 72 % of households wanted at least one species planted suitable for making building poles, and 16 % wanted at least one species suitable for timber. These uses combined account for half the rural trees that grow around farmlands and households (Scherr 1997). In the Leyte province of the Philippines 81 % of households grow rural trees on their land, mainly to produce timber for construction purposes (Cedamon et al. 2005). In countries such as Mexico, *Brosimum alicastrum*, although quite a soft wood, has many uses, including low cost furniture, beehives, packing crates and tool handles (Peters and Pardo-Tejeda 1982).

Fruit and Seeds

Fruit trees in developed countries are generally intensively managed, and rely heavily on inputs of agrochemicals and fertilizers. Fruit production from rural trees is not as structured as traditional fruit tree cultivation yet still provide important economic benefits. In developing countries i.e. Brazil, fruit plantations cover 2.3 million ha, with citrus tree plantings alone covering one million hectares (Anon 2002). Some 2.7 million ha of orchards are harvested in Iran with an annual production of *ca.* 16.5 million t that include crops such as the Persian walnut, muskmelon, lime (fruit) and pomegranate. Traditional home garden fruit trees are grown along the edges of terraces, bunds or alleys in association with farm crops such as rice, tobacco and vegetables. These kinds of systems are found throughout heavily populated areas within Africa, Asia and Australia, Caribbean, Central America, and South America. Up to two-thirds of the rural tree species used by farmers in Bangladesh are fruit or food trees (Mehl 1991). In Bukina Faso, the landlocked country in west Africa, the desert date palm (*Balanites aegyptiaca*) produces an estimated 100–150 kg of fruit per mature tree each year; Indian jujube (*Ziziphus mauritiana*) 80–130 kg; and tamarind 150–200 kg (Boffa 2000a). In Sumatra, Indonesia small holders clear an area of land to plant hill padi. After the harvest of rice, fruit tree seedlings are planted within young coffee or pepper vines. In this way, farmers harvest coffee, pepper in the medium term and fruit could be harvested 30–50 years later (de Foresta and Michon 1994; Bouamrane 1996).

Flowers

Essential oils, honey and other apiculture products are derived from rural trees. These products can constitute a financial income for many under-developed countries. Apiculture is popular in rural Mozambique, where Acacia (*Acacia* spp.) and Miombo (*Brachystegia*) are the most common honey trees. Some 20,000 traditional apiculturists in this region produce an average 360,000 kg per year of honey and 60,000 kg per year of wax. Modern apiculturists produce an estimated 20,000 kg per year of honey, and 8,000 kg of wax (Mange and Nakala 1999). Wild honey gathering from a wide number of rural tree species produces 20,000 t of honey in India annually, twice the amount from managed beekeeping systems. In Mirpur, North Pakistan, traditional apiculturists manage 7,000 colonies which produce 70 t of honey mainly from *Zizyphus jujuba, Zizyphus sativus, Adhatoda vasica, Acacia, Arabica*, and *Robinia pseudoacacia* flowers (Ahmad et al. 2007). In Europe, the main rural tree species producing unifloral honeys are Sweet Chestnut, *Eucalyptus* and *Robinia*, although Hawthorn also has great apicultural importance in Spain (Oddo et al. 2004). China is a major honey-producing country and the common rural Ivy tree, (*Schefflera octophylla*), provides an important winter source of nectar in apiculture, producing 10–15 kg per colony (Youhua 1997). The Tupelo or River lime (*Nyssa ogeche*), grows naturally along three major rivers and their tributaries in the US state of Florida, with the Tupelo honey produced from their flowers being highly desired (Ciesla 2002).

Sap

Beverages made from sap are popular, such as palm wine, made from the rural tree species *Borassus aethiopum*, *Hyphaene coraiacea* and *Phoenix reclinata*. Palm wine production in Mozambique, with daily production figures as high as 20 L per day, is a year-round activity for households living along major roads (Mange and Nakala 1999). In the Bassila region of Benin, palm wine-making and the related distilled product is the only tree-linked enterprise which can cover an individual's subsistence costs (Boffa 2000a; Obire and Putheti 2010). Indigenous tribes in North America tap sap from Maple trees; *Acer saccharum* is frequently tapped for its sap, and is usually boiled-down to make syrup (Ciesla 2002). An alternative to maple syrup is made from birch sap, containing a higher content of fructose than sucrose. The extraction of birch sap from birch trees is a common practice in Korea, Finland, Japan, and Ukraine where it is a valuable drink in its raw form (Cielsa 2002) as well as a highly valued product in British Colombia (Byrne and Dow 2006). Birch beer is made from sap harvested from trees in the southern Appalachian Mountains, in eastern North America (Ciesla 2002). Birch sap also has medicinal value, and is proven to strengthen the immune system and be an effective diuretic (Zevin et al. 1997). The sap of the Kitul palm (*Caryota urens*) in Sri Lanka is an important food source for local communities; being boiled down to make a sugar-like substance called "Jaggery" (Caron 1995). The Coconut palm (*Cocos nucifera*), produces palmyra sap which is popular in Sri Lanka, after it is fermented and distilled to make a common beverage named "Arrack" (Albala 2011). Natural gum is a valuable resource from several gum and resin producing trees such as *Eucalyptus* spp. producing three times the income at a household level than crop production (Tadesse et al. 2007). The sap from the Blood Tree, (*Harunga madagascariensis*), in East Africa is used to make an orange dye for use in households of local communities and by local fishermen (Weiss 1979). Mangrove tree species have high tannin content, meaning industrial use of their sap as a black dye is common in production of Polynesian Tapa cloth (Tomlinson 1994). Atlantic poison oak, (*Toxicodendron pubescens*), has been recognized as a culturally significant plant in USA, as the sap can be harvested and used as a shoe polish, and a permanent ink (Casey and Wynia 2010). The sap of *Ficus spp.* are all well known in Ecuador for their use in candle making, as are other tree resins such as *Hevea brasiliensis*, and *Mauria spp.* (Pohle and Reindhart 2004). Resins obtained from the bark of *Copaifera demeusi* are used as a varnish, and are a popular source of income in rural areas of the Belgian Congo (Anderson 1995).

Rural Trees—A Underrated Environmental Resource

An African adage states that the "Earth is not a legacy we inherit from our parents, but rather a loan from our children". Rural trees play a leading role in meeting resource conservation and management needs (Anon 2002).

Soil and Water

World-wide the rate of soil degradation now outstrips soil regeneration. In Canada, soil degradation costs $US 1 billion a year, while in India soil erosion affects 25–30 % of the total land under cultivation. Without conservation measures, the total area of cropland in developing countries in Asia, Africa, and Latin America would shrink by 544 million ha over the long term due to soil erosion and degradation. The role of rural trees in halting the advance of desert land, checking wind and water erosion, facilitating the percolation of rainwater, and enhancing agricultural production is widely accepted. In dense or thin stands, rural tree line plantings, singly or in hedgerows, preserve soil organic matter and boost soil fertility (Roose and Boli 1999). Farmers in Niger exploit the nitrogen fixing *Faidherbia albida* for this purpose as well as providing shade and shelter for humans and animals. Other important tree species that serve similar purposes include *Combretum glutinosum*, *Guiera senegalensis* and *Adansonia digitata.* In countries such as Egypt, Iraq and Libya, windbreaks have a substantially positive impact on production yields (Anon 1986). Rural tree planting has been an old tradition in many Near East countries. Among the cultivated trees, poplars occupy prominence in Turkey, Syria and Iraq. *Casuarina* spp. predominate on the irrigated land of Egypt whilst *Eucalyptus* spp. are grown widely in the rain-fed areas of the region. More recently, some acacias (e.g. *A. saligna*) have been grown for sand dune stabilization. The area of land in China vulnerable to desertification is dwindling by about 128,300 ha per year due to planting efforts using primarily poplars, eucalyptus, tamarind and pines. Similarly, windbreaks planted in Mauritania stabilize sand dunes, reduce the advance of the desert, and combat drought (Salem 1991). Oases in Iraq intercrop a top story of palms, with an understory of fruit trees preventing wind erosion of soil. Farmers in the mountainous areas of Iran leave 20–100 trees per hectare standing on farmland to ensure soil stabilization and crop protection. For the same reasons, Afghan farmers plant and grow mulberry trees, poplars, and eucalypts and fruit trees along the boundaries of fields and along irrigation canals (El-Lakany 1993). *Acacia colei* trees are pruned once a year and their branches spread over potential farming land to increase soil fertility enhancement. In addition, mulches made solely from *A. colei* are used as a source of organic matter throughout the Sudano-Sahelian region of Africa. Rural trees, regenerating spontaneously and/or planted to maintain or extend tree cover, are of great benefit in watershed management, reducing soil degradation and controlling desertification.

Biological Diversity

Rural tree line-plantings are a source of biological richness for ecological preservation, bio-conservation, water purification, and storm protection. Rural trees growing in line-plantings, clumps or woodlands can have a role in conservation-oriented water, biomass and soil fertility management (Anon 2002). Where established, tree plantings can replace mechanical approaches to soil and water protection, conservation and restoration. Likewise rural trees can provide a valuable phytogenetic resource that can be utilized to sustainably increase food supplies and address environmental issues such as climate change (Anon 2002). Within the United Kingdom for example native oaks, *Quercus* spp., are of very high importance to wildlife. They are unique hosts of at least 284 species of foliar feeding insects, as well as many wood decay, leaf litter and seed feeding invertebrates supporting predators, parasites and hyperparasites, which rely on those mentioned above (Kennedy and Southwood 1984). Areas of rural tree cover can support a more substantial and diversified biomass than forests. Tree savannah of the Serengeti plains in Tanzania provides shade, protection, and shelter for mammals and birds. Hedgerows, windbreaks and woodlands on farmland constitute refuges for wild plants and animals, forming islands of biodiversity and biological corridors. Woody species grow and thrive at the foot of tree rows, their seeds having been distributed by animals, in a further contribution to biodiversity. The existence of many animal species has been safeguarded by the planting, maintenance or restoration of hedges. Trees lining rivers and streams are a source of biological enhancement, providing spawning-beds for fish and shellfish, as their shade acts to limit the development of aquatic flora, thus reducing eutrophication. Rural tree linings also function as biological corridors for terrestrial wildlife (Anon 2002). Standing dead wood is of particular importance to bats and birdlife, providing roosting and nesting sites. Such sites include tree cavities and loose bark, including that of stumps. Dead wood characteristics define bat use as a specific roosting or nesting site. Home gardens in many countries offer refuges for rare plant and tree species contributing to the biological spectrum. This is also true of agroforests, with their high densities and range of woody species, which render environmental status comparable to those of a forest.

Rural Trees as Water Purification Systems

Rural trees planted on slopes, as buffers, hedgerows and windbreaks, can purify surface and groundwater, industrial and residential effluents, downwashes from power lines, storm waters, agricultural runoffs, diluted sludges, and even radionuclide-contaminated solutions (Anon 2002). Rural trees with rapid-growth root systems and tolerance to flooding such as willows and poplars can remove toxic metals such as copper, cadmium, chromium, nickel, lead, and zinc as well as low levels of radioactive contaminants from streams. Importantly rural trees grow on land of marginal quality and possess long life-spans with little or no maintenance costs. A young tree buffer can remove an average of 0.38 g nitrogen per day per square metre of land; equivalent to 38 times greater than a grassy meadow (Balent 1996). The restoration of a riparian buffer 50 m wide along 120 km of the Garonne river in France can, during the growing season, remove 5.6 t per day of nitrogen that otherwise is deposited within local waterways.

Climate

Rural trees can act as both reservoirs and potential sources of carbon. The role of tropical forest ecosystems in carbon storage and release is recognized in the global context in the regulation of atmospheric carbon, and subsequent reduction of greenhouse gases. The impact of rural trees on increasing sequestered carbon has received little study but will become increasingly meaningful as carbon dioxide (CO_2) emissions world-wide increase annually (Kongsager et al. 2012; Nepal et al. 2012).

Economic Benefits of Rural Trees

It is hard to assess the economic contribution of rural trees at the local level and even more complex at national and international levels. Wood and non-wood products from rural trees only classify as statistics when traded on official, national or export markets. This has the effect of masking the economic impact of the rural tree resource. It is evident however that the flowering period for orchards and arboreta are becoming tourist attractions in developed countries and can provide substantial boosts to local rural economies (Dixon 2012).

Tree autumnal colours are a tourist attraction to visitors to North America, particularly since the 1950s in New England (Ryden 2001). The colours are described as a phenomenon that attracts thousands of visitors that travel to view this forest which has increased in size quite significantly post-war. Westonbirt arboretum is the best example of tourist attraction for autumn colours within the United Kingdom, with wide varieties of Maple, Beech and Cherry (Mitchell 1968).

Rural Tree Management

Rural trees in developed countries are managed with specific strategies designed to ensure trees do not become structurally unsound and a danger to the public. Likewise in developing countries some form of management system usually exists where in other countries rural trees are left unmanaged. Irrespective of the appropriate management system, maintaining a stable environment around rural trees is critical in delaying the transition from maturity to decline. Management programmes should be proactive rather than reactive. Treatments should be applied preventively to maintain tree health rather than remedial once decline begins. Properly performed cultural practices, including pruning, soil and root system management and timely pest management can increase the longevity of rural trees. Within certain countries, such as India or Vietnam for example, woodlands are protected at the community level with the area divided into small blocks or compartments with each allotted to local inhabitants.

Inspections

Periodically inspecting trees for structural defects, including dead and broken limbs, cracks, split crotches, decay and root defects, is essential to prevent failures that could lead to personal injury or property damage. Improved guidelines for evaluating defective trees and communicating risk to the consumer have been developed (Hayes 1997). Inspections also reveal early symptoms of stress that may be corrected before decline occurs. Tree structure and health inspections should be performed annually and after major storms.

Tree Protection

Domestic animals such as goats and cows will gnaw the bark of trees such as elm or beech. Cows, sheep and goats will eat the foliage and twigs of trees of all sizes. With young trees this is usually fatal. Wild animals such as rabbits, hares, grey squirrel, and deer can damage trees by eating the leading shoots and side branches, gnawing the bark, damaging fruit and eating seeds and buds. Regular inspections and maintenance programmes can arrest some forms of injury especially in the early years, such as the removal of a dead or damaged branch. However, in most cases fencing or tree guards are used.

Conservation Dead Wooding

Whilst the removal of dead wood is an important part of ensuring tree health and reducing the potential for failure, dead wood provides an important ecological resource. The dead wood of different species has varying importance due to both the plant's longevity and its own, but all have some ecological value. For example oak wood decays slowly and the trees are long-lived whilst birch wood decays rapidly and the tree has a shorter life-span. The size and positioning of dead wood; twigs to trunks, on the ground to in the canopy, change the resources and niches rural trees can provide. Dead limbs, holes, decay columns, stumps and fallen dead wood of varying size and even dead wood that falls into watercourses, can support different communities and have different ecological effects (Ferris-Kaan et al. 1993). The niches and their resources are utilized by birds, bats, invertebrates, fungi and the species which consume them (Ferris-Kaan et al. 1993; Bunnell and Houde 2010).

The hazards to human health presented by dead wood such as branch failure can lead to management plans that heavily restrict the amount of dead wood left on the tree. Conservation dead wooding is a compromise between safety and ecological benefit where only extremely hazardous dead wood is removed and the remainder left for utilization by wildlife. Branch snags up to 300 mm are left, providing less hazardous standing dead wood. Large limbs which represent the greatest danger by failure are removed, being removed at the point where their diameter exceeds 50 mm. Another option for creating such naturalistic branch ends is 'coronet' cutting, which mimics nature's own breakages in the wood. Any dead wood removed from a tree need not be removed from the site itself if viable and left to decay naturally for further ecological benefits.

The Future of Rural Trees

Trends in rural tree development differ between underdeveloped, developing and developed countries. For example rural trees within Europe located within hedgerows, fruit-tree meadows and riparian buffers have increased in number due to conservation and rehabilitation efforts. These trends can be linked to changes in agricultural economies, and needs of urban infrastructure for wood and non-wood forest products, which have in turn driven up the economic value of rural trees. Population growth within urban environments has created a situation where a larger number of tree types and formations need to be conserved in rural areas to meet potential economic challenges. Contrary to this, with the expansion of agricultural systems in tropical countries, land clearing is often followed by a phase in which trees supply wood and non-wood products as an intrinsic part of the newly implemented production systems. Agricultural development within developing industrialized countries, with its stages of mechanization, irrigation, ever-larger plots and farms and land consolidation has been responsible for the removal of many trees from the rural landscape (Anon 2002).

References

Aguilar S, Condit R (2001) Use of native tree species by an hispanic community in Panama. Econ Bot 55(2):223–235

Ahmad F, Joshi SR, Gurung MB (2007) Beekeeping for rural development. ICIMOD, Nepal

Albala K (2011) Food cultures of the world encyclopedia: four volumes. ABC-CLIO, California

Alexandre DY, Lescure JP, Bied-Charreton M, Fotsing JM (1999) Contribution à l'état des connaissances sur les arbres hors forêt (TOF). IRD-FAO, France, p 185

Anderson A (1995) Recovery and utilization of tree extractives. Econ Bot 9(2):108–140

Anon (1986) Brise-vent et rideaux abris avec référence particulière aux zones sèches, FAO Conservation Guide 15. FAO, Rome, p 385

Anon (1999) State of the World's Forests. FAO, Rome, p 154

Anon (2002) Trees outside forests. A key factor in integrated urban and rural management. Conservation Guide 35. FAO, Rome, pp 1–59

Anon (2004) Producing spalted wood. Forest Products Laboratory, Madison. Available via USDA. Forest Service (FS). http://www.fpl.fs.fed.us/documnts/techline/producing-spalted-wood.pdf. Accessed 15 Nov 2012

Arbonnier M (2000) Arbres, arbustes et lianes des zones sèches d'Afrique de l'Ouest. CIRAD-MNHN-UICN, Paris, p 541

Arnold JEM (1996) Economic factors in farmer adoption of forest product activities. In: Leakey RRB, Melnyk M, Vantomme P (eds) Domestication and commercialization of non-timber forest products in agroforestry systems. Proceedings of an international conference held in Nairobi, Kenya Feb 1996, vol 9 Non-Wood Forest Products. FAO, Rome

Auclair D, Prinsley R, Davis S (2000) Trees on farms in industrialized countries: silvicultural, environmental and economic issues. Proceedings of XXI IUFRO World Congress, Kuala Lumpur, Malaysia, August 2000, p 776

Balent G (1996) La forêt paysanne et l'aménagement de l'espace rural. Etudes et Recherches sur les Systèmes Agraires et le Développement 29:267

Bernus E (1980) L'arbre dans le Nomad's land. In: Pélissier P (1980) L'arbre en Afrique tropicale-La fonction et le signe. Cahiers ORSTOM des Sciences Humaines 17(3–4):171–176

Boffa JM (2000a) Les parcs agroforestiers en Afrique de l'Ouest: clés de la conservation et d'une gestion durable. Unasylva 51(200):11–17

Boffa JM (2000b) Les parcs agroforestiers en Afrique subsaharienne. FAO Conservation Guide 34. FAO, Rome, p 258

Bouamrane M (1996) A season of gold-putting a value on harvests from Indonesian agroforests. Agrofor Today 8(1):8–10

Bunnell FL, Houde I (2010) Down wood and biodiversity-implications to forest practices. Environ Rev 18:397–421

Byrne W, Dow K (2006) Market and product development for birch timber and non-timber products: current status and potential in British Colombia. Royal Roads University, Victoria

Caron CM (1995) The role of non-timber tree products in household food procurement strategies: profile of a Sri Lankan village. Agrofor Syst 32:99–117

Casey PA, Wynia RL (2010) Culturally significant plants. United States Department of Agriculture, Kansas

Cedamon ED, Ematage NF, Suh J, Herbohn JL, Harrison SR, Mangaoang EO (2005) Present tree planting and management activities in four rural communities in Leyte province, the Philippines. Ann Trop Res 27(1):19–34

Ciesla WM (2002) Non wood forest products from temperate broad-leaved trees. FAO, Rome

Clément CR (1989) The potential use of the pejibaye palm in agroforestry systems. Agrofor Syst 7(3):201–212

Dallimore W (1913) Minor Agricultural Industries II. The culture of early flowers in Cornwall and the Scilly Islands. Bull Misc Inf 1913(5):171–177 (Royal Gardens, Kew)

Dixon GR (2012) Apple phenomenon (Cider). Horticulture week, 4th May 2012, pp 33–34

El-Lakany MH (1993) Forestry policies in the near east region: analysis and synthesis. In: Anon (ed) FAO Forestry Paper Series, vol 111. FAO, Rome

de Foresta H, Michon G (1994) Agroforests in Indonesia: where ecology and economy meet. Agrofor Today 6(4):12–13

Ferris-Kaan R, Lonsdale D, Winter T (1993) The conservation management of deadwood in forests, Research Information Note 241, Forestry Commission. http://217.205.94.38/pdf/RIN241.pdf/$FILE/RIN241.pdf. Accessed 13 Aug 2012

Gakou M, Force JE, McLaughlin WJ (1994) Non-timber forest products in rural Mali: a study of villager use. Agrofor syst 28:213–226

Gebreegziabher Z (2007) Household fuel consumption and resource use in rural-urban Ethiopia. Wageningen University, Wageningen

Gilmour DA (1997) Rearranging trees in the landscape in the middle hills of Nepal. In: Arnold JEM, Dewees PA (eds) Farms, trees and farmers: responses to agricultural intensification. Earthscan, London, p 21

Gouwakinnou GN, Lykke AM, Assogbadjo AE, Sinsin B (2011) Local knowledge, pattern and diversity of use of Sclerocarya birrea. J Ethnobiol Ethnomed 7:1–9

Hayes SC (1997) Behavioral epistemology includes nonverbal knowing. In: Hayes LJ, Ghezzi PM (eds) Investigations in behavioral epistemology. Context Press, Reno, pp 35–43

Ilamurugu S (2005) Neem for sustainable agriculture. In: Bandopadhyay A, Sundaram KV, Moni M, Kundu PS, Mrityunjay M (2005) Sustainable agriculture. Northern Book Centre, New Dehli, p 223

Kennedy CEJ, Southwood TRE (1984) The number of species of insects associated with British trees: a re-analysis. J Anim Ecol 53:455–478

Kongsager R, Napier J, Mertz O (2012) The carbon sequestration potential of tree crop plantations. Mitig Adapt Strategies Glob Chang 18:1–17

Leuze E (1966) Die Scilly-Inseln (The Isles of Scilly). Erdkunde 20(2):93–103

Mange DP, Nakala MO (1999) Country brief on non-wood forest products, Republic of Mozambique. Data collection and analysis for sustainable forest management in ACP countries. FAO-EC Partnership Programme. Available via FAO. http://www.fao.org/forestry. Accessed 5 Sep 2012

Mehl CB (1991) Trees and farms in Asia: an analysis of farm and village forest use practices in South and Southeast Asia multipurpose tree species network research series, report n°18. US Agency for International Development, Washington DC, p 83

Mitchell AF (1968) Westonbirt in colour. H. M. Stationery Office, UK

Mohanta TK, Patra JK, Rath SK, Pal DK, Thatoi HN (2007) Evaluation of antimicrobial activity and phytochemical screening of oils and nuts of Semicarpus anacardium L.F. Sci Res Essay 2:486–490

Mustafa MM, Teklehaimanot Z, Haruni AKO (2002) Traditional uses of perennial homestead garden plants in Bangladesh. For Trees Liveli 12:235–256

Mutchnick PA, McCarthy BC (1997) An ethnobotanical analysis of the tree species common to the subtropical moist forests of the Petén, Guatemala. Econ bot 51:158–183

Namsa ND, Mandal M, Tangjang S, Mandal SC (2011) Ethnobotany of the Monpa ethnic group at Arunachal Pradesh, India. J Ethnobiol Ethnomed 7(31):1–14

Nepal P, Ince PJ, Skog KE, Chang SJ (2012) Projection of US forest sector carbon sequestration under US and global timber market and wood energy consumption scenarios, 2010–2060. Biomass Bioenerg 45:251–264

Obire O, Putheti RR (2010) The oil Palm Tree: a renewable energy in poverty eradication in developing countries. Drug intervention Today 2:34–41

Oddo LP, Piana L, Bogdanov S, Bentabol A, Gotsiou P, Kerkvliet J, Martin P, Morlot M, Ortiz-Valbuena A, Ruoff K, Von DOheK (2004) Botanical species giving unifloral honey in Europe. Apidologie 35:82–93

Parsons JJ (1962) The acorn-hog economy of the oak woodlands of southwestern Spain. Geogr Rev 52(2):211–235

Peters CM, Pardo-Tejeda E (1982) Brosimum alicastrum (Moraceae): uses and potential in Mexico. Econ bot 36(2):166–175

Pohle P, Reindhart S (2004) Indigenous knowledge of plants and their utilisation among the Shuar of the lower tropical mountain forest in southern Ecuador. Lyonia 7(2):134–149

Prance G, Nesbitt M (2005) The cultural history of flowers. Routledge, Abingdon

Quine CP, Sharpe AL (1997) Evaluation of exposure and the effectiveness of shelterbelts on the Western and Northern Isles of Scotland. Scott For 51(4):210–216

Ribeiro A, Romeiras MM, Tavares J, Faira MT (2010) Ethnobotanical survey in Canhane village, district of Massingir, Mozambique: medicinal plants and traditional knowledge. J Ethnobiol Ethnomed 6:1–15

Roberts MJ (2000) Edible and medicinal flowers. Spearhead, Claremont

Roose E, Boli Z (1999) L'influence de l'homme sur l'érosion, vol 1. Centre IRD, Montpellier. Bulletin du Réseau Erosion 19:608

Ryden KC (2001) Landscapes with figures: nature & culture in New England. University of Iowa Press, Iowa

Salem BB (1991) Combattre et prévenir l'érosion éolienne dans les régions arides. Unasylva 2(164):33–39

Salem BB, Van Nao T (1981) Fuelwood production in traditional farming systems. Unasylva 33:13–18

Scharma S (1996) Landscape and Memory. Fontana Press (Harper Collins), London

Scherr SJ (1997) Meeting household needs: farmer tree-growing strategies in western Kenya. In: Arnold JEM, Dewees PA (1997) Farms, trees and farmers: responses to agricultural intensification. Earthscan, London

Shackleton SE, Shackleton CM, Netshiluvhi TR, Geach BS, Ballance A, Fairbanks DHK (2002) Use patterns and value of savanna resources in three rural villages in South Africa. Econ Bot 56(2):130–146

Somarriba E (1990) Sustainable timber production from uneven-aged shade stands of Cordia alliodora in small coffee farms. Agrofor Syst 10(3):253–263

Spore (2000) La recherche au secours d'un arbre à trésor. Spore 88:9

Tadesse W, Desalegn G, Alia R (2007) Natural gum and resin bearing species in Ethiopia and their potential applications. Invest Agrar Sist Recur For 16(3):211–221

Taylor P (2008) Gardens of Britain and Ireland. Dorling Kindersley, London

Tenkir E (2011) Trees, people and the built environment. In: Proceedings of the urban trees research conference, Institute of Chartered Foresters and Forestry Commission 13–14th April 2011

Tomlinson PB (1994) The botany of mangroves. Cambridge University, Cambridge

Turnbull JW (1999) Eucalypt plantations. New For 17:37–52

Vabi M (1996) Eliciting community knowledge about uses of trees through participatory rural appraisal methods: examples from Cameroon and Central African Republic. Rural Dev For Netw Pap 19e:30–37

Watkins C (1998) A solemn and gloomy umbrage: changing interpretations of the ancient oaks of Sherwood forest. In: Watkins C (ed) European woods and forests: studies in cultural history. CAB International, Wallingford

Weiss EA (1979) Some indigenous plants used domestically by East African fishermen. Econ bot 33(1):35–51

Yazzie D, VanderJagt DJ, Pastuszyn A, Okolo A, Glew H (1994) The amino acid and mineral content of Baobab (Adansonia digitata L.) leaves. J Food Compos Anal 7(3):189–193

YouHua Z (1997) Ivy Tree: a major nectar plant in winter for Apis cerana in South China. Bee World 78:128–130

Zevin IV, Altman N, Zevin LV (1997) A russian herbal: traditional remedies for health and healing. Healing Arts Press, Vermont

Chapter 22
Management of Sports Turf and Amenity Grasslands

David E. Aldous, Alan Hunter, Peter M. Martin, Panayiotis A. Nektarios and Keith W. McAuliffe

Abstract Grasslands are significantly important vegetation resources that occupy about one-quarter of the earths vegetative surface. Whereas sports turf has been defined as a surface layer of vegetation, consisting of earth and a dense stand of grasses and roots, amenity grasslands have been defined as grasslands which provide recreational, functional, or aesthetic value, but which is not used primarily for agricultural production. Sports turf and amenity grassland management can vary in terms of the grass assets, function, composition, use and level of intensity and their annual cost of maintenance. The resources required to maintain the level of playing surface performance are described in terms of mowing, fertilizer application, irrigation, aeration, topdressing and plant protection procedures, as well as a means of effectively measuring player and playing surface performance.

Keywords Turf management · Amenity grassland management · Sports turf · Playing quality · Playing surface performance

---

Professor David E. Aldous – deceased 1st November 2013

A. Hunter (✉)
College of Life Sciences, School of Agriculture and Food Science,
Agriculture & Food Science Centre, University College Dublin, Belfield, Dublin 4, Ireland
e-mail: alan.hunter@ucd.ie

D. E. Aldous
School of Land, Crop and Food Science, The University of Queensland,
Gatton Campus, Lawes, Queensland 4343, Australia

P. M. Martin
Amenity Horticulture Research Unit, University of Sydney Plant Breeding Institute,
107 Cobbitty Road, Cobbitty, NSW 2570, Australia
e-mail: peter.martin@sydney.edu.au

P. A. Nektarios
Department of Crop Science, Lab. of Floriculture and Landscape Architecture,
Agricultural University of Athens, Iera Odos 75, 11855 Athens, Greece
e-mail: pan@aua.gr

K. W. McAuliffe
Sports Turf Research Institute, 52 Raby Esplanade, Ormiston, QLD 4160, Australia
e-mail: keithm.sti@gmail.com

G. R. Dixon, D. E. Aldous (eds.), *Horticulture: Plants for People and Places, Volume 2*,
DOI 10.1007/978-94-017-8581-5_22,

Introduction

Approximately one-third of the world's vegetative cover consists of grass (*Poaceae*) dominated ecosystems (Jacobs et al. 1999) and approximately 80% of this area is managed amenity grassland plant communities in western countries. The *Natural Environmental Research Council (NERC)* in Great Britain *defined amenity* grassland as "all grassland which has recreational, functional, or aesthetic value, but which is not used primarily for agricultural production" (NERC 1977; Rorison and Hunt 1980). This inclusive definition covers the spectrum from intensively managed sports turf through lawns of all types to low-input but managed semi-natural and natural grasslands. Sports turf assets range from elite stadiums on fully engineered growing-medium profiles with annual turf maintenance costs (including turf replacement) that exceed AU$ 200,000/ha to suburban sports fields on natural soil with maintenance costs ranging from AU$ 14,500 to AU$ 18,600/ha per annum (Anon 2007). Regardless of its intensity, the management of sports turf and lawns aims to maintain a grass monoculture suitable for regular active or passive use (Beard 1973), whereas in contrast the management of natural and semi-natural grasslands aims to maintain a species rich community of grasses and non-grasses for aesthetic, biodiversity and conservation purposes (NERC 1977; Groves 1990; Marshall 1994). Despite some overlap, the underlying aims and the management practices applicable to regularly mown amenity grasslands as against natural and semi-natural amenity grasslands are very different and both areas have developed an extensive research literature. However, because of limitations of space the primary focus of this review will be sports turf and regularly mown amenity grasslands.

In economic terms, the sports turf and amenity grassland industry can be categorized into four sectors, namely, the physical asset or facility; the manufacturing and supply sector (e.g. equipment, seed, fertilizer); the servicing sector (e.g. designers and consultants, distributors and retailers, soil and water testing laboratories) and the research, education and training sector (community colleges and/or universities, public and private research, extension and outreach agencies) (Beard 1973; Aldous et al. 2007), all of which add considerable value to the gross domestic product (GDP) of many developed and developing countries. Hence a comprehensive view of sports turf and amenity grasslands needs to include the elements of design, construction, establishment, management and performance assessment of these assets. Considerable communal benefits flow from the upkeep of amenity grasslands, including provision of a healthy environment for humans, protection of biological diversity, reduced global warming and amelioration of the fundamental life-support systems of air, soil and water (Beard and Green 1994; Higginson and McMaugh 2008).

Grass Types and Seasonal Growth Cycles

The family *Poaceae* contains approximately 785 genera and 10,000 species (Watson and Dallwitz 1992). Of these, only about 25 species are generally regarded as quality turfgrasses i.e. grasses having of desirable characteristics required in sports turf or lawns, although many more species find use as plant cover in lower-input amenity situations or in special habitats such as heavy shade or salt affected land (Beard 1973; Duble 2001; Christians 2003; Aldous and Chivers 2002). An interesting feature of the distribution of the main turfgrasses (and many of the associated weeds) is the extent to which they have become naturalised in suitable climatic regions in countries far removed from their native habitats. By the mid-nineteenth century many turf grasses indigenous to Asia, Europe and Africa had been introduced to North and South America, Australasia and other parts of the world either by the deliberate action of some early settlers, or accidentally imported as seeds contained in livestock feed, packing around household goods or amongst the ballast of ships. For case-studies, see Warnke (2003) on *Agrostis stolonifera* and Sauer (1972) on the *Stenotaphrum* species.

In physiological terms, the species of the grass family fall into two distinct groups, those characterised by the C_3 photosynthetic pathway and those characterised by the C_4 pathway. This well-known subdivision is of fundamental importance in relation to the adaptation and productivity of the grasses. One of the most obvious differences is in temperature adaptation; C_3 grasses having an optimum temperature for growth of circa 18 °C and a minimum circa 4 °C, while for C_4 species the corresponding figures are more than 30 °C and 10 to 12 °C. In the natural distribution of grasses, C_3 species dominate in the cool-temperate regions while C_4 species dominate in the tropical areas. In areas intermediate between these two climatic types, species from both groups are prominent, but with different seasonal growth pattern governed by the time of occurrence of appropriate temperature conditions together with adequate solar radiation. Beard (1973), Watson (1994) and Wilson (1999) provide detailed descriptions of the morphological, ecological and physiological characteristics of the C_3 (cool-temperate) and C_4 (tropical) turfgrasses.

In Great Britain, northern Europe and the cool humid parts of the northern United States, C_3 turfgrasses such as *Lolium perenne* and *Agrostis stolonifera* achieve their best growth in the summer months when temperatures are optimal and solar radiation is at its maximum, while winter production is severely limited or even reduced to zero by low temperatures and minimal solar radiation. On the other hand, these same two grasses grown in a warm-temperate area such as Brisbane, Australia, achieve their maximum growth in spring and autumn when temperature and solar radiation conditions are favourable, are moderately productive in the mild winters (but somewhat solar radiation limited) and are least productive at the height of summer because temperature conditions are distinctly supra-optimal for them. As might be expected, C_4 turfgrasses such as *Cynodon dactylon* and *Stenotaphrum secundatum* grow actively when temperatures are sufficiently high and water is not limiting, meaning that in lower-altitude equatorial climates year-round growth is typical, while in warm-temperate areas there is a distinctly seasonal pattern with growth slowing in the autumn and ceasing in the winter as the grass goes into dormancy.

As a result of the differential sensitivity to temperature of the two physiological groups of grasses, in warm-temperate regions such as the southern United States, parts of South America and much of Australia, as well as areas with a Mediterranean climatic pattern, under conditions of adequate water, C_4 turfgrasses provide good playing surface conditions over spring and summer, but are generally inadequate over winter, whereas C_3 grasses provide a good playing surface in autumn, winter and much of spring but lose vigour and have difficulty recovering from wear during summer. In areas such as these, not surprisingly, it has long been the custom to refer to the two groups as "warm season grasses" and "cool season grasses". If, however, we consider a cool-temperate maritime region with short summers and rarely a daily maximum above 22 °C, it is apparent that the only cool season turfgrasses would flourish during the warm season of that locality. The lack of logic in this terminology when applied on a global basis is apparent and it would be preferable to use the C_3 and C_4 descriptors instead, but gaining general acceptance for this approach could be a slow process, because as Watson (1994) pointed out that the "cool season"/"warm season" terminology, which originated in the USA, has been widely adopted in turf science literature.

The main grass species that account for greater than 90 % of the turf managed worldwide (Martin 2004) are as follows:

- C_3 (cool season) grasses, *Agrostis stolonifera, Agrostis capillaris, Festuca arundinacea, Festuca rubra* subsp.*rubra, Lolium perenne* and *Poa pratensis.*
- C_4 (warm season) grasses, *Axonpus affinis, Cynodon dactylon* (inclusive of the interspecific hybrids with *C. transvaalensis*), *Pennisetum clandestinum, Paspalum notatum, Stenotaphrum secundatum* and *Zoysia japonica.*

All of the listed C_3 species and three of the C_4 species (*Cynodon dactylon, Stenotaphrum secundatum* and *Zoysia japonica*) have experienced at least 150 years of domestication during which time many ecotypes have been selected and in the last 50 years a considerable number of cultivars of almost all the species listed have originated from hybridisation programs. There are, however, other species, such as *Festuca ovina, Digitara didactyla* Willd. and *Eremochloa ophiuroides* (Munro) Hack. that receive little attention from plant breeders but have for many decades occupied small but steady market niches. In addition, relative newcomers for commercial use such as *Dactylocterium australe* (R.Br.) Beauv. *Paspalum vaginatum* Sw., *Sporobolus virginicus* (L.) Kunth., *Microlaenia stipoides* (Labill.) R.Br. and *Buchloe dactyloides* (Nutt.) Columbus have proved to be useful genera for special environments including inundated and salt affected land (Duncan and Carrow 2000; Loch et al. 2006; Marcum 2008) and low light environments (Dudeck and Peacock 1992), leading to a small but growing demand for planting material.

Beard (1973), Craigie (1994), Martin (2004) and Turgeon (2008) provide important criteria in selecting and assessing sports turf and amenity grassland species on merit (Table 22.1). Criteria common to all the lists are the need for rapid establishment, persistence, fitness for purpose and good appearance. Additional criteria for sports turf species include high seed yield in the traditional seed-growing districts and tolerance of the pesticides commonly used in turf management (Martin 2004). For traditional mown lawns in urban public space, Hitchmough (1990) provided recommendations for species that could be established at a relatively low cost and yet tolerate use for a range of informal and formal recreational activities.

Table 22.1 Environmental adaptations of some common cool and warm-season turfgrasses. (Source: Aldous 1999)

Turfgrass	Temperature		Sun/Shade		Drainage potential		Water requirements		Soil fertility		Preferred soil pH	
	Cool-season	Warm-season	Sun	Shade	Good	Poor	High	Low	High	Low	pH 6–8	pH 5–7
Agrostis capillaris	*		*		*	*	*		*			*
Agrostis stolonifera	*		*			*	*		*			*
Agrostis gigantea	*		*			*	*		*			*
Agrostis canina	*		*			*	*		*			*
Cynodon dactylon		*	*		*			*	*			*
Buchloe dactyloides		*	*		*		*	*		*	*	
Axonopus affinis		*		*						*		*
Digitaria didactyla		*	*		*	*		*	*			*
Pennisetum clandestinum		*	*		*			*		*		*
Microlaena stipoides		*			*			*			*	
Eremochloa ophiuroides		*		*	*			*		*		*
Poa pratensis	*		*		*		*		*		*	
Festuca ovina and other	*		*		*			*		*		*
Stenotaphrum secundatum		*		*	*		*		*			*
Festuca arundinacea	*		*	*	*			*		*		*
Zoysia spp.		*		*		*		*		*		*

Establishment of Sports Turf and Amenity Grassland

Establishment of sports turf and lawns involves the preparation of the site, the sowing of the grass seed or the laying of turf and the management of the site until the grass is sufficiently strong to pass into normal use.

Site Preparation The nature of the site preparation depends on whether a fully-constructed profile is to be installed on an impermeable sub-base (e.g. a major stadium pitch on a concrete base or a golf green on a compacted clay base) or a natural soil profile is to be used, albeit with some modification. It is not the purpose of this review to go into the essential engineering and soil mechanics aspects of fully-constructed profile preparation to meet specified performance criteria, but interested readers are directed to the exhaustive and practical treatment of the subject in Chaps. 3 and 4 of Adams and Gibbs (1994) and the brief but pertinent presentation of Neylan et al. (1999). Similarly, the essentials of preparing a site on a natural soil profile with or without significant modification can be found in the same sources. More specific information on the design and construction of sports fields, golf courses, lawn bowling greens, lawn tennis courts and cricket grounds will be found in Chaps. 6 to 8 of Adams and Gibbs (1994) and Chaps. 14 to 19 of Aldous (1999) with horse racing tracks covered in Field (1994). If the profile is to be based on the perched water table principle, as in the United States Golf Association's recommended method for construction of golf putting greens, particular attention must be paid to the physical properties of the various layers in the profile. Sub-surface drainage pipes, power sources and irrigation systems are usually installed during the construction process rather than disrupting the finished surface at a later stage. The key considerations in site preparation include clearing and rubbish removal, elimination of persistent perennial weeds, preparation of the sub-grade, installation of drainage works, irrigation and power systems and application of fertilizer and conditioners (Beard 1973; Liebao and Aldous 1999; Christians 2003; Turgeon 2008). Final levelling can be achieved by dragging light-weight rakes or levellers over the area. For small irregularities a weighted roller can be used to remove depressions, as well as assist in consolidation of the seedbed.

Establishment of the Grass

Assuming that the preparation has resulted in a smooth and properly graded surface composed of soil, sand or growing medium of appropriate biological, chemical and physical properties, the next step is to establish the grass. The earliest method of establishing a lawn involved the cutting of turf in pieces (sods) about 300 mm × 300 mm × 50 mm from closely grazed swards on commons, river banks or the shores of estuaries and laying these down on a well-prepared bare and level area. The demand for lawns' turf grew so rapidly following the invention of the mechanical lawn mower in the 1830s that these natural turf sources were soon overtaxed and enterprising seed companies were quick to demonstrate in Great Britain, north-western Europe and the north-eastern United States of America that lawns could be successfully established from seed of the same species of *Lolium, Agrostis, Festuca* and *Poa* found in the natural turfs. Thus, arose the custom in the predominately C_3 grass regions of the world for sports turf and lawns to be established from seed. Because of the lack of understanding of the importance of ecotypes and varieties in turf, the seed supplied until the mid-twentieth century, although generally correctly labelled as to species, frequently contained a high proportion of stemmy, early-flowering forms. It is for this reason that many high-quality golf and bowls facilities continued to use turves or plugs taken from old established greens of superior uniformity and fineness well into the 1960s. In the warmer parts of the world where C_4 grasses predominate, the relative ease with which the turf species can be established from turves or sprigs and the rapidity with which the swards regenerate after turf-stripping has meant that vegetative methods of establishment have remained the preferred method for the majority of the important species to the present time.

Establishment by Seed and Vegetative Methods Seed mixtures, blends and single cultivars are used in establishing sports turf from seed. As defined by Beard (1973) seed mixtures are composed of two or more cultivars/varieties from different grass species; blends are composed of two or more cultivars/varieties from the same species. For the C_3 grasses, conventional wisdom holds that mixtures are desirable because the various components are able to demonstrate improved performance at different times within the growing season. For example, Eade (1990) states that a typical seed mixture may include turf-type *Lolium perenne* and *Poa pratensis* to the extent of 70% by weight in the mixture together with fine *Festuca* or *Agrostis* as 'filler' grasses, which add density and variety to the mixture. A similar case is made for the advantages of blends. However, innumerable research studies as well as simple observation show that a grass community is a dynamic entity responsive to weather and management conditions and as such the proportions of the constituents rapidly drift from those sown and that many species or varieties are soon reduced to trace levels. In contrast, other C_3 grass species, such as *Agrostis stolonifera*, are normally sown alone as a single cultivar.

Although a number of C_4 turfgrasses, including *Pennisetum clandestinum, Buchloe dactyloides* and *Cynodon dactylon* (L.) can be established from seed on a practical scale, for most of them vegetative establishment is the preferred method. Some *Cynodon* hybrids, such as Santa Ana, Tifgreen (328) and Tifdwarf are sterile and cannot produce seed, so their establishment is limited to vegetative propagation. Vegetative establishment involves the practices (in increasing order of cost) of stolonizing, sprigging or chaffing, plugging and turf laying using turves or rolls. The latter is a popular vegetative source which can be purchased washed or unwashed from the sod farm. The commercial sod should be delivered fresh, or kept in the shade if a delay is expected and the area cultivated to seed bed quality in advance to ensure rapid transplant rooting. Commercial sod is laid out in a brick-work pattern and consolidated with a light roller. New sod should show new root growth within 7–10 days and should be fit for use within weeks.

Table 22.2 Sowing and stolonizing rates of common C_3 and C_4 turfgrasses

C_3 turfgrasses	Initial sowing (g m^{-2})	Oversowing[a] (g m^{-2})
Agrostis spp.	7	5
Poa pratensis	15	10
Festuca arundinacea (turf type)	40	30
Festuca rubra subsp. *commutata*	30	15
Festuca rubra subsp. *rubra*	25	15
Lolium perenne (turf type)	35	25
C_4 turfgrasses		Stolonizing rate[b] (kg.100 m^2)
Cynodon dactylon	12	12
Cynodon hybrids	n.a.	12
Stenotaphrum secundatum	n.c.a	25
Axonopus affinis	10	12
Pennisetum clandsetinum	1.5	10
Digitaria didactyla	12	20
Zoysia spp.	8	50
Paspalum spp.	3	12

n.a. Seed not available

n.c.a. Seed not commercially available

[a] Seeding into an existing turf surface, usually with temporary C_3 turfgrasses to provide green, active lawn grass growth during the dormancy of the existing lawn

[b] Sprigged material, such as stolons, rhizomes or tillers used to establish a turfgrass area by planting in furrows or small holes. Chaffing is practised by spreading shredded turf pieces over a slightly moistened, prepared site, then rolling the stolons and rhizomes into the surface

For a detailed presentation of the underlying principles and the practical aspects of establishment by either seed or vegetative methods, the reader is referred to the relevant chapters in standard text-books including those of Beard (1973), Adams and Gibbs (1994), Emmons (2000) and Christians (2003). These sources cover seedbed preparation, selection of quality seed or sod, optimum time of planting, depth of planting, planting techniques, protection from birds and ants, after planting care, disease management in the planting bed and weed control during establishment. Typical seeding and/or stolonising rates for turf species popular in Australia are given in Table 22.2.

It will be noted that the seeding rates, when converted to kilograms per hectare (kg/ha), are very high compared with the rates used in pasture establishment. Thus, with *Lolium perenne* (perennial ryegrass) the rate mentioned above is equivalent to 350 kg/ha, approximately one hundred times the rate used in pasture work. In terms of seed numbers, with *L. perenne* at an average weight of 2.0 mg per seed, there would be 1.75 seeds/cm^2 at the sowing rate stated above. These high rates are used in turf work to achieve a rapid cover on the ground, to assist in weed suppression and to help develop a high density sward that can go into sporting use in as little as 3 months from sowing. Under these conditions competition between the seedlings is intense, so that when seed mixtures are used species that are slow to emerge or have small seedlings are rapidly suppressed. This was well demonstrated by Crocker and Martin (1964) in outdoor experiments with *Lolium perenne/Festuca pratensis* mixtures in which after 10 weeks, seedlings of *F. pratensis* growing in competition with *L. perenne* were half the size of the seedlings in a *F. pratensis* monoculture.

Failure to achieve a satisfactory emergence rate is unfortunately a common occurrence with the small seeded grass species used in turf. Under commercial conditions time is usually of the essence in establishment projects, so if a poor emergence results a prompt decision needs to be made in regard to re-sowing. Note that at many of the major stadiums over-seeding is carried out post-match throughout the season. To assist in knowing how long to wait before deciding to re-sow, it is useful to have some idea of the time taken from sowing to emergence for the common turf grasses. Assuming an appropriate soil temperature for the species concerned, the following may be taken as a guide. *Lolium multiflorum*, *Lolium perenne*, *Agrostis capillaris*, *Agrostis stolonifera* and *Festuca arundinacea* have a rapid germination rate and take 8–10 days to emerge, whereas *Festuca rubra* subsp. *commutata*, *Festuca rubra* subsp. *rubra*, *Poa pratenis*, *Cynodon dactylon* and *Digitara didactyla* have a medium germination rate and should emerge in 14–16 days, while *Poa trivialis* and *Zoysia japonica* are slow germinators that take some 21–25 days to emerge.

Post-Sowing or Sodding Care and Growing-In

Seeded surfaces. Frequent light irrigation to keep the surface damp (but not saturated) is necessary at establishment until the seedlings are at least 30 mm high or at the 3–4 leaf stage of growth. Basal fertilizer applied at sowing time will keep the turf species growing for 4–6 weeks, but subject to a soil test, after that time regular small applications are necessary to ensure continued growth and active tillering. The first mowing should not be carried out until the sward is at least 30 mm high and should be very lenient. After this the mowing height is reduced gradually to the recommended height of cut over the next few months. Establishing the correct mowing height and frequency at this time will develop strong turf by promoting a good balance between the shoots (leaves) and the roots (Aldous 1999; Emmons 2000; Christians 2003) and stimulating the adaptive morphological response necessary to achieve high tiller density and short internode lengths (Beard 1973). Clippings may be returned to the seed bed, but if there is extremely heavy growth of grass or weed inflorescences with seed it is advisable to remove the clippings. If possible do not apply herbicides during the first 3 to 4 months, as juvenile turf is very sensitive to herbicide damage. Minimise traffic on the newly-established stand and when it is judged sufficiently mature adopt a cautious approach about bringing it into play. Months of work can be destroyed in a few hours if brought into play prematurely.

Vegetatively-Established Surfaces In the case of commercial sod, or other vegetative forms of establishment, irrigation needs to be applied as soon as the grass has been installed, with repeat waterings 2 to 3 times during the day, depending on the weather conditions. The development of small white roots protruding from the underside of the sod will indicate when the watering frequency can be reduced. Mowing and fertilizer requirements are similar to the sowing of seed.

Fig. 22.1 The main mechanised mower types include the cylinder or reel (*left*), rotary (*middle*), mulch and flail mowers (*right*). *Rotary mowers* use a rotating blade or blades to cut vegetation with mulching technology introduced for use with rotary mowers

Management Strategies for Sports Turf and Lawns

Mowing Regime

Mowing accounts for more than 75 % of the labour costs in maintaining sports turf and amenity lawns. Despite the apparent simplicity of the operation, poor mowing, as will be explained in a later paragraph, can lead to a rapid deterioration of the sward.

For sports turf the main mechanised mower types include the cylinder or reel, rotary, mulch and flail mowers (Fig. 22.1). The cylinder or reel mowers, hand or power driven, use a scissors action to cut the leaf blade and if properly adjusted provide the highest quality cut. The rotary mower, which cuts by impact using horizontally moving high-speed blades, is better suited for rough grass areas and the coarser grasses. Its close relative the mulch mower can be used where the clippings are turned into instant mulch and directed back into the grass surface. A modification to rotary mowing is the hover mower, where the sucking in of air and pressing it against the edge of the hood, lifts the machine making it float under a rotary cutting blade. Hover mowers have a place on sloping lawn or inaccessible areas. The flail mower has a horizontally-aligned shaft from which hang vertical rotating blades and finds purpose on rough areas and long grass or on undulating or rocky ground such as highway banks, roadsides or orchard areas.

Improved turfgrass appearance and performance are the results of adhering to preferred mowing heights, choosing the right frequency pattern, using judicious clipping management and selecting appropriate mowing equipment (Aldous 1990). No single mowing height is correct for all grass species as there are differences in growth habits and response to water, fertiliser and mowing regimes. Each sports turf grass has a preferred range (Table 22.3). Continual mowing below these preferred heights can result in a gradual thinning of the grass cover and reduction in the extent of the root system as well as increasing the susceptibility to weed, insect, disease and moss invasion. Mowing consistently above the preferred mowing height can lead to an increase in thatch and open, clumpy turf. Scalping, or excessive removal

Table 22.3 Preferred mowing heights for some common cool and warm season turfgrasses. (Source: Beard 1973; Aldous 1990; Chivers and Aldous 2005)

Designated cutting height	Mowing height (mm)	Turfgrass species
Very close	5–12	*Agrostis stolonifera, Agrostis canina* subsp. *canina,* hybrid *Cynodon* turf grasses, *Zoysia matrella*
Close	12–25	*Agrostis capillaris, Poa annua, Cynodon dactylon, Digitaria didactyla, Lolium* spp. (turf type), *Poa pratensis* (improved types), *Festuca arundinacea* (dwarf types), other *Zoysia* spp.
Medium	25–50	*Buchloe dactyloides, Festuca rubra, Eremochloa ophiuroides, Axonopus affinis* and *A, compressus, Penisetum clandestinum, Poa pratensis, Lolium perenne* and *L. multiflorum, Festuca arundinacea* (medium cvs.), *Festuca pratensis, Stenotaphrum secundatum, Microlaena stipoides*
High	35–75	*Paspalum notatum, Festuca arundinacea, Stenotaphrum secundatum*
Very high	75–100	*Poa canadensis, Bromus inermis*

of leaf tissue, not only reduces the amount of leaf available for photosynthesis, but also tends to decrease root and rhizome development and minimise the ability of the turf to withstand climatic or agronomic stress. With very closely-cut fine grasses such as *Agrostis stolonifera* on golf greens, surprisingly small alterations in cutting height have a large influence on leaf area and hence photosynthetic potential. In an experiment reported by Bell and Danneberger (1999), raising the height of cut from 0.125 inch (3.18 mm) to 0.141 inch (3.58 mm) increased the leaf area and photosynthetic potential by 13 % and by 25 % when increased to 0.156 inch (3.96 mm). In a mixture or blend of grasses, the preferred mowing height should favour the predominant or most useful grass species. Mowing heights 1.0–2.0 mm above the preferred height can assist the turf to survive better as well as enhance its tolerance to heat stress and traffic.

Mowing frequency will be determined by the growth rate and the preferred mowing height, rather than by following a fixed time interval. Increasing the mowing frequency can increase shoot density and reduce shoot growth, carbohydrate reserves, rooting and rhizome development (Beard 1973). No more than 30–35 % of the green leaf should be removed at any one mowing. Maintaining consistent mowing patterns and turning the mower in the same place can lead to localised soil compaction, invite weed invasion and cause permanent thinning of the turf. In managing a bowling green the greenkeeper can choose from three mowing patterns; a single cut, where the green is mown once, a double cut, where they return down the previous cut and cutting twice, where the green is cut in two different directions. The most successful and recommended method is double cutting as it produces a finer leafed grass, smoother surface and distributes the wear over the surface.

Chemical growth inhibitors of sports turf, when applied at the recommended times, can assist in reducing the mowing frequency during periods of maximum growth (Anon 2005). Plant growth retardants work by either inhibiting cell division and differentiation in the meristematic regions of the grass, or suppressing cell elongation and as a result suppressing turfgrass growth and seed head development. Although the cost of using growth regulators limits broad acre use, there are specific turf systems, such as shaded turf in major stadiums, where they are important management tools.

Supplemental Fertilization and Nutrition

Some 16 chemical elements are essential for plant growth and development and these nutrients are cycled and re-cycled through the grass production system. The relevance of the cycle is twofold; if clippings are not returned, supplemental fertilizer needs to be provided and secondly, if soil microbial activity is reduced through poor nutrition, thatch levels may increase that in turn lead to management problems. In the nutrient cycle there has to be a balance between the nutrients lost from the soil (output) and nutrients applied (input) for good grass growth, with the storage capacity being limited by the nutrient holding capacity of the soil.

Commercially-available nitrogen carriers (fertilizers) are grouped into whether they are inorganic or organic and whether they are natural or synthetically manufactured (Table 22.4). Most of the inorganic carriers are highly soluble in water, odourless, have a short release time and can be lost or leached out if heavily watered. The natural, organic carriers, such as animal by-products (bone meal, animal manures and wastes) are generally lower in nitrogen content and take some 4 to 8 weeks to release the nitrogen to the grass (Beard 1973). Recent developments in fertilizer technology have produced certain controlled release products. Thus the resin coated Multicote®, where the nutrient release is via osmosis, provides generally good control for up to 9 months, the polymer/sulphur coated products such as Poly S® generally provides good control for up to 14 weeks, depending on temperature and rainfall. With organic compounds like isobutylidenediurea (IBDU) and methylene urea the release of the contained nitrogen is slow and dependent on environmental conditions.

Commercial fertilizers can be applied dry in powered or granulated form, while many can also be applied in water. Dry fertilizer can be broadcast by hand through a centrifugal-type or gravity fed fertiliser distributor and spreader (Fig. 22.2). For larger areas tractor-drawn or self-propelled spreaders are used.

Fertilizer application contributes to turf colour, density, uniformity and growth rate. The timing of fertilizer applications depends on the grass species, the quality of turf desired, release rate of the fertilizer, budget and the capacity of the soil type to retain the nutrients. A basic rule in fertilizer use is that the grass must be actively growing at the time of application, otherwise there is the potential for the material to be lost and/or the environment polluted. Fertilizer rate will be influenced by the turfgrass species, the level of quality desired, intensity of use, the growing season and the environment and cultural conditions. With the exception of nitrogen

Table 22.4 Examples of simple and complete fertilizers, showing percentages of nitrogen (N), Phosphorus (P) and potassium (K)

Examples of principal simple fertilizers				
Nitrogenous fertilizers	Fertilizer	Nitrogen (%)	Phosphorus (%)	Potassium (%)
Inorganic	Ammonium nitrate	33	0	0
	Nitrate of soda	16	0	0
	Sulphate of Ammonia	21	0	0
Organic	Dried blood	12–14	0.2–2	0.4–0.8
	Blood & bone	4–5	5	0
	Animal wastes	7–10	2–6	0.4–0.6
	Raw bone meal	3–5	15–27	0
	Activated sewerage sludge	4–7	4–6	0.4–0.7
	Poultry manure	4–8	1–2	0.8–1.6
	Cow manure	0.2–2.7	0.01–0.3	0.06–2.1
	Sheep manure	1–3	0.1–1	0.3–1.5
Manufactured organics	Urea formaldehyde	38	0	0
	Isobutylidene diurea	31	0	0
	Sulphur-coated ureas	32–37	0	0
	Urea	46	0	0
Phosphatic fertilizers	Concentrated superphosphate	0	38	0
	Superphosphate	0	22	0
Potash fertilizers	Muriate of potash	0	0	50
	Sulphate of potash	0	0	60
Examples of some complete fertilizers				
	Pivot ®	8	11	10
	Summermix ®	17	0.6	9
	Wintermix ®	12	3	18

Dried blood, hoof and horn, meat and bone, animal wastes not available in Europe because of Bovine spongiform encephalopathy (BSE) or 'mad cow disease'. Even if they were, very strict environment rules apply

Fig. 22.2 Granular fertilizer application can be carried out by drop-type (gravity) spreaders or rotary-type (centrifugal) spreaders

fertilizers, where the local manager generally judges the need for applications by the colour of the turf, the rest of the fertilizer program should be based on regular and properly interpreted soil tests or preferably tissue tests in the interests of economy and sustainability. For best nutritional results soil amendments such as compost, and inorganic amendments such as lime, dolomite and gypsum, need to be used in conjunction with the fertilizer program (Adams and Gibbs 1994).

Water-soluble fertilizers can increase the salt concentration around the roots of the grass and in turn reduce water availability to the plant, with the result that the grass may wilt and growth rate is reduced. In severe cases turf death can occur. Fertilizer burn can also occur with leaf contact, leading to spotting and browning. The adverse effects are most marked with lush turf growth and in hot, low humidity weather. To avoid these problems it is important to consider the salt index of the fertilizer. This index expresses the osmotic effect of a given weight of the fertilizer in terms of the amount of pure sodium chloride needed to produce the same osmotic effect (Beard 1973; Harivandi 1999). Values for common fertilizers range from a low of 8 for superphosphate to a high of 116 for muriate of potash. A list of the salt index values of common fertilizers and the precautions to be adopted when applying material with values above 20, is given by Beard (1973).

Irrigation

Recent surveys carried out on the turfgrass industry in United States of America and in Australia have both identified decreasing availability of quality water at reasonable cost as the most important turf issue for both large scale and small scale users (Beard and Kenna 2008b; Aldous et al. 2007). This highlights the need to maximise the efficient use of the available water. It has long been known that C_4 turfgrasses are more efficient in using water to make a unit of dry matter under similar solar radiation conditions than C_3 species. C_4 grasses typically have a moderate transpiration rate under summer conditions and good drought tolerance, whilst in comparison C_3 grasses have high transpiration rates coupled with fair to poor drought tolerance (Beard 1989; Beard 2002). This creates the opportunity in suitable climatic areas to reduce irrigation water requirement by growing C_4 grasses rather than C_3 grasses and further reducing irrigation requirement by exploiting the greater drought tolerance of the C_4 grasses.

The Irrigation Program

Types of irrigation systems for sports turf include, sub-surface drip, soak hoses, big gun sprinklers, pop-up rotary sprinklers and hand move sprinklers or pipes. Turf farms may opt to use other forms of irrigation system such as travelling booms or central pivot irrigation systems (Fig 22.3). An irrigation program is built around four basic components; how much water should be applied, the optimum frequency

Fig. 22.3 Sprinkler systems can be based on overhead sprinklers or sprays, installed on permanent risers or buried underground where the sprinklers rise when the water pressure rises. Centre pivot irrigation is a form of overhead irrigation where the irrigation infrastructure is mounted on wheeled towers and the sprinklers run the length of the frame

Fig. 22.4 A pressure gauge allows for the accurate checking of the sprinkler or spray head operating pressure

of application, the rate of water application and at the time of day water should be applied for greatest irrigation efficiency (Connellan 1999). Pressure testing assesses whether the sprinkler system is performing to the manufacturers specifications (Fig. 22.4).

Amount of Water to Apply The amount of water that can be applied without resulting in drainage loss will depend on the depth of the root zone, the available water-holding capacity of the soil and the amount of water present when irrigation is commenced. Soil water levels can be checked physically by soil sampling or by using portable instruments such as a tensiometer, moisture meter or infrared thermometer (Connellan 1999). Furthermore, soil type impacts on available water content (Table 22.5).

Table 22.5 Available water values for selected soil types. (Source: Connellan 1999)

Soil type	Available water (mm h^{-1})
Sand	60
Fine sand	90
Sandy loam	110
Loam	170
Silt loam	170
Clay loam	165
Clay	140

Optimum Frequency of Application Irrigation frequency will be determined by the climatic conditions, the type of turf, the soil type and the available moisture within the root zone. Generally newly-established turf, with limited rooting depth benefits from light, frequent applications, whereas established turf requires infrequent and deep watering. The matrix in Table 22.6 shows the impact of soil texture and root depth on desirable watering frequency in warm weather.

Rate of Water Application It is important to apply irrigation water at a rate below the soil infiltration rate (surface irrigation systems are an exception to this). If water application rate exceeds infiltration rate water will pond and in turn runoff or seep past the root zone via any cracks or fissures (termed by-pass flow). Most modern pop-up sprinkler systems are designed to apply water at a rate of less than 20 mm/hr, which is generally well below the soil infiltration rate.

Time of Day to Irrigate Some operators choose to irrigate in the late evening or early morning because water loss from evapotranspiration is minimal, a better water pressure is generally available and often wind speeds are lower. Others prefer early morning watering as this allows the surface to dry off fairly rapidly, reducing the incidence of summer fungal diseases. Turf surfaces watered in the late afternoon or early evening can remain damp all night thereby encouraging disease if nights are warm.

Evenness of Watering For the best grass growth response, water should be applied as uniformly as possible over the turf surface. This is a function of good irrigation design and practical management to cope with fluctuating water pressure, wind direction and slope. Periodic objective measurement of the uniformity of water distribution is an essential management tool.

Water Source and Quality In recent years repeated dry conditions in many countries have forced turf managers using potable water to explore alternative irrigation water sources. These non-potable waters may be of good quality for irrigation, but frequently have high concentrations of soluble salts, bicarbonates, toxic elements such as boron or chlorine, or other contaminants such as bathroom and laundry effluent and other gritty particles. Before using such sources thorough testing and specialist advice must be sought to consider both short-and long-term implications of using these types of waters (Harivandi 1999). Similarly storage ponds or dam water should be free of algae, weed seed and other foreign debris.

Table 22.6 Frequency of watering turf based on soil type and root zone depth

Soil texture	Depth of lawn roots (cms)		
	10	20	30
Fine-medium sands (open and well drained)	2 × daily	Daily	Daily
Sandy loams (loose and friable)	Daily	3 × per week	2 × per week
Loams-clays (heavy and difficult to work)	3 × per week	2 × per week	Weekly

Fig. 22.5 Cultivation (aeration) is a mechanical process that uses metal hollow-tine or solid-tine corers, that pierce the ground to replace soil air and improve drainage. A major contributor to poor aeration is soil compaction caused by machinery or foot traffic

Renovation

Renovation is a process of restoring the turf beyond that achievable through routine cultural practices, but without complete tillage of the soil (Turgeon 2008) and as such results in improved condition of the playing surface and a restored growing environment (Neylan et al. 1999). In temperate and warm-temperate climate areas renovation is usually carried out over summer for C_4 grasses and early autumn or spring for C_3 grasses, with recovery taking some 6–8 weeks in summer to up to 10–12 weeks with the same grass in spring.

Compaction relief can be carried out with mechanical hollow-tine corers, solid-tining, deep grooving, slicing, vertical cutting and spiking, all of which open the soil surface to improve soil air, water percolation and fertiliser distribution in the soil profile (Emmons 2000; Christians 2003) (Fig. 22.5). Deeper compaction relief can be achieved with a borer drill, deep penetrating spiker or vibra-moling units, which not only relieve surface compaction but promote water infiltration.

Renovation also involves the removal of excessive thatch, a layer that not only buffers the surface variation in soil temperature, but assists in cushioning the surface when in play. However, thatch has a down side in that it can become hydrophobic in nature, resulting in dry patches where water will not penetrate. Thatch also provides a suitable microclimate for disease spores and insect larva and reduces the effectiveness of many herbicides because they are absorbed in the layer. Dethatching with fixed, flail and spring-toothed mowers as well as groovers and slicers, is generally carried out just prior to or during the active growth period of the sward (Fig. 22.6).

Fig. 22.6 Scarification or de-thatching is a mechanical process whereby metal blades, tines or prongs remove organic material from the playing surface

Another renovation process is that of topdressing, which is used in conjunction with these mechanical surface treatments to prevent a layering effect from occurring and to dilute the thatch and aid its decomposition (Fig. 22.7). The amount of topdressing material applied will vary with the grass species, temperature, purpose of the topdressing and condition of the grass surface. Light, frequent applications of top dressing, as often as monthly to three or four times per year and is continued over an extended period, are considered more beneficial than heavy, infrequent applications. Frequent, light soil topdressing to a thickness of 3.0–5.0 mm can help dilute thatch and can assist in the stimulation of vegetative material such as stolons and plugs. Topdressing is distributed evenly over the surface mechanically and the material thoroughly worked in with a heavy broom or a steel mat or rakes and levelled with screeds to smooth out the surface (Aldous 1999; Emmons 2000; Christians 2003).

Sports pitches may also be renovated by a procedure known as fraise mowing, or planning, which involves setting specifically made machinery to a depth where it just removes the surface verdure including thatch, monocotyledonous and dicotyledonous weeds along with a very small quantity of rootzone material (Fig. 22.8). In Europe and the United Kingdom fraise mowing removes organic debris and shallow rooted *Poa annua* leaving most of the perennial ryegrass crowns intact. In this situation, the crown and root system of established grasses remain intact and quickly recovers to form a new sward. The practice is particularly effective for rhizomatous grasses such as Bermuda grass. Alternatively, the machine can be set to remove a deeper layer (up to 50 mm); a procedure that effectively removes all vegetation and surface compacted rootzone, while simultaneously re-levelling the pitch surface. In this situation, the surface is restored either by re-turfing or seeding, or by regeneration of the rhizome system of certain grasses. Irrespective of the depth of material that is removed, the debris is deposited into trailers travelling alongside via a conveyor belt attached to the machine. It represents a very rapid method for surface restoration and is particularly suitable for sand constructed football pitches.

Fig. 22.7 Topdressing involves the even application of sand or soil and is worked into the surface by means of a drag mat. It can improve infiltration rates and assist with the biological control of thatch. Light topdressings termed "dusting" is often undertaken on golf greens to prepare for tournaments

Fig. 22.8 Fraise mowing is suitable for sports fields, cricket pitches and tennis courts that contain a lower percentage of *Poa annua* and a higher percentage of other cool-season perennial turfgrasses

Weeds, Pests and Diseases of Sport Turf and Amenity Grasslands

Weeds of Sports Turf and Lawns Accurate identification of weeds is basic to devising correct control measures. In many countries there are local or national guides for the identification of turf weeds, for example McCarthy and Everist (2008) for the United States of America. In addition there is an important on-line Global Compendium of Weeds (2011) that contains, amongst other things, reference material for weeds of sports turf and lawns found in many countries. Weeds by their nature compete with the more desirable grass species for sunlight, nutrients,

Fig. 22.9 Wear protective clothing and equipment mentioned on the label. Read the label in regard to health and safety concerns in spray application

water and space and reduce the quality and competiveness nature of the grass surface. In turf they tend to be most frequent in spots where the desirable grass species has been weakened by adverse biotic or abiotic conditions. Wind, water, foot traffic, machinery and materials used for soil modification, renovation and topdressing are the usual forms of dispersal.

Weed control mechanisms need to take into account the status of their life cycle (annual, biennial or perennial), season in which growth occurs and the method of reproduction (Shildrick 1990). With perennial weed control a number of strategies need to be considered that could involve one or more of the following methods: mechanical (hand pulling, tillage, smothering, mowing), physical (barriers mulches, light exclusion), physical energy (thermal or heat radiation, flame weeding, soil solarisation), biological control (classic biological control, mycoherbicides, companion planting) and natural chemical control with allelopathic compounds and microbial phytotoxins (Morgan 1989) or other natural products such as liquid borax and corn gluten meal (Christians 2003). Busey (2003) undertook a review of the cultural management of weeds in C_3 turfgrasses and found that a small number of cultural factors, such as mowing height, high rates of N fertilization and the use of cultivars and species well adapted to tolerate biotic and environmental stresses can strongly influence weed colonization.

Herbicidal control involves chemicals that kill or alter the normal growth of weeds. Herbicides are classified into two different groups based on the nature of their activity; pre-emergence and post-emergence. Pre-emergence herbicides are applied to established turf before the weeds have emerged. Systemic herbicides are absorbed by the leaf and are then translocated throughout the plant. Pre-emergence herbicides are applied just before the desired grass seed is sown or before the weeds emerge from the seed bed. Post-emergence herbicides are applied after the emergence of a specific weed. For small areas of turf the hose-end applicators work very effectively for the control of many weed species. Larger areas require application by means of a knapsack sprayer or a tractor or mower-mounted pressure sprayer (Anon 2013) (Fig. 22.9).

Insect Pests of Sports Turf and Lawns

Fermanian et al. (2003), Watschke et al. (1994), Brandenburg and Villani (1995) and Vittum et al. (1999) provide a comprehensive coverage of the major insect pests of sports turf and lawns. The most important insect groups and their feeding patterns are as follows: sub-surface feeder pests (white grubs and beetles, weevils, mole crickets), sub-surface chewing feeder pests (sod webworm, armyworm and cutworm, grasshoppers, leaf bugs), nuisance pests (ants, wasps and bees, centipedes and millipedes, mosquitoes, ticks and fleas), sub-surface sucking pests (aphids, leafhoppers, scale insects, spider and clover mites, spittlebugs) and beneficial insects such as lady beetles, lacewings and parasitic wasps. Insects, as well as weeds, find their way into turf areas through contaminated topsoil or in some cases poor planting material and can be transported by drainage and irrigation water, infected topsoil, vegetative plant parts (sprigs, plugs, turf rolls) and equipment. Many insects fly or crawl in from surrounding areas.

Preventative measures that suppress the occurrence of destructive turfgrass pests include sanitation (such as the selection of clean seed, sod, or topsoil), reducing the sources of inoculum, crop rotation and cultivation or the cleaning up of areas where insects may overwinter or build up resistance, as well as making greater use of resistant grass cultivars where available. Maintaining strong, vigorous grass, by means of a correct and well-managed cultural programme will help to outgrow small insect populations and assist in turf recovery from insect damage. Another way of controlling selected insect populations is through biological agents that affect part of the insect's life cycle and so reduce future populations. There is also a range of soil fungi and parasitic wasps that can help to control certain scarab larvae, as well as a range of insecticidal controls by means of sprays, dusts, granules or fumigants of various types. More recently insecticides derived from plant extracts have become available that may, when objectively tested, prove effective against some turf pests. In many countries the local Department of Agriculture (or equivalent) provides services for the identification of insect pests and recommendations on their management or control (Aldous and Brereton 1999).

Parasitic nematodes and certain free-living but phytophagous nematodes are serious problems in some turf areas, particularly those with moist, mild-to-warm (18–28 °C) summer climates and well aerated soils. Nematodes can be dispersed on soil attached to equipment, in irrigation and drainage waters, in imported topsoil, or by transfer of vegetative material. Nematode injury to turfgrasses takes many forms and often exhibits itself as a general lack of thriftiness rather than any definite symptoms. Specialist assistance is usually required for the identification of nematodes in or on plants and in soil. Nematode control can be achieved through effective sanitation, pre-planting hot water treatment, the use of resistant turf cultivars (where available) and a range of chemicals. For details see Aldous and Brereton (1999).

Diseases of Sports Turf and Lawns

Smiley et al. (2005) and Toshikazu and Beard (2002) provide a thorough coverage of the occurrence, diagnosis and control of diseases that affect sports turf and the more intensively managed amenity grasslands such as lawns. The causal agents of disease can either be biotic agents (bacteria, fungi, mycoplasmas, nematodes and viruses), or abiotic agents (a wide range including pesticides, animal urine, chemical salts and chemical spills, extremes of temperature, lightning, soil compaction, mower scalping and abrasion injury). The field of turf pathology has proved to be one of the most complex in plant pathology, especially in relation to the soil-borne diseases. In broad terms, disease management strategies translate into the use of good sanitation procedures, greater use of disease resistant plant cultivars and operating best practice in cultural management. Management practices and environmental factors necessary to maintain turf have a decided effect on disease incidence and development. Most cultural practices exert an indirect effect on disease development by modifying the micro-environment that in turn affect disease incidence. Similarly, increasing soil fertility to improve turf vigour can in some cases (but by no means all) increase tolerance of disease. Apart from a small number of familiar diseases with obvious and consistent symptoms, a turf manager confronted with disease problems should always seek the advice of a properly qualified specialist in turf pathology before attempting any remedial measures.

In view of the multitude of weeds, insect pests, diseases and other biological problems afflicting managed turf, it is not surprising to find that the industry is a significant user of chemically based control measures. It therefore seems appropriate to conclude this section of the Chapter with a few comments on Integrated Pest Management (IPM). Leslie (1994) defined this emerging strategy as one which uses multiple control mechanisms in a compatible manner to prevent pest levels from reaching unacceptable economic or aesthetic levels, while minimizing their effects on humans and the environment. All in the turf industry are encouraged to adopt this approach as an active investment in the sustainable future of the industry and the communities which they serve.

Performance Assessment of Sports Playing Surfaces

Management of turf grass for sports use is directed to the production of an aesthetically pleasing and functional playing surface that encourages skilful play and minimises the risk of injury to the participants (Baker 1999). The general visual assessment of turf aesthetics using criteria such as colour, density and uniformity is adequately dealt with in all standard text-books on turf science and will not be further discussed here. This section will instead focus on the more recently-developed objective approaches to the assessment of the playing quality or performance of the surfaces prepared for a range of sports and on the related issue of links between the nature of the surface and the risk of injury.

Playing quality is a function of the physical properties of the immediate turfgrass surface layer and the physical properties of the underlying soil or growing medium. As summarised by Neylan et al. (1999) the physical properties of the turfgrass surface layer are dependent on factors such as the grass species, climatic and weather conditions, cultural practices and past usage patterns, while the properties of the underlying soil are determined by relatively stable factors such as texture, porosity, reinforcing materials and short term variables, in particular soil moisture content and reinforcement of the soil by root development, which interact with the other factors to determine the deformability of the surface and the overall strength and hardness of the profile. It is obvious that the properties discussed above will influence foot traction, surface hardness and ball response. There has been a tendency to place more emphasis on the soil conditions rather than turf surface in relation to player injuries, but Bell et al. (1985) Canaway and Baker (1993) have stressed the importance of the "agronomic factor" in assessing injury potential. More recently, Baker (1999); Orchard et al. (2005) and Chivers (2008) have suggested grass species and cutting height are important components of this agronomic risk factor.

Although the appearance or aesthetics of the surface are important in achieving high public approval ratings, particularly for televised coverage of events, it is the performance of the playing surface that matters most. The key components of playing performance quality include the ball-surface interaction (racing excepted) and the player-surface (or horse-surface) interaction are covered by Bell et al. (1985) and Baker and Canaway (1993). The indicators used to quantify both of these types of interaction involve measurement of the forces that the player or ball applies on the surface. These forces have horizontal and vertical components as well as at times a circular component, the degree of torque, for example, the twisting action of the foot while it is in contact with the ground (Bell et al. 1985). In horse racing these forces are expressed differently and include the forces of deceleration, support and propulsion.

In the player-surface interaction the principal indicators are traction, surface hardness and surface evenness. Inadequate traction may lead to a greater number of falls or potential collisions between players and increase the risk of injury, whereas excessive traction may increase the incidence of knee and ankle injuries as has been observed with some artificial playing surfaces (McNitt et al. 1996). The principal tools currently used to measure traction include the pendulum test, the studded disc (Fig. 22.10) or the towed sledge, whilst surface hardness is best measured by either the penetrometer, the Clegg Impact Soil Tester (Fig. 22.11), the Berlin and Stuttgart Artificial Athlete (Baker 1999) or the studded disc technique (Canaway et al. 1990). Surface evenness, another player-grass surface characteristic, can be measured by means of a straight edge or moving rods in a frame (McClements and Baker 1994a, b). More recently a device, the Trueness meter, has been developed to accurately gauge the levelness and smoothness of golf greens (pers comm. STRI). For a review of the devices used to assess racetrack conditions, see Neylan and Stubbs (1997).

With the ball/surface interaction, the main indicators are ball rebound, spin and roll (Baker 1999). The main method used to measure ball rebound is the vertical bounce test (ball bounce/rebound). Ball roll methodology ranges from measuring the distance

Fig. 22.10 The traction apparatus measures the rotational force required to initiate movement of a studded disc

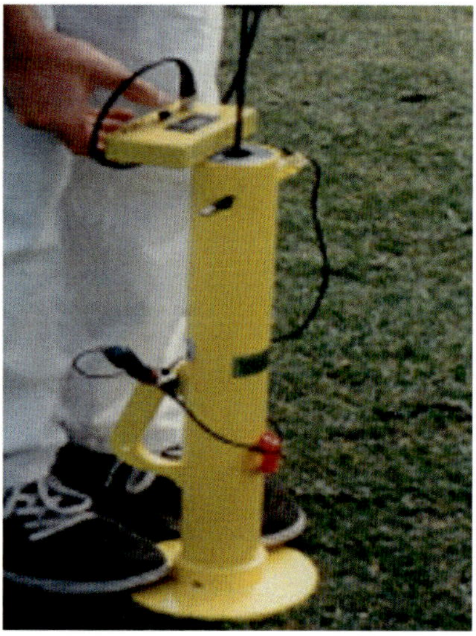

Fig. 22.11 The Clegg Impact Tester utilizes a test mass (either 0.5 kg or 2.25 kg are used) released down a tube guide where an accelerometer is used to record peak deceleration during impact with the turf

rolled after release of the ball down a standard ramp (Fig. 22.12), for example the stimpmeter used in golf (Radko 1980), the deceleration of the ball using electronic timing gates and, in the case of lawn bowls, measurement of the time the bowl travels a set distance (Canaway et al. 1990). For ball spin a modified baseball/cricket bowling machine has been used, while ball impact is measured with a stroboscopic instrument that produces a succession of photographic images from which changes in angle, velocity and spin can be calculated (Baker and Canaway 1993).

Fig. 22.12 Release of the ball down a standard ramp assesses the ball roll properties including distance rolled

Regardless of the test method used, it is important to ensure results correlate with player opinion (Baker and Canaway 1993; Aldous et al. 2005; Chivers 2008). Players have been interviewed on their perceptions of the physical properties of a playing surface, not only in regard to grip, but also in regard issues of safety while running or falling on different types of surfaces under various conditions. The results of these interviews have influenced the setting of standards for some sports and in at least one football code have led to changes in the approach to surface preparation. One interesting outcome of the interviews with Australian Football League players was the establishment of a link between low surface resilience and increased player fatigue (Chivers et al. 2005; Orchard et al. 2005).

The majority of sports using natural or synthetic turf surfaces have developed a system of performance standards for defining the required performance of the surface. For example, for a competition event to be held with sports such as hockey, lawn bowls or tennis the playing surface must conform to standards endorsed by the international body. Recognised testing houses accredited by the international sports bodies are assigned the task of testing and signing off any surface wanting to be awarded competition status.

Performance requirements have been established for soccer pitches Canaway et al. (1990), rugby (McClements and Baker 1994b), hockey (McClements and Baker 1994a), lawn bowls (Bell and Holmes 1988), golf greens (Baker et al. 1996) and Australian Football League (AFL) football (Aldous and Chivers 2003; Chivers 2008). For the sport of cricket, various efforts have been made over the years to develop meaningful standards for how a cricket pitch should perform (McAuliffe and Gibbs 1993). With cricket the ball-surface interaction is more complex (and arguably more critical) than with other sports and as such it has been challenging to find testing devices that offer accurate correlation with player assessment of pitch performance. At this point in time the ICC (international body governing cricket)

Table 22.7 Importance of a range of performance indicators for different sports played on grassed surfaces (adapted from Baker 1999)

Sporting surface	1	2	3	4	5	6	7	8	9	10	11	12	13
Ball/surface interaction													
Ball rebound	C	C	A	B	C	C	C	C	C	C	C	A	A
Ball spin	C	C	A	C	C	C	A	B	C	C	C	C	A
Ball roll	C	A	C	A	A	C	A	B	A	C	C	A	C
Ball impact	C	C	A	B	C	C	A	B	C	C	C	C	A
Player impact/surface interaction													
Traction	A	C	B	B	C	A	C	B	B	A	A	A	A
Surface hardness	B	C	B	B	B	A	C	C	B	B	B	A	B
Surface evenness	B	A	A	A	A	C	A	B	A	C	B	B	A

A very important, *B* important, *C* less, or no importance
1 baseball, *2* lawn bowling green, *3* cricket pitch, *4* cricket outfield, *5* croquet, *6* football, *7* golf green, *8* golf fairways, *9* hockey, *10* lacrosse, *11* racecourse, *12* soccer, *13* lawn tennis

has not formally adopted a system of performance standards, instead relying on past performance or trialling to assess pitch performance.

Agronomic factors have also influenced playing quality with traction which has been shown to be highly correlated with the percentage of grass cover (Bell and Holmes 1988) and the morphology of the grass used in the surface (tufted habit vs. stoloniferous habit) (Orchard et al. 2005; Chivers 2008). Safety is also influenced by the interaction between the surface and the type of studs, cleats or spikes used on players footwear (Livesay et al. 2006).

A convenient summary of the performance indicators for a wide range of sporting activities is set out in Table 22.7.

Maintaining the playing quality of sports turf is of paramount importance not only to achieve a high standard of play but also to minimise the extent of different types of player injuries and to provide for an increased degree of player safety. Because playing quality is dependent on the properties of the grass surface in conjunction with the properties of the underlying soil or growing medium, it is clear that delivery of safe, high performance grounds cannot rely on enlightened agronomic practices alone but must begin with the optimisation of the soil system before a single blade of grass is planted.

In recent times, the Performance Assessment of Sports Surfaces (PASS) approach has been used to measure and assess the performance of sports fields in Australia and New Zealand. The PASS approach provides an objective means of recording playing quality and safety of the sports turf surface, which in turn enables benchmarking to be undertaken in relation to past results or in accordance to other national databases or recommended national guidelines. The parameters, methods of testing and required limits are found in Table 22.8. These features identify whether or not a sports field is achieving its optimal use and the limits to its field performance (McAuliffe 2012).

Table 22.8 Parameters, method of testing and limits in determining playing quality of sports fields. (Adapted from McAuliffe 2012)

Parameter	Method of test	Limits
Surface gradient	Digital inclinometer or topographic survey information	Recommended maximum gradient in any direction of 1 : 70; up to 1 in 50 could be used for lower grade
Surface levelness (mm)	3 m straight edge and/or 5 m string line	< 100 mm variation under a 5 m string line; < 25 mm over 3 m straight edge
Smoothness/Trip index (for smoothness of ball roll & safety)	1 m straight edge and wedge	No turfy grass plants > 15 mm above turf canopy. No divots deeper than 20 mm
Ground cover (both total % cover and turf type)	0.25 m^2 quadrate	Total turf cover to be > 98 % start of season and >than 85 % mid-season. No bare patches > than 200 mm @ start of season and < 10 bare areas > 200 mm @ per field mid-season
Weed content (%)	Quadrate	< 20 % grass weeds; < 5 % broad leaf weeds
Rooting depth (mm)	Use of ruler after core sampling	Effective rooting depth greater than 125 mm
Sward height (mm)	Floating disk	Hockey—8 mm to 20 mm Cricket-(excluding wicket)—10 mm to 35 mm Soccer—20 mm to 40 mm AFL—20 mm to 50 mm Rugby Union/League—40 mm to 55 mm
Surface hardness (player/surface interaction)	2.25 kg Clegg Hammer for player/surface interaction	Clegg hammer readings between 50 to 130 gravities (first drop)
Surface stability	Going stick and/or digital cone penetrometer	Going stick index value between 10 to 14.5. Penetration resistance < 2500 kpa
Traction (Nm)	Use of studded disk apparatus	Test values to fall between 30 to 50 Nm
Infiltration rate (mm/hr)	Ring infiltrometer	> 25 mm/hr for general purpose grounds

Conclusions

Sport and amenity grasslands play a significant part in improving the environmental, social, economic and health benefits of society. Although the maintenance and management of sport and amenity grasslands can vary in terms of grass selection, function, intensity of use and the cost of maintenance, the effective measurement of player and playing surface performance can greatly influence the health and safety of the modern playing surface.

References

Adams WA, Gibbs RJ (1994) Natural turf for sport and amenity: science and practice. CAB International, Cambridge, 404 p

Aldous DE (1990) Mowing principles and practices for effective turfgrass management. In: Trevor T, Aldous DE (eds) Proceedings of management of amenity and sports turf. Victorian Region of the Royal Australian Institute of Parks and Recreation, Melbourne, 28–30 March, pp 116–128

Aldous DE (ed) (1999) International turf management handbook. Inkata—A Division of Butterworth Heinemann, Melbourne (ISBN 0 7506 8954 4 347)

Aldous DE, Brereton JS (1999) Turfgrass plant health and protection. In: Aldous DE (ed) International turf management handbook, Chap 10. Butterworth-Heinemann, Melbourne, pp 173–194

Aldous DE, Chivers IH (2002) Sports turf and amenity grasses: a manual for use and identification. Landlinks, Collingwood, 152 p

Aldous DE, Chivers IH (2003) Scoping study to performance test Australian league turfgrass surfaces to ACL injuries. Horticulture Australia, project number TU 01007, July, 21 p

Aldous DE, Chivers IH, Kerr R (2005) Player perceptions of Australian football league grass surfaces. Int Turfgrass Soc Res J 10:318–326

Aldous DE, Haydu JJ, Satterthwaite LN (2007) Economic analysis of the Australian turfgrass industry. HAL Project No TU06004

Anon (2005) Plant growth regulators in the turfgrass industry. http://www.msu.edu/~babbageb/pgr.htm. Accessed 15 April 2012

Anon (2007) GHD report strategies for managing sports surfaces in a drier climate, July. http://ebookbrowse.com/ghd-report-strategies-for-managing-sports-surfaces-in-a-drier-climate-july-2007-pdf-d365197104. Accessed 2 Feb 2012

Anon (2013) Herbicides. http://en.wikipedia.org/wiki/Herbicide. Accessed 20 Feb 2013

Baker SW (1999) The playing quality of turfgrass sports surfaces. International turf management handbook, Chap 14. Butterworth-Heinemann, Melbourne, pp. 231–244

Baker SW, Canaway PM (1993) Concepts of playing quality: criteria and measurement. Int Turfgrass Soc Res J 7:172–181

Baker SW, Hind PD, Lodge TA, Hunt JW, Binns DJ (1996) A survey of golf greens in Great Britain. IV playing quality. J Sports Turf Res Inst 71:9–24

Beard JB (1973) Turfgrass: science and culture. Prentice-Hall, Englewood Cliffs

Beard JB (1989) Turfgrass water stress: drought resistance components, physiological mechanisms and species diversity. In: Takatoh H (ed) Proceedings of the 6th international turfgrass ressearch conference. Japanese Society of Turfgrass Science, Tokyo, pp 23–28

Beard JB (2002) Turf management for golf courses. Ann Arbor, Chelsea, 793 p

Beard JB, Green RJ (1994) The role of turfgrass in environmental protection and their benefits to humans. J Environ Qual 223:542–460

Beard JB, Kenna MP (2008) Water issues facing the turfgrass industry. USGA Green Sect Rec 46(6):9–17

Bell GE, Danneberger TK (1999) Temporal shade on creeping bentgrass turf. Crop Sci 39:1142–1146

Bell MJ, Holmes G (1988) The playing quality of association football pitches. J Sports Turf Res Inst 64:19–47

Bell MJ, Baker SW, Canaway PM (1985) Playing quality of sports surfaces: a review. J Sports Turf Res Inst 61:26–45

Brandenburg RL, Villani MG (1995) The handbook of turfgrass insect pests. The Entomological Society of America, Lanham, 140 p

Busey P (2003) Cultural management of weeds in turfgrasses: a review. Crop Sci 43:1899–1911

Canaway PM, Baker SW (1993) Soil and turf properties governing playing quality. Int Turfgrass Soc Res J 7:192–200

Canaway PM, Bell MJ, Holmes G, Baker SW (1990) Standards for the playing quality of natural turf for association football. In: Schmidt RC, Hoerner EF, Milner EM, Morehouse CA (eds)

Natural and artificial playing fields: characteristics and safety features. American Society for Testing and Materials, Philadelphia, pp 29–47
Christians NE (2003) Fundamentals of turfgrass management, 2nd edn. Wiley, Hoboken, 368 p
Chivers IH (2008) The development and equipment and techniques for the objective assessment of turf. PhD dissertation. Faculty of land and food resources. The University of Melbourne, March, p 709
Chivers IH, Aldous DE, Orchard J (2005) The relationship of australian football grass surfaces to anterior cruciate ligament injury. Int Turfgrass Soc Res J 10:327–332
Connellan GJ (1999) Turfgrass irrigation.In: Aldous DE (ed) International turf management handbook. Chap. 7. Butterworth-Heinemann, Melbourne, pp 119–138
Craigie V (1994) Grass and forb selection characteristics for open space. In: Aldous DE, Arthur T (eds) Trends in sports turf and amenity grasslands. Proceeding of royal Australian institute of parks and recreation, Melbourne, August, pp 5–12
Crocker RL, Martin PM (1964) Competition between perennial ryegrass and meadow fescue under field plot conditions. J Br Grassl Soc 19:27–29
Duble RL (2001) Turfgrasses: their management and use in the southern zone, 2nd edn. Texas A & M University, College Station, 323 p (ISBN 1-58544-161-9)
Dudeck AE, Peacock CH (1992) Shade and turfgrass culture. American Society of Agronomy, Madison
Duncan RR, Carrow RN (2000) Seashore paspalum-the environmental turfgrass. Ann Arbor, Chelsea
Eade M (1990) Mixtures and blends for the establishment of turf from seed. In: Trevor T, Aldous DE (eds) Proceedings of management of amenity and sports turf. Victorian Region of the Royal Australian Institute of Parks and Recreation, March, pp 46–49
Emmons RD (2000) Turfgrass science and management, 3rd edn. Delmar Thomson Learning, Albany, 528 p
Fermanian TW, Shurtleff MC, Randell R, Wilkinson HT, Nixon PL (2003) Controlling turfgrass pests. Pearson Education, New Jersey, p 654
Field TRO (1994) Horse racing tracks. In: Adams WA, Gibbs, RJ (eds) Natural turf for sport and amenity science and practice. CAB International, Wallingford, pp 329–353
Global Compendium of Weeds (2011) http://www.hear.org/gcw/index.html. Accessed 15 April 2012
Groves R (1990) Native grassland species in revegetation programs. In: Trevor T, Aldous DE (eds) Proceedings of management of amenity and sports turf. Victorian region of the royal Australian institute of parks and recreation, March, pp 4–12
Harivandi MA (1999) Interpreting turfgrass irrigation water test results. University of California Publication 8009. http://ucanr.org/freepubs/docs/8009.pdf. Accessed 19 May 2012
Higginson FR, McMaugh PE (2008) Environmental, social, economic and health benefits of turfgrass: a literature review. Project TU07034, horticulture Australia limited, Sydney, 65 p
Hitchmough J (1990) Options for amenity grassland management. In: Trevor T, Aldous DE (eds) Proceedings of management of amenity and sports turf. Victorian region of the royal Australian institute of parks and recreation. 28–30 March, pp 58–67
Jacobs BF, Kingston JD, Jacobs LL (1999) The origin of grass-dominated ecosystems. Annals of the missouri botanical gardens, vol 86. Missouri Botanical Garden Press, USA, pp 590–643
Liebao H, Aldous DE (1999) Turfgrass establishment, revegetation and renovation. In: Aldous DE (ed) International turf management handbook, Chap 5. Butterworth-Heinemann, Melbourne, pp 81–100
Leslie AR (1994) Handbook of integrated pest management for turf and ornamentals. Lewis, Boca Raton, 660 p
Livesay GA, Reda DR, Naumann EA (2006) Peak torque and rotational stiffness developed at the shoe-surface interface—the effect of shoe type and playing surface. Am J Sports Med 34(3):415–422
Loch DS, Poulter RE, Roche MB, Carson CJ, Lees TW, O'Brien I, Durant CR (2006) TU02005: amenity grasses for salt affected parks in coastal Australia. Final project report for horticulture Australia Ltd, p 93

Marcum KB (2008) Relative salinity tolerance of turfgrass species and cultivars. In: Pessarakli M (ed) Handbook of turfgrass management and physiology. CRC, New York, pp 389–406
Marshall EJP (1994) Amenity grass for non-sport use. In: Adams WA, Gibbs RJ (eds) Natural turf for sport and amenity. CAB International, Wallingford, pp 354–376
Martin PM (2004) The potential of native grasses for use as managed turf. In: Proceedings of the fourth international crop science congress, Brisbane. http://www.cropscience.org.au/icsc2 004/symposia/2/3/2136_martin.htm. Accessed 30 May 2012
McAuliffe KW (2012) A national system for the performance testing and assessment of sports fields in Australia. Ifpra World July, pp 6–9
McAuliffe KW, Gibbs RJ (1993) A national approach to the performance testing of cricket grounds and lawn bowling greens. Int Turf grass Soc Res J 7:222–230
McCarthy LB, Everest JW (2008) Colour atlas of turfgrass weeds: a guide to weed identification and control strategies, 2nd edn. Wiley, Hoboken, 423 p
McClements I, Baker SW (1994a) The playing quality of natural turf hockey pitches. J Sports Turf Res Inst 70:13–28
McClements I, Baker SW (1994b) The playing quality of natural turf rugby pitches. J Sports Turf Res Inst 70:29–43
McNitt AS, Waddington DV, Middour RO (1996) Traction measurement on natural turf. In: Hoerner EF (ed) safety in American football, ASTM STP 1305, American society for testing and materials, pp 145–155
Morgan WC (1989) Alternatives to herbicides. Plant Prot Q 4(1):33–36
NERC (1977) Amenity grasslands—the needs for research, natural environment research council, publication series "C", No. 19. Natural Environment Research Council, London
Neylan JJ, Stubbs A (1997) Review of devices currently available for assessing racetrack conditions. Rural industries research and development corporation. Project Number PTP-5A, Kingston, 31 p
Neylan JJ, McGeary DJ, Robinson MR (1999) Prescription surface development: sports field and arena management. In: Aldous DE (ed) International turf management handbook, Chap 17. Butterworth-Heinemann, Melbourne, pp 281–298
Orchard J, Chivers IH, Aldous DE, Bennett K, Seward H (2005) Rye grass is associated with fewer non-contact anterior cruciate ligament injuries than Bermuda grass. Br J Sports Med 39(10):704–709. (ISSN: 1473–0480)
Radko AM (1980) The USGA Stimpmeter for measuring the speed of putting green. Beard JB (ed) Proceedings of the third international turfgrass research conference ASA and CSSA. International Turfgrass Society and ASA and CSSA, Madison
Rorison IH, Hunt R (1980) Amenity grassland: an ecological perspective. Wiley, Hoboken, 261 p
Sauer JD (1972) Revision of *Stenotaphrum* (Gramineae: Paniceae) with attention to its historical geography. Brittonia 24:202–222
Shildrick JP (1990) Weed control in sports turf and intensively managed amenity grassland. In: Hance RJ, Holly K (eds) Weed control handbook: principles, 8th edn. Wiley, Hoboken, pp 407–430
Smiley RW, Dernoeden PH, Clarke BB (2005) Compendium of turfgrass diseases. The American Phytopathological Society, USA, 167 p
Toshikazu T, Beard JB (1997) Colour atlas of turfgrass diseases. Ann Arbor, Chelseam, 245 p
Turgeon AJ (2008) Turfgrass Management, 8th edn. Pearson Prentice Hall, Upper Saddle River, 436 p
Vittum PJ, Villani MG, Tashiro H (1999) Turfgrass insects of the United States and Canada, 2nd edn. Comstock, Ithaca
Watschke TL, Dernoeden PH, Shetlar D (1994) Managing turgrass pests. Lewis, Boca Raton, 361 p
Watson L, Dallwitz MJ (1992 onwards) The grass genera of the world: descriptions, illustrations, identification and information retrieval; including synonyms, morphology, anatomy, physiology, phytochemistry, cytology, classification, pathogens, world and local distribution and references. http://delta-intkey.com. Accessed 10 April 2012

Watson JR (1994) Warm season turfgrasses. In: Adams WA, Gibbs RJ (eds) Natural turf for sport and amenity Chap 11. CAB International, Wallingford

Warnke SE (2003) Creeping bentgrass. In: Casler MD, Duncan RR (eds) Turfgrass biology, genetics and breeding. Wiley, Hoboken, pp 175–185

Wilson JR (1999) Turfgrass growth and physiology. In: Aldous DE (ed) International turf management handbook. Chap. 3. Butterworth-Heinemann, Melbourne pp 49–63

Chapter 23
Interior Landscapes

Ross W. F. Cameron

Abstract Interior landscapes (plantscapes) can vary in scale from an individual pot plant, to representations of entire plant communities housed within large glasshouses or atria. Inclusion within buildings brings many challenges to plants (inappropriate irradiance, temperature, humidity, irrigation, aerial environment, nutrition), but potentially many benefits to the humans who occupy those buildings (higher humidity, reduced aerial pollutants, psychological benefits, enhanced 'sense of place' and improved interior décor). This chapter reviews the role, history, benefits, installation and management of interior plants and outlines their contribution to human society. The desire to include plants within our built structures has traditionally been through an appreciation of their aesthetics (colour, form, texture and line) and our fascination with their diversity and cultivation. Only recently though, have we begun to understand that exposure to plants in our everyday activities, such as during our office work or recreational activities, may provide us with a number of additional benefits linked with environmental improvements to the interior space as well as positive impacts on our health and well-being.

Keywords Air quality · House plant · Interior design · Ornamental · Plantscape · Well-being

Introduction

Plants in interior landscapes can range from a single spider plant (*Chlorophytum comosum*) placed on a domestic bathroom windowsill, through to depictions of natural habitats or entire plant communities housed in elaborate, technically-sophisticated atria. They are used to furnish and improve the aesthetics of domestic homes, offices, schools, leisure centres, shopping malls, hotels, cafes and restaurants as well as provide distinctive iconic landscapes in themselves e.g. 'The Great Glasshouse' at the National Botanic Gardens of Wales (UK) or the Tropical and Mediterranean Biomes at the Eden Project (UK). Interior planted landscapes or 'plantscapes' are

R. W. F. Cameron (✉)
Department of Landscape, The University of Sheffield, Western Bank,
Sheffield, South Yorkshire S10 2TN, UK
e-mail: r.w.cameron@sheffield.ac.uk

G. R. Dixon, D. E. Aldous (eds.), *Horticulture: Plants for People and Places, Volume 2*,
DOI 10.1007/978-94-017-8581-5_23, © Springer Science+Business Media Dordrecht 2014

Fig. 23.1 The Winter Gardens in Sheffield, UK. The glasshouse structure has been designed to accommodate large plant specimens, as well as provide a strong architectural form in its own right. The winter gardens act as an entrance to an Art Gallery, a thoroughfare from one city civic square to another and a retail forum with shops, bars and cafes

employed to convey a certain brand or image (Steele 1992), for example tall, imposing plant specimens set within a building foyer to emphasis grandure and denote the aspirational attitude of the parent company. In contrast, soft-textured plants positioned within a restaurant to provide a relaxing and welcoming ambience. Plants are used to enforce the interior design style. Palms (e.g. *Howea forsteriana* or *Veitchia merrillii*) and ferns (e.g. *Blechnum gibbum* or *Nephrolepis exaltata*) may complement the wicker furniture of an 1900s CE 'Edwardian theme', or *Spathiphyllum wallisii* reflect the elegance of art deco, whereas the strong bold forms of *Yucca elephantipes* and *Sansevieria trifasciata* are combined with geometric, metallic planters to promote a modern, 'minimalist' style. Scale and complexity vary too, from individual pot plants; through to themes employing a number of identical potted specimens, for example placed regularly along a corridor; up to large specialised planters integral to the building's framework. At the grandest scale extensive 'landscapes' complete with trees, specimen shrubs, bamboos, epiphytes, climbers and ground cover plants can be grown under glass or some other translucent material (Fig. 23.1). Even within this last category the context can vary from conservatories attached to domestic homes, traditional orangeries, botanic glasshouses such as the Temperate Glasshouse at Royal Botanic Gardens, Kew, UK or large scale biomes, e.g. as proposed at Riyadh (Saudi Arabia). The plans for the Riyadh biome incorporate a section of rainforest, a 65 m high waterfall, hotel and retail complex all of which will be housed under one glass roof.

Despite their wide use, the retailing of interior plants tends to divide into two distinct markets—those for the domestic market (house-plants) and those used in offices, lobbies and a range of other public or private spaces. Indeed, in the latter case the plants may not actually be sold *per se*, but rented from an interior landscaping company.

The reasons for adopting interior plants are not dissimilar to those of the outdoor landscape. Well-designed interior plantscapes not only add to the aesthetics of interior design, but help provide privacy, filter sunlight or reduce glare. They can provide focal points, block unsightly objects, absorb or deflect noise and direct pedestrian traffic flows. They also provide a readily available opportunity to engage with living objects and promote a sense of nature. Increasingly, the wider benefits attributed to plants are used in their justification for inclusion within buildings. This includes positive influences on air quality, therapeutic aspects and ability to modify atmospheric conditions (e.g. raise humidity and provide thermal cooling).

History

The practice of keeping plants within buildings is thought to date back at least 3000 years, although it is probably more accurate to describe the earliest 'house plants' as transitional plants i.e. grown in pots and located at the interface between a building and the outdoors. There is some evidence for containers being used in China to hold plants, perhaps as long as 2000 BCE (Chen et al. 2005). Remnants of terracotta pots have been found in the Minoan palace of Knossos (1900 BCE) and Egyptian royalty are thought to have grown Frankincense (*Boswellia sacra*) in temples as early as 1500 BCE (Horwood 2007). Perhaps the first true interior plants were those grown by the Romans. The Emperor Tiberius (42 BCE–37 CE) grew melons within a "speculari-um", where light was transmitted through small fragments of translucent mica "Lapis Specularis" (Paris and Janick 2008) although irradiance transmission levels were likely to be considerably less than conventional glass. Nevertheless, it is also supposed that the plants themselves may have been grown on movable barrows and placed outside during the day to maximise photosynthesis. In ancient times, interior plants were probably used for culinary and medicinal purposes primarily, although they may also have been located in temples and palaces from a symbolic perspective. The first evidence for glass itself being used to help cultivate plants dates back to 290 CE.

In Europe, there appears to have been little advancement in the use of interior plants through the dark ages (500–1000 CE), although the semi-enclosed spaces of monastery cloisters, arguably may have housed climber plants such as roses (*Rosa*) and honeysuckle (*Lonicera*) (Nichols 2003). By the 1250 CE however, Arabic influences were beginning to make their mark as noblemen returning from wars in the Middle-East (Crusades) brought new plants and new techniques back with them. This included the use of growing plants in pots. This is thought to have been accomplished by glazed ceramic pots, as the unglazed clay flower pot was not in common use until 1600 CE (Currie 1993).

Italy is also considered the location for the 're-invention' of the glasshouse in the 1200s CE, with rudimentary glasshouses *giardini botanici* (botanical gardens) used to house exotic plants brought back from the tropics by explorers. These glasshouses were later adopted in the Netherlands and England, although there were on-going problems in maintaining adequate temperatures at night and during winter.

By 1700–1800 CE, both flowering and fruiting plants grown in pots were becoming commonplace in the houses of the aristocracy (being more popular than cut flowers) (Woudstra 2000). *Citrus* and *Malus* spp. were grown as dwarf potted stock and brought in at banquets to have their fruit picked fresh from the tree. Flowering plants were used to improve the aesthetics of the house and provide colour and perfume throughout the growing season—e.g. *Galanthus*, *Narcissus*, *Tulipa* followed by *Viola, Lilium, Philadelphus*, *Rosa, Lonicera* and then *Aster, Helichrysum* and *Viburnum* (Woudstra 2000). The common name of Chimney flower was attributed to *Campanula pyramidalis*, due to its placement during summer in the temporarily redundant fireplaces of large, estate houses. Plants were often grown in plain earthenware pots in the garden, but then inserted into larger more decorative pots when positioned *in situ* within the house.

Glasshouses (greenhouses/orangeries) began appearing within the vicinity of the grand estate houses of Europe in the 1700s CE. This corresponded to a period of greater social stability where wealthy citizens had time and the financial resources to invest in plant cultivation, other than that previously associated with subsistence agriculture. The size/sophistication of these glasshouses increased as technology advanced through better quality glass and improved construction techniques; the glasshouse at the Palace of Versailles being state of the art in terms of elaboration and size (150 × 13 × 14 m, lbh). In parallel, this was an era of exploration and trade with new plant species being brought into Europe from warmer climates. Imported seeds and young plants needed to be carefully cultivated and this required the development of protected growing structures. Many plant species were exploited to provide a greater variety of food crops and for economic gain e.g. cultivation and transplantation of plantation crops, but there was also a heighted scientific interest in plants in general, and an increasing interest in gardening.

The late 1700s CE saw the development of small glasshouses within botanic gardens to house the tropical species and this lead on to the development of iconic structures in the 1800s CE such as the Great Conservatory at Chatsworth, UK (1837), the Tropical Palm House at the Royal Botanic Gardens, Edinburgh, UK (1858) and the Palm (1844) and Temperate Houses (1863) at the Royal Botanic Gardens, Kew, UK. Some of these structures resulted in the extensive use of curvilinear glass for the first time. (Fig. 23.2). Such structures helped educate the public and stimulate interest in tropical plants. The invention of the Wardian case in 1833 (a small glass terrarium) also made transportation on long sea voyages more feasible.

House architecture also changed in the 1800s with many of the new villas possessing recessed bay windows and sun-porches. Larger windows became popular in Britain after the repealing of the Glass Excise Act (1845 CE) and this, combined with the manufacture of less expensive and better quality glass, promoted windows with greater light transmission and hence more conducive to plant growth.

Fumes from coal fires and gas lamps though, were still restricting the cultivation of some of the more sensitive plant species. In Victorian Britain, lifestyle changes also meant that the burgeoning middle-class had more leisure time and the cultivation of the pot plant took off as a trend. This was fuelled by a combination of culture—the Victorians hankered after contact with nature and the elegance provided by indoor plants was the epitome of 'good taste'; at the same time new and exciting

Fig. 23.2 The Glass Pavilion (built 1838) at Sheffield Botanic Garden was one of the first examples where curvilinear glass panes were used to good effect to maximise light transmission and strengthen the roof structure

plant species were being discovered and retailed through specialist nurseries and catalogues. Many of the species that are popular as houseplants today came to the fore during the Victorian era, including *Aspidistra, Abutilon, Jasminum, Ficus, Fuchsia, Citrus, Heliotropium*, as well as various ferns and palms.

The next significant phase for plant use within buildings rose with the advent of office work in the post-1950s CE. The increasing numbers of women working in offices after World War II is thought to be a contributing factor to greater incidences of plants (and cut flowers) being used to improve the aesthetics of the office environment. Plants were more formally embraced within the Bürolandschaft or 'office-landscape' concept to create designated spaces and a degree of privacy within an open-plan office structure. Unlike the traditional offices there were no stationary partitions, walls, or private offices to inhibit communication among workers. Similar approaches were adopted across Europe and North America in the 1970s and 1980s CE. The use of plants become commonplace in open plan offices both to provide structure within the interior and a degree of seclusion to the individual workplace. Plants, however, were not restricted to the office and are now employed in the reception areas of offices/hotels etc. as well as in communal 'break out' areas for employees, such as café bars etc. Increasingly, in high profile developments the plantscape is designed into the architecture of the building, rather than being left as an 'add-on' at the end of the process. The plant range has also expanded with time, with at least 1000 different plant species (covering 100 genera) now being represented (Chen et al. 2005).

Why Use Plants in Buildings?

Plants enhance the interior aesthetic, helping to provide a pleasant, tranquil environment where people can work or relax. They are used to provide a design statement and hence reflect current fashions and trends. In the business sector they help re-enforce a certain image or provide a connotation about a company and its

Fig. 23.3 The strong architectural form of *Yucca elephantipes* (*left*) is used effectively to provide interest beside a reception desk. The container adds to the contrasts in form and colour. Form also plays an important role in this mixed composition (*right*) but so does geometrical 'line' with upright lines associated with the cacti providing contrast to the directional lines of the path, walls and glass panes

aspirations. Particularly, in large retail and shopping areas plants help give identity to the site through providing a 'sense of place' that would be difficult to achieve through other media. Indeed, in an environment increasingly dominated by a limited range of retail chain stores, with their 'universal' brands, careful plant design and selection can provide an element of uniqueness to any given site. In a similar manner, international air travellers are exposed to airport terminals that tend to be largely uniform in character, but by incorporating large displays of orchids, Singapore's Changi airport promotes itself as a distinct, 'place apart' and one that reflects the nature of the host country; the orchid being the national flower. Some companies use similar symbolism in their plantscapes to highlight their national or cultural connections. The plant choice itself influences the ambience of the setting, with form and line being key components (Fig. 23.3). *Ficus benjamina*, *Schefflera sctinophylla* "Amate" and *Howea forsteriana* provide relaxing, flowing lines, whereas the sharp ascending forms of *Sansevieria trifasciata*, *Dracaena draco* and *Aechmea fasciata* deliver a dynamic, dramatic display. Alternatively, more rounded, but large-leaved, strongly textured species such as *Monstera deliciosa*, *Fatsia japonica* and *Ficus elastica* help promote feelings of strength, reliance and power.

In atria, (and at a smaller scale, conservatories) plants help blur the transition from indoor and outdoor and link the two. In large, open plan spaces, such as shopping malls, stations and airports, plantscapes are orientated to channel pedestrian traffic towards important landmarks, such as check-in desks, entrances and exits, shopping or food facilities (Steele 1992). In these situations plants may need to act as signage or reference points and hence be bold and easily visible. Humans often find such large open spaces intimidating in themselves and landscape plants provide a familiar feature and create more intimate space (Davidson 2000).

The use of a limited range of species or forms is often advocated too to provide some harmony to the design, and avoid a cluttered and irregular appearance to the interior space. Practical and managerial issues need to be considered carefully. Plant choice should reflect busy, public open spaces, and soft flexible foliage is to be preferred to spikes and sharp or serrated leaves that people may brush against. Low irradiance may mean coloured leaf forms are either ineffective of lose their colour (red and purple leaf types turning green in low light); species with variegated leaves may be preferable in these situations where there is an ambition to provide variety. It may also be advisable to exclude species that regularly flower as unsightly dead flowers require frequent deadheading, and fallen flower heads/fruits need to be swept up on a regular basis. Fruit too, may be attractive to young children but be potentially toxic e.g. *Solanum pseudocapsicum* (Der Marderosian 1976).

Interior plants may provide some economic advantage through their aesthetics and associated ambience by attracting people into restaurants, shops and hotels. Wolf (2002) argues plants create shop interiors which are directly more favourable for 'retail activity'. Well-planted interiors have been linked with reduced absenteeism in the workforce with economic implications for the employing company (Fjeld 2000; Bringslimark et al. 2007; see also Sect. 4).

Plants are often quoted too as attenuating noise in offices, but there is little in the peer reviewed literature to give details of this. Certainly, plants are likely to reflect and absorb sound waves (Costa and James 1999 as cited in Tarran et al. 2007) but so too are soft furnishings, and whether plants provide an added advantage is a moot point. Certainly they may help provide a greater intimacy within an interior environment, and give the perception that an area is quieter.

Health and Well-Being

The benefits of green space on human health and well-being has become increasingly documented (e.g. Grinde and Patil 2009; St Leger 2003; Kuo and Sullivan 2001; van den Berg et al. 2010; see Chap. 26). Detailed understanding, however, of how factors such as context, scale of landscape, diversity of features etc. might affect responses, remain elusive (Cameron et al. 2012). Therefore, research findings on the impact of house plants and interior landscapes are particularly relevant as they can provide some of the more extreme scenarios; for example, can the introduction of a limited number of natural elements such as a plants, really affect the well-being of the occupants of a room? There is evidence to suggest this is indeed the case. Moreover, the presence of interior plants has been linked with a number of positive factors, namely:

- Increased attentiveness or mental skills (Lohr et al. 1996; Shibata and Suzuki 2001, 2002; Kim and Mattson 2002; Khan et al. 2005)
- Positive mood and creativity (Isen 1993; Larsen et al. 1998)
- Faster reaction time (Lohr et al. 1996)
- Lower systolic blood pressure (Lohr et al. 1996; Park and Mattson 2008)
- Lower electrodermal activity (Park et al. 2004)
- Less fatigue and physical discomfort symptoms (dry throat, coughing, dry skin) (Asaumi et al. 1995; Fjeld et al. 1998; Fjeld 2000, 2002; Park and Mattson 2008)
- Increased social interaction and engagement in hospital patients with psychological health problems (Talbott et al. 1976)
- Stress recovery or feeling of calmness (Kim and Mattson 2002; Shibata and Suzuki 2004; Chang and Chen 2005; Dijkstra et al. 2008; Park and Mattson 2008, 2009)
- Higher pain tolerance (Lohr and Pearson-Mims 2000; Park et al. 2004; Park and Mattson 2009)
- Higher ZIPERS values (Zuckerman's Inventory of Personal Emotional Reactions Score) (Lohr and Pearson-Mims 2000)
- A more attractive working or living environment (Larsen et al. 1998; Adachi et al. 2000; Fjeld 2002; Lohr and Pearson-Mims 2000; Park and Mattson 2009).

In a review of plants and their influence on human productivity Bakker and van der Voordt (2010) sum up the most dramatic positive effects of introducing plants as:

- 12 % increase of in shop sales (Wolf 2002)
- 12 % increase in response speed during simple recognition tests (Lohr et al.1996)
- 23 % reduction in symptomatic physical complaints within a population of 51 office employees (Fjeld 1998)
- 25 % decrease in health complaints among 48 employees of a hospital X-ray department (Fjeld 2000)

Increasing the proportion of plants within view is also thought to enhance the positive effects, with greater reduction in anxiety and more positive responses reported (Han 2009).

Within this context, those studies that are located in hospitals are particularly valuable as routine physiological measurements are a requirement for monitoring patient recuperation. Therefore, research activities involving assessments of human physiological reactions, rarely cause additional anxiety, which otherwise might arise under a different set of circumstances. Although, it should be recognised that responses in hospital patients may not always be analogous to that of fully healthy participants. In a comprehensive study, Park and Mattson (2008) showed that patients recovering from abdominal surgery, had lower blood pressure and heart rate, significantly less need for analgesics, and recorded lower ratings of pain, anxiety and fatigue when plants and flowers were introduced into their hospital room (Fig. 23.4). Patients also reported more positive feelings and higher satisfaction levels with their rooms compared to patients without plants. Despite these positive factors associated with interior plants, some medical authorities now actively

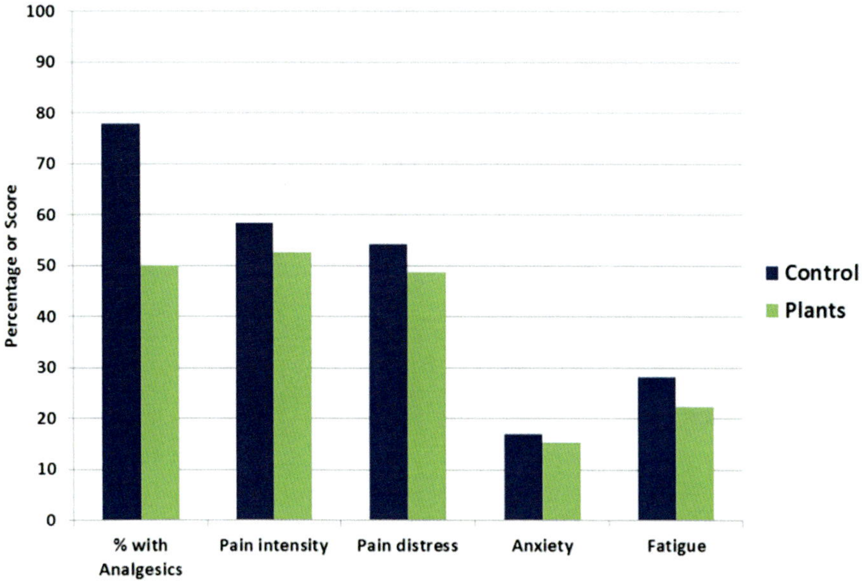

Fig. 23.4 Hospital patient recovery when plants present (plants) or absent (control) from hospital rooms during recuperation. Data represents percentage of patients taking pain killers (analgesics) 3 days after surgery and scored for self-reported feelings of pain (intensity and distress), anxiety and fatigue following surgery. (From Park and Mattson 2008)

discourage the placement of plants and cut flowers in hospital wards due to concerns related to cleanliness, perceived risks from hospital acquired infections and increased workload for nurses (e.g. watering plants).

Using simulations of office environments Chang and Chen (2005) suggest that both external views of nature and interior plantings can reduce reported anxiety levels and blood volume pulse rate, compared with scenarios that omitted these elements. Dijkstra et al. (2008) using photo images of a hospital room, with and without plants, suggest the stress-reducing effects associated with plant-based images, was mediated through an increased perceived attractiveness of the room. They conclude that by making hospital environments more attractive (and not only with natural objects such as plants) they can contribute to the health and well-being of patients.

In contrast, a number of studies showed no positive relationships. Plants had no impact on work satisfaction (Shoemaker et al. 1992) nor task preparation and performance with women (but had a positive effect on men) when carrying out sorting and word association tasks (Shibata and Suzuki 2002). Larsen et al. (1998) found the presence of plants improved mood scores in participants but they also seemed to reduce concentration levels.

In a survey of 385 Norwegian office workers, Bringslimark et al. (2007) gave an indication that plants affected absenteeism due to illness in a small, albeit statistically reliable way. The distance an employee was situated from the plant, however, altered the response. Plants within view of the workstation and the presence of the employees own plants reduced absenteeism, but plants at closer proximity increased it. This somewhat ambiguous situation reflects other research using controlled settings to evaluate the effects of plants on psychophysiological stress, (Chang and Chen 2005; Coleman and Mattson 1995; Kim and Mattson 2002; Liu et al. 2003; Lohr et al. 1996). The results relating to the location of plants and who owns them may also reflect deeper issues about employees having ownership over their workspace. Workspaces that allow the individual to express some element of their personality and interests also provide positive outcomes (e.g. well-being and productivity; Knight and Haslam 2010), so perhaps then plants chosen by employees to reflect their individuality provide greater benefits than 'corporately imposed' specimens placed within the vicinity of the workstation. Random sampling in the USA via e-mail also gave marginally significant positive responses in terms of job satisfaction attributed to both plants and presence of windows within office environments (Dravigne et al. 2008).

In contrast to many previous studies where scenarios were recreated over the short-term under semi-controlled laboratory conditions, Han (2009), studied student responses in real classrooms over longer time intervals. This involved live plants being placed at the back of school classrooms (i.e. not in the direct line of vision or only peripheral to the students' viewpoint for much of the time). Results showed that there were significant differences in the immediate preference, comfort and feeling of friendliness in rooms with plants, but no differences over longer time periods. There was no significant difference either on academic performance, but pupils in rooms with plants present were shown to have significantly fewer sick leave hours and fewer punishment records (Han 2009).

In their review of plants and human productivity, Bakker and van der Voordt (2010) hypothesise that variation in results could relate to the fact that impacts vary based on different types of productivity, that there is a lack of standardised methods to compare responses (a viewpoint widely held) and that little account of the plant type and health status are included in the methodologies. Plant quantity, form, size and species may have an effect, and this area is certainly under researched and worthy of further exploration. On the other hand the implied criticism that studies were potentially flawed due to the plants used being in a sub-optimal state of health (plant health status not being reported in the methodologies) seems less credible, not least because significant numbers of the researchers were also experienced horticulturalists.

If plants do have a positive impact on well-being, then the mechanisms behind this have also intrigued researchers in the field. Wilson (1984) argued that humans have an innate requirement for nature (Biophilia) due to the fact that so much of our evolutionary adaptation took place in forests and grasslands. Deviations from these evolutionary environments, i.e. working in factories and offices rather than forests and savannahs suggest mismatches, and that negative mismatches (discords) can cause stress in humans. Analogies are often made to zoo animals, in which if many of their natural environmental conditions are not replicated, certain species demonstrate discordant behaviour; pacing, self-harm or lack of mating success. Some researchers argue that a disconnection with nature can be a source of psychological and sociological problems in humans too, at least for certain individuals who may be particularly predisposed to such factors (individual peculiarities and cultural factors may influence the degree of discord experienced or their susceptibility towards the negative effects (Ulrich et al. 1991; Kaplan 1995; Hartig et al.2003; Ottosson and Grahn 2005)).

If the role of plants can be categorically proved to enhance mood, improve productivity and reduce absenteeism through sickness, the economic implications could be significant. For example absence from work cost UK business £16.8 billion in 2010. Job dissatisfaction and low morale can be prominent factors in short-term absence and could account for as much as 15% of all reported sickness absence. By providing more amenable, stimulating and perhaps healthier working environments through using plantscapes more widely, then savings could be in the region for £4.3 billion pa. (e.g. based on the Fjeld 2000 data).

Employers can tackle this most easily by re-examining people management policies and the working environment, to see what can be done to improve staff productivity and well-being. If companies with the worst absence rates could meet average levels, the UK economy would be £1.9 billion better off (Anon 2003).

The health and well-being benefits of interior plants has not been lost on the interior plant landscaping industry, and many companies promote the use of plants indoors due to these considerations as well as aesthetic virtues. In a number of high-profile news cases where journalists have highlighted the amount of money politicians have 'squandered' on decorating their offices with plants, the industry has not been slow to demonstrate that the use of plants may actually save money through improved productivity and reduced employees absenteeism (e.g. Saner 2012).

Indoor Air Quality

Air quality and the effect of aerial pollutants has been a concern for health officials since the 1950s. It is often assumed that air quality is worse outdoors, but levels can actually be higher indoors (Baek et al. 1997). Atmospheric nitrous oxides (NO_x), sulphur oxides (SO_x), carbon monoxide (CO) and dioxide (CO_2), ammonia (NH_3) and organic gases can diffuse indoors and can be augmented by other aerial pollutants generated indoors. Indoor-generated pollutants include particulate matter (dust), but also volatile organic compounds (VOCs), of which 300 have been identified but include compounds such as formaldehyde, toluene, benzene and xylene. VOCs have been closely associated with 'sick building syndrome' where building occupants suffer ailments included headaches, nausea, dizziness, respiratory problems, dry throat and eyes, and complain of loss of concentration. VOCs are commonly generated from office features/chemicals including furniture, printers, laminated finishings and the detergents, paints, varnish and polishes used within the building. Poor ventilation can enhance levels of common gases such as CO_2 and even this has been associated with reduced workplace productivity (Seppänen and Fisk 2006) and mental ability (Shaughnessy et al. 2006).

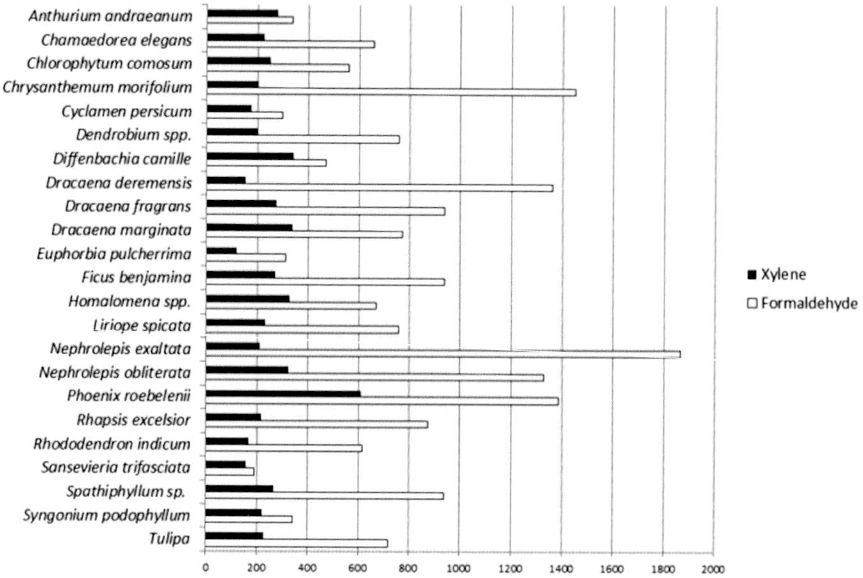

Fig. 23.5 Removal (μg hr^{-1}) of volatile organic compounds (xylene—solid, formaldehyde—open) by different plant species (From Wolverton and Wolverton 1993)

Interior plants are thought to be a positive force by both removing aerial particulate matter by adsorbing dust onto their surfaces and by absorbing and metabolising a range of gaseous pollutants. Plants have been cited as reducing levels of NO_2 (Coward et al. 1996), SO_2 (Lee and Sim 1999) particulate matter (Lohr and Pearson-Mimms 2000) and VOCs (Wolverton and Wolverton 1993; Dingle et al. 2000; Orwell et al. 2004; Tarran et al. 2007) from interior environments. Removal of aerial pollutants have been researched through the use of biomass filters, e.g. using moss species and associated microflora (Darlington et al. 2001) as well as pot plants. Research has indicated that some pot plants may be more effective at removing pollutants than others (Fig. 23.5), although the relative effectiveness can vary depending on the pollutant. It is clear that plants absorb some of the gases and metabolise them directly—CO_2 of course being essential for photosynthesis. Tarran et al. (2007) found that by placing just three plants in an office environment CO_2 levels were reduced by 25 % (and 10 % in an air conditioned building), but these plants were also effective at removing any CO present (86–92 % reduction).

After further careful study however, it became evident that much of the benefits derived from the plants in terms of air quality was through the action of associated micro-organisms (Wolverton and Wolverton 1993; Tarran et al. 2007), primarily in the rhizosphere (roots and growing media), but also the phyllosphere (leaves and stems). For example, De Kempeneer et al. (2004) showed that by enhancing the population of *Pseudomonas putida* bacteria on leaves of *Rhododendron indica*, levels of airborne toluene could be removed much more rapidly compared to leaves with only a background level of micro-flora. The same effect was not apparent, however, when the *Pseudomonas* strain was placed on an artificial surface, suggesting the role of the plant was important albeit via the secondary mechanism as a host to the bacteria.

Although studies often show significant reductions of air borne VOCs when plants are present, the rates at which plants and micro-organisms metabolize these compounds depends on the growing conditions, such as irradiance levels, temperature, humidity and both the nutrient and water status of the growing medium. Some of the reactions appear to be concentration dependant too, with metabolism decreasing when background levels fall, thus plants will not necessarily eliminate the VOCs completely from a room.

In contrast to the aforementioned data, some interior plants are thought to contribute to respiratory related disorders in humans (e.g. asthma, rhinitis and laryngitis), notably *Ficus benjamina* (Axelsson et al. 1985) and *Spathiphyllum wallisii* (Kanerva et al. 1995), although sensitivity seems to be maximised in people who have regular contact with the plant, or work within a very close proximity to them, i.e. recorded cases centre around plantscape maintenance workers, rather than office employees.

Irradiance

The interior environment does not replicate the 'natural habitat' for most plant species. In nature, ferns and bryophytes may cope with low light (irradiance) levels when colonising cave mouths or deep gullies, but in general the lack of light within buildings is the significant factor affecting plant growth and development. Light is not only needed for photosynthesis, but also regulates plant development (photomorphogenesis), chlorophyll, other pigment synthesis, stomatal behaviour, leaf temperature, rate of transpiration and mineral uptake. It is often considered that house plants placed more than 1.5–2 m from a north facing window or perhaps 3 m from a southern aspect one will begin to experience sub-optimal irradiance. This 'rule of thumb' though will vary with latitude, season and plant species.

Light as the human eye perceives it is measures as lux (lumens per m^{-2}), but plants require certain specific wavelengths more than others, particularly absorbing energy in the red and blue spectra for photosynthesis. Therefore, what is more critical to them is the amount of irradiance within the photosynthetically active region (PAR, i.e. 400–700 nm) and this is measured as µmol photons $m^{-2}\,s^{-1}$. Lux can be converted to µmol $m^{-2}\,s^{-1}$, but the convertion factors can vary depending on the light source, i.e. will be different from daylight (multiply lux by 0.018) compared to say 'coolwhite' fluorescent tube lighting (multiply lux by 0.013). It is thought that most plants will not survive below 400 lx, with shade tolerant species coping with 400–750 lx, medium light requiring species 750–1500 lx, and high light 1500–2500 lx. To put these values into context, on a clear sunny day outdoors light can exceed 50,000 lx, whereas a typical office environment may only be 500 lx.

Fig. 23.6 The impact of coloured foliage *Codiaeum variegatum* (*left*) and *Coleus x hybridus* (*right*) is accentuated at higher interior light levels, even when the species themselves may be understorey plants in their natural habitat

The amount of natural light entering a building, will depend on the design of the building, and in many cases supplementary light is required to provide the necessary irradiance for plant growth. The amount of supplementary light required will be determined by plant choice—large, tall growing specimens may need higher irradiance (e.g. >2500 lx), as they will shade-out their own lower leaves and any other plants planted below them (Lockwood 2000). In contrast, tropical species adapted to life on the forest floor may only require 500 lx. Flowering plants often need higher levels of irradiance to stimulate new shoot growth and to provide enough energy to support the developing blooms. Coloured foliage too, is often best expressed when light levels remain high (Fig. 23.6). Supplementary light can be provided via a number of different lamp types with varying intensities, light spectra and energy use efficiencies, but it is important to remember that for many situations humans will also be experiencing (requiring) the light produced and this should be taken into account when specifying the type of light required. Artificial illumination can be provided by incandescent, fluorescent, halogen, metal halide (high intensity discharge) lamps or through combinations of these, with increasing interest be evident in the application of light emitting diodes (LED) or other forms of energy efficient lighting. Although artificial lighting can add to the overall design effect when placed at low level and directed upwards, this should be avoided when attempting to grow plants effectively. Providing irradiance intensity is optimal, illumination durations in line with office working hours are usually sufficient to provide daily requirements of PAR (8–14 h being ideal).

In the more iconic plantscapes the environment may well have been designed to accommodate plants, including the incorporation of atria, conservatories and courtyards into the building design. These have the advantage of providing roof glass and hence greater natural light, although factors such as aspect, degree of shading from adjacent walls or buildings, depth and width of atria need to be taken in to consideration. A deep narrow atrium, for example, with a small ground area, but surrounded by many storeys of built structure, may still not provide enough light to facilitate plant requirements.

It is also important to recognise that on multi-storey plantings, trailing plants and balcony plantings may also intercept the light and increase the shade experienced by plants at the base. Although much of the emphasis in plantscape design is about optimising light penetration (especially to account for lower irradiance levels in winter), scenarios involving excessive light should not be disregarded. For those tropical 'forest floor' species adapted to mid or lower irradiances, direct summer sunlight and high temperatures (e.g. when combined with low humidity) can be damaging, and provision is required for venting of the glasshouse structures or providing additional shading. Such shading can be accomplished through automated blinds, or even the use of electrically controlled glass ('electrochromatic glass' or 'smart-glass') that can change its colour or transparency.

Acclimatization

In the early days of the large corporate plantscapes, it became evident that the leaves of plants often became chlorotic and abscissed shortly after being transplanted from the nursery to the interior plantscape. Subsequently, this was linked to a lack of light adaptation in the existing leaves, and that plants were exposed to too severe a transition from relatively high light (and temperature) conditions within a nursery glasshouse (plants are also shipped in from outdoor nurseries located in warmer climates, where natural irradiance levels are also higher) to the relatively low light conditions of an interior space. Today, interior plants are placed through an acclimation procedure where light levels are gradually reduced down by artificially shading the plants during the production phase. This usually takes 6–8 weeks, with the light intensity being reduced by 50 % on 2–3 separate occasions over this period. This acclimation process involves both physiological and anatomical changes in the plant leaf, e.g. chlorophyll being rearranged within the grana to provide greater exposure to light, altered nitrogen metabolism, reduced but more efficient photosynthetic capacity, the presence of smaller or less frequent stomata, the development of larger, but thinner leaves, and altered root to shoot ratios (Rodríguez-Calcerrada et al. 2008). In addition to reducing irradiance intensity, acclimation can be aided by less frequent irrigation and lower nutrient supply (Chen et al. 2005).

Temperature, Humidity and Aerial Pollutants

Human thermal comfort is considered to be optimal at 21°C during the day and 18°C at night and the choice of interior plant species tends to reflect the temperatures typical of modern buildings. The absence of winter chilling within buildings can be problematic to temperate species that are permanently placed and require a period of cold to break dormancy and complete their physiological development. In essence, most plants used are from the tropical, sub-tropical or warm temperate regions. These themselves, however, can be divided into classifications based on their biomes. Those derived from the humid tropics may require 30°C during the day, but 15–18°C at night. In contrast, dry tropical desert species may tolerate higher day-time temperatures, but are adapted to sub-zero night temperatures. Warmer temperate or sub-tropical species prefer 20–25°C during the day, with 10–12°C at night. Additionally, certain species from cooler temperate zones may prefer temperatures that are adjusted based on the seasons with cooler regimes during the winter (e.g. 12–15°C day-time temperatures compared to 18–20°C during summer).

Draughts and direct placement close to open windows, heaters or air conditioning units should be avoided. Not only will these aspects affect temperature, but may decrease humidity or cause excessive movement of the foliage, resulting in necrotic lesions on leaves. Plants with fine foliage can be particularly susceptible to desiccation under such circumstances.

Relative humidity is the volume of moisture within the air as a percentage of the total moisture the air could potentially hold at any given temperature and pressure. During the nursery stage when plants are grown in glasshouses they often experience relative humidity of 85–90%. Office environments are usually closer to 45–50% relative humidity although domestic properties can be<25%. As with irradiance a period of acclimatization is recommended by progressively lowering the relative humidity before moving plants from the glasshouse nursery. Plant species also vary in their optimum humidity with species from arid climates e.g. cacti requiring lower values than those from moist tropics. Bamboos and ferns and other fine leaved plants may require 70% relative humidity once *in situ*, but most broadleaved species tolerate the humidity typical of offices (i.e. 45–50%). Where humidity is too low, hand misting can be employed to wet the leaves and raise the localized humidity, but this is too labour intensive for large plantscapes. Here, mist or fog nozzles can be fitted into the design and automated to counteract low humidities. Current systems use mains water and compressed air to create the mist or fog, and standing storage tanks of water should be avoided to minimise risks due to *Legionella pneumophilia* (Legionnaires' disease). Ideally nozzles should be located to direct the spray into the plant canopy and away from seating areas, paths and electrical equipment. Apart from hazards associated with puddling water on the hard standing areas, hissing noises issuing from the vegetation are not always appreciated by an unsuspecting public!

Aerial pollutants and gaseous fumes can be problematic to interior plantscapes (Ingels 2009). Industrial cleaning products, paints, varnishes, preservatives etc. which may release ammonia, chlorine and volatile hydrocarbons are often phytotoxic.

Ethylene released as a by-product of hydrocarbon combustion can cause epinasty and chlorosis, even at relatively low concentrations. Effects seem to be worse when there are prolonged exposure periods and where there is limited air movement, or the plants are in confined spaces where pollutants can build up. In swimming pools and leisure centres, problems associated with chlorinated water can be minimised, by ensuing the plants are far enough away from the pool to avoid direct splashing, and that there is forced air movement and ventilation to help reduce the background levels of chlorine in the air. Particulate matter (dust) also needs to be removed from foliage on a regular basis to maintain photosynthetic capacity, and this can be done by washing down the foliage periodically, or cleaning large leaves with a fine cloth (Ingels 2009).

The Root Environment

The growing medium within plantscapes needs to provide anchorage, supply water, air and nutrients and look aesthetically pleasing. Root development is required not only to access water and nutrients but also to synthesis endogenous hormones that regulate the growth and physiological development of the entire plant. Growing media and the water it contains can be the key factor affecting the weight loading of the plantscape, and architects need to account for this when designing buildings or anticipating the placement of planters on floors above ground level. As excessive weight can be problematic the growing media should be a light as possible, whilst still providing sufficient bulk density (e.g. 0.15–0.75 g cm^{-3}) to anchor the plants and retain moisture. Both loam and loamless (e.g. peat based) media have been used for interior landscapes (Lockwood 2000). Inert materials, such as clay leca are also used more widely than would be the norm in exterior landscapes. Leca is expanded clay formulated into round pellets and is used within hydroculture systems, where its honeycomb micro-structure facilitates both moisture retention and aeration. Hydroculture involves the placement of leca or some other inert material (e.g. perlite, rockwool, glass beads) within a planter which is then part filled with water or a weak nutrient solution. Water then rises and is accessed by the plant roots via capillary action.

In large plantscapes, such as the Eden Project, UK, the soils have been 'manufactured' to suite the plants' needs (e.g. mimicking the conditions of a tropical rainforest, whilst accounting for the fact that soil depth and drainage characteristics may be compromised by the location (a disused quarry). After initial research, a blend of mineral materials and green compost was selected (Russell and Best 2006).

Where conventional growing media are used then more care is required on the levels of irrigation supplied. The old adage that house plants usually fail due to over-watering or under-watering is based on sound principles. The overall strategy is to provide regular, consistent levels of irrigation, without letting the medium become saturated or hydrophobic. A number of irrigation systems are used in practice within interior landscapes (Table 23.1).

Table 23.1 Various forms of irrigation systems employed in plantscapes

System	Details	Advantages	Disadvantages
Hand watering	Usually at a domestic level via watering can or similar	Opportunity to regularly monitor plants Good for plants with special irrigation requirements, e.g. orchids, bromeliads	Still evident in some commercial situations, but cumbersome, labour intensive, and risk of water spillage in offices High potential for inaccurate irrigation
Manual watering via hose pipe	Some atrium situations with a range of taps provided in convenient locations Alternatively, maintenance crew may arrive with water bowser Extension systems available to water aerial planters/hanging baskets and other inaccessible plants	Water individual specimens within the display	Difficult to move hoses/bowser around Risk of pedestrian tripping or water spillage
Self-watering planters	For group planters/containers Employs a sub-surface reservoir, which irrigates the medium via wicks or other capillary mechanisms. The reservoir is topped up periodically via a feeder pipe	System is self-regulating	Certain depth of container required to ensure capillary action is effective Difficult to tell if the plant is being irrigated correctly as surface media is dry Roots can encircle the reservoir, making plant replacement difficult
Automated drip lines	Drip irrigation for container plantings or display gardens. Can use seep hose to wet entire substrate or discrete drip emitters around the base of individual plants Often sophisticated control systems, automatically irrigating after set thresholds for light integrals, evapotranspiration demand or media moisture availability	Pipes can be hidden under the substrate	Roots & silt can block drainage holes Drippers prone to blockages If nutrients are supplied via irrigation, get build-up of salts at the dripper as the moisture evaporates With discrete dripper type can get moist 'cone' within the substrate, with limited lateral movement of moisture
Precision micro-irrigation	Deliver minimal volumes at regular intervals	Each emitter can be adjusted for flow rate thereby meeting the needs of individual plants Small volumes reduce the likelihood of water run-off, when the medium is initially dry	Takes time for each emitter to be adjusted to match the demands of individual plants

Pest and Pathogen Management

Much of ornamental horticulture has had to deal recently with withdrawal of chemical pesticides, due to increased regulations based on human health and environmental concerns, and the costs of developing new products for relatively small and specific markets. These pressures are not new however, for the interior landscape industry, where close contact between humans and plants has tended to rely traditionally on more 'benign' chemical compounds (detergents or spot treatment with alcohols or oils), or more innovative approaches such as the use of biocontrol agents (e.g. the introduction of predatory/parisitoid insects, mites, nematodes, fungi etc.) (Steiner and Elliot 1987). In some countries the use of chemical control products are regulated e.g. in the UK by the Chemicals Regulations Directorate. Common plant pests are now regularly controlled by biocontrol means, for example *Trialeurodes vaporariorum* (glasshouse whitefly) by *Encarsia formosa*, *Tetranychus urticae* (glasshouse red spider mite) by *Phytoseiulus persimilis*, *Planococcus citri* (Citrus mealybug) by the *Leptomastix dactylopii* wasp and *Coccus hesperidium* (soft brown scale insect) by *Metaphycus alberti* wasps. Biocontrol agents are less common for disease control, and cultural factors are the primary tool for controlling pathogens within plantscapes. Minimising stress through careful regulation of irradiance, humidity and irrigation is key to reducing the chances of infection and pathogen spread (Lockwood 2000). For example, root pathogens such as *Pythium*, *Phythophthora* and *Rhizoctonia* can be more prevalent after plants have been over-watered or roots have been damaged by an excessive accumulation of fertilizer salts. High salt levels can cause desiccation and necrosis of tissues, encouraging subsequent infection from grey mold (*Botrytis cinerea*). As a last resort, plants can be removed from the interior landscape and placed back in a glasshouse or quarantine area, where more virulent pesticides may be applied. Plants can then be restored to full-health, before being placed back in the plantscape (assuming the original cause of the disease/pathogen in the interior landscape has been remedied in the meantime).

Managing the Interior Landscape

The majority of plantscapes are designed, installed and maintained by specialised plantscape companies. Many of these companies provide a 'complete service', employing a range of personnel that can include architects, landscape architects, interior designers, horticulturalists and irrigation technicians. Some too may have their own production nursery whereas others buy in plant specimens from specialist supply nurseries. The production of interior landscape plants is a whole industry in itself (See Chap. 9). Centres of production occur in Florida & California, USA, the Netherlands, Belgium, China, Thailand, Sri Lanka, Singapore, Costa Rica, Honduras and Brazil; many nurseries now specialising in the propagation of young plants (via tissue culture or cuttings) or importing these from other countries to grow-on, 'finish-off' and acclimatize the crop. As Chen et al. (2005) states

> Today someone living in Poland may be watering a *Diffenbachia* cultivar that was propagated in China, finished in the USA and then sold via the Aalsmeer auction in the Netherlands. The plant itself may be a hybrid of species collected in Brazil and Colombia, but bred in the UK.

The arrangements and contracts between plantscape companies and the office/building owners (clients) vary. At one end of the spectrum the company may simply sell or rent plants to the client, to the other end where a complete consultation, design, installation and service package is included. In the latter circumstances the plantscape company will visit the proposed site taking account of infrastructure (site access, weight loading of floors, suitability of locations—irradiance and photoperiods, draughts, seating arrangements etc.) and interior design (existing or desired) before providing a planting design to suit the client's needs (including ancillary aspects such as proposed container/pot design, colour etc.). Once a plan is agreed the company will install and maintain the landscape, usually offering a service whereby landscape technicians will visit the planting on a regular basis to water, clean, prune, fertilize and inspect plants for pathogens, disease or stress effects. Often the plants are rented rather than owned by the client and the plantscape company will replace damaged, diseased or oversized plants as part of the contract. It is good practice for the plantscape company to insure itself against any damage or other liabilities potentially caused by its personnel, when working within a client's building. The personnel need training, not only in the horticultural skills required, but also ensuring a 'high customer service' and being able to respond in a rapid and professional manner to the client's needs.

Work schedules vary depending on number of plants/size of area (Anon 2000), but a typical schedule for an office complex housing 30 plants might encompass:

1. Notification of service technician(s) arrival/departure (or identification of regular time set)
2. Inspection of substrate moisture status and plants watered as required, (e.g. fortnightly, if semi-automated irrigation system employed)
3. Inspection of substrate and containers—cleaning and topping up substrate with surface pebbles/leca or other mulch materials as appropriate (monthly)
4. Liquid fertilizer applied (e.g. monthly)
5. Foliage cleaned (monthly) and re-polished with leaf shine wax (bi-annually)
6. Dead and dying leaves removed (fortnightly)
7. Inspected for pests and pathogens (monthly). Plants treated with biological control agents as required
8. Pruned to remove unsightly or badly positioned branches (annually)
9. Failing or excessively large plants replace free of charge (as required)
10. Office visited by member of plantscape management team (bi-annually)

Environmental factors or maintenance arrangements are not always appropriate for the installation of live plants, and artificial plants are also used in interior landscapes. A number of plantscape companies will also offer quite sophisticated artificial interior plants, made of silk or other high quality materials to mimic their living counterparts. Such 'specimens' may still need some, albeit less frequent, maintenance, e.g. removing dust from the 'leaves'.

Conclusions and Future Directions

Over half of the world's population now live in towns or cities, and as a result urban areas cover c.2.8 % of the world's land surface (Seto et al. 2011). Increasingly, engagement with a 'green world' for a large proportion of the world's citizens is going to be manifest through urban green space, with interior plantscapes becoming an increasingly important component of this. Most of the previous advances in the interior plantscape sector have been interlinked with changes in lifestyle and habits, with alterations in transportation, architecture, technology, demographics and economics acting as catalysts. There is no reason to assume the future will be different. Plants in, around and on buildings will have a vital role in ensuring that future generations appreciate the role of vegetation and help facilitate an enjoyment and understanding of nature, albeit perhaps in a highly-stylised manner. As such, horticulturalists, architects and engineers need to work together in future to ensure city landscapes progressively turn from 'grey' to 'green', for a wide variety of environmental, social, health and aesthetic reasons. Our understanding of where *interior* plantscapes fit precisely into these benefits remains incomplete. Nevertheless, if the evidence for the beneficial effects of interior plants (e.g. on psychological health) is substantiated and the quantifiable aspects determined more fully, then the use of interior plants need to be implemented more widely. Currently few school classrooms and university lecture theatres have great numbers of plants (if any), and many hospital environments bar them completely based on concerns over hygiene. Similarly, their full potential to enhance the retail environment of individual department stores and shops still seems at best, under-utilised. The interior landscape sector is an important component of horticulture, with the industry evolving and adapting to meet changing fashions and ideologies. One would envisage that the required drive for a more sustainable, low carbon future, combined with the desire for better quality urban lifestyle, would result in a greater prominence for interior plantscapes. Futuristic ideas such as sky gardens, may mean that whole sections of multi-storey buildings are dedicated to accessible green space, with both exterior and interior landscapes playing a significant role. Indeed modern technologies, may see the blurring of these distinctions between locations as new materials provide opportunities for the development of semi-protected or temporary protection structures and features; conceivably a necessary requirement for even conventional 'hardy' plants when grown on an exposed terrace 100 m above ground level! Perhaps, once again, the pot plant may assume its original transition role between interior and exterior environments. In tandem with these developments and challenges, the industry itself needs to continue to evolve and innovate to ensure the construction and maintenance of future plantscapes meets the demands set out by new objectives (low energy consumption, efficient use of water and other resources, quantifiable environmental and health benefits and increased urban density).

References

Adachi M, Rohde CLE, Kendle AD (2000) Effects of floral and foliage displays on human emotions. HortTechnology 10:59–63

Anon (2000) Interior Plant Maintenance Schedule, Section 12850, BMS Basic Master Specification, Alberta Infrastructure, California. www.infrastructurealbertaca/Content/doctype486//12850bdoc

Anon (2003) In sickness and in health. Business Voice Magazine, Confederation of British Industry, June 2003, p 11

Asaumi H, Nishina H, Nakamura H, Matsui Y, Hashimoto Y (1995) Effect of ornamental foliage plants on visual fatigue caused by visual display terminal operation. J Shita 7:138–143

Axelsson G, Skedinger M, Zetterströ O (1985) Allergy to weeping fig—a new occupational disease. Allergy 40:461–464

Baek S-O, Kim Y-S, Perry R (1997) Indoor air quality in homes, offices and restaurants in Korean urban areas—indoor/outdoor relationships. Atmos Environ 31:529–544

Bakker I, van der Voordt T (2010) The influence of plants on productivity: a critical assessment of research findings and test methods. Facilities 28:416–439

Bringslimark T, Hartig T, Patil GG (2007) Psychological benefits of indoor plants in workplaces: putting experimental results into context. HortScience 42:581–587

Cameron RW, Blanusa T, Taylor JE, Salisbury A, Halstead AJ, Henricot B, Thompson K (2012) The domestic garden—its contribution to urban green infrastructure. Urban For Urban Green 11:129–137

Chang C-Y, Chen P-K (2005) Human responses to window views and indoor plants in the workplace. HortScience 40:1354–1359

Chen J, McConnell DB, Henny RJ, Norman, DJ (2005) The foliage plant industry. In: Janick J (ed) Horticultural reviews, vol. 31. John Wiley & Sons, Oxford

Coleman CK, Mattson RH (1995) Influences of foliage plants on human stress during thermal biofeedback training. HortTechnology 5:137–140

Coward M, Ross D, Coward S, Cayless S, Raw G (1996) Pilot study to assess the impact of green plants on NO_2 levels in homes. Building Research Establishment Note N154/96. Watford, UK

Currie CK (1993) The archaeology of the flowerpot in England and Wales c. 1650–1950. Garden Hist 21:227–246

Darlington AB, Dat JF, Dixon MA (2001) The biofiltration of indoor air: air flux and temperature influences the removal of toluene, ethylbenzene and xylene. Environ Sci Tech 35:240–246

Davidson J (2000) '…The world was getting smaller': women, agoraphobia and bodily boundaries. Area 32:31–40

De Kempeneer L, Sercu B, Vanbrabant W, Van Langenhove H, Verstraete W (2004) Bioaugmentation of the phyllosphere for the removal of toluene from indoor air. App Microbiol Biot 64:284–288

Der Marderosian AH, Giller FB, Roia FC (1976) Phytochemical and toxicological screening of household ornamental plants potentially toxic to humans I. J Toxicol Environ Health 1:939–953

Dingle P, Tapsell P, Hus S (2000) Reducing formaldehyde exposure in office environments using plants. Bull Environ Contam Toxicol 64:302–308

Dijkstra K, Pieterse ME, Pruyn A (2008) Stress-reducing effects of indoor plants in the built healthcare environment: the mediating role of perceived attractiveness. Prev Med 47:279–283

Dravigne A, Waliczek TM, Lineberger RD, Zajicek JM (2008) The effect of live plants and window views of green spaces on employee perceptions of job satisfaction. HortScience 43:183–187
Fjeld T, Veiersted B, Sandvik L, Riise G, Levy F (1998) The effect of indoor foliage plants on health and discomfort symptoms among office workers. Indoor Built Environ 7:204–209
Fjeld T (2000) The effect of interior planting on health and discomfort among workers and school children. HortTechnology 10:46–52
Fjeld T (2002) The effects of plants and artificial daylight on the well-being and health of office workers, school children and health-care personnel. Proceedings of International Plants for People Symposium, Floriade, Amsterdam
Grinde B, Patil GG (2009) Biophilia: does visual contact with nature impact on health and well-being? Int J Environ Res Pub Health 6:2332–2343
Han KT (2009) Influence of limitedly visible leafy indoor plants on the psychology, behavior, and health of students at a junior high school in Taiwan. Environ Behav 41:658–692
Hartig T, Evans GW, Jamner LD, Davis DS, Garling T (2003) Tracking restoration in natural and urban field settings. J Environ Psychol 23:109–123
Horwood C (2007) Potted history: the story of plants in the home. Frances Lincoln Ltd, London
Ingels JE (2009) Interior plantscaping. In: Landscaping principles and practices, 7th edn. Delmar, New York
Isen A (1993) Positive affect and decision making. In: Lewis M, Haviland JM (eds) Handbook of Emotions. Guilford, New York
Kanerva L, Mäkinen-Kiljunen S, Kiistala R, Granlund H (1995) Occupational allergy caused by spathe flower (*Spathiphyllum wallisii*) Allergy 50:174–178
Kaplan S (1995) The restorative benefits of nature: toward an integrative framework. J Environ Psychol 15:162–182
Khan AR, Younis A, Riaz A, Abbas MM (2005) Effect of interior plantscaping on indoor academic environment. J Agric Res 43:235–242
Kim E, Mattson RH (2002) Stress recovery effects of viewing red-flowering geraniums. J Therap Hort 13:4–12
Knight C, Haslam SA (2010) Your place or mine? Organizational identification and comfort as mediators of relationships between the managerial control of workspace and employees' satisfaction and well-being. Bri J Manage 21:717–735
Kuo FE, Sullivan WC (2001) Environment and crime in the inner city: does vegetation reduce crime? Environ Behav 33:343–367
Larsen L, Adams J, Deal B, Kweon B-S, Tyler E (1998) Plants in the workplace: the effects of plant density on productivity, attitudes, and perceptions. Environ Behav 30:261–281
Lee J-H, Sim W-K (1999) Biological absorption of SO_2 by Korean native indoor species. In: Burchett MD (ed) Towards a new millennium in people–plant relationships. Contributions from International People–Plant Symposium, Sydney, pp 101–108
Liu M, Kim E, Mattson RH (2003) Physiological and emotional influences of cut flower arrangement and lavender fragrance on university students. J Therap Hortic 14:18–27
Lockwood SL (2000) Interior planting: a guide to plantscapes in work and leisure spaces. Gower Publishing Ltd, Aldershot
Lohr VI, Pearson-Mims CH (2000) Physical discomfort may be reduced in the presence of interior plants. Horttechnology 10:53–58
Lohr VI, Pearson-Mims CH, Goodwin GK (1996) Interior plants may improve worker productivity and reduce stress in a windowless environment. J Environ Hort 14:97–100
Nichols RS (2003) English pleasure gardens, 2nd edn. David R Godine, New Hampshire
Orwell RL, Wood RL, Tarran J, Torpy F, Burchett MD (2004) Removal of benzene by the indoor plant/substrate microcosm and implications for air quality. Water Air Soil Poll 157:193–207
Ottosson J, Grahn P (2005) A comparison of leisure time spent in a garden with leisure time spent indoors: on measures of restoration in residents in geriatric care. Landsc Res 30:23–55
Paris HS, Janick J (2008) What the Roman emperor Tiberius grew in his greenhouse. In: Pitrat M (ed) Cucurbitaceae. INRA, Avignon, pp 33–41

Park S-H, Mattson RH (2008) Effects of flowering and foliage plants in hospital rooms on patients recovering from abdominal surgery. HortTechnology 18:563–568

Park S-H, Mattson RH (2009) Therapeutic influences of plants in hospital rooms on surgical recovery. HortScience 44:1–4

Park S-H, Mattson RH, Kim E (2004) Pain tolerance effects of ornamental plants in a simulated hospital patient room. Acta Hortic 639:241–247

Rodríguez-Calcerrada J, Reich PB, Rosenqvist E, Pardos JA, Cano FJ, Aranda I (2008) Leaf physiological versus morphological acclimation to high-light exposure at different stages of foliar development in oak. Tree Physiol 28:761–771

Russell S, Best L (2006) Setting the standards for compost. Biocycle Int J Comp Org Recycl June 2006:53–57

Saner E (2012) Pot plants in the office: good or bad, The Guardian April 12, 2012 http://www.guardiancouk/lifeandstyle/shortcuts/2012/apr/03/pot-plants-in-the-office

Seppänen OA, Fisk W (2006) Some quantitative relations between indoor environmental quality and work performance or health. Heat Vent Air-Cond Ref Res 12:957–973

Seto KC, Fragkias M, Güneralp B, Reilly MK (2011) A meta-analysis of global urban land expansion. PLoS ONE 6:e23777

Shaughnessy RJ, Haverinen-Shaughnessy U, Nevalainen A, Moschandreas D (2006) A preliminary study on the association between ventilation rates in classrooms and student performance. Indoor Air 16:465–468

Shibata S, Suzuki N (2001) Effects of indoor foliage plants on subjects' recovery from mental fatigue. N Am J Psychol 3:385–396

Shibata S, Suzuki N (2002) Effects of the foliage plant on task performance and mood. J Environ Psychol 22:265–272

Shibata S, Suzuki N (2004) Effects of an indoor plant on creative task performance and mood. Scand J Psychol 45:373–381

Shoemaker CA, Randall K, Relf D, Geller ES (1992) Relationships between plants, behavior and attitudes in an office environment. Horttechnology 2:205–206

Steele J (1992) Interior landscape dictionary (Landscape architecture). John Wiley & Sons, New Jersey

Steiner MY, Elliot DP (1987) Biological pest management for interior plantscapes, 2nd edn. Alberta Environmental Centre, Edmonton

St Leger L (2003) Health and nature—new challenges for health promotion. Health Prom Int 18:173–175

Talbott JA, Stern D, Ross J, Gillen C (1976) Flowering plants as a therapeutic/environmental agent in a psychiatric hospital. HortScience 11:365–366

Tarran J, Torpy F, Burchett M (2007) Use of living pot-plants to cleanse indoor air—research review. Proceedings of 6th international conference on indoor air quality, ventilation & energy conservation—sustainable built environment, vol 3. Sendai, Japan, pp 249–256

Ulrich RS, Simons RF, Losito BD, Fiorito E, Miles MA, Zelson M (1991) Stress recovery during exposure to natural and urban environments. J Environ Psychol 11:201–230

Van den Berg AE, Maas J, Verheij RA, Groenewegen PP (2010) Green space as a buffer between stressful life events and health. Soc Sci Med 70:1203–1210

Wilson EO (1984) Biophilia. Harvard University Press, Cambridge

Wolf KL (2002) Retail and urban nature: creating a consumer habitat. Proceedings of international plants for people symposium, Floriade, Amsterdam, Netherlands

Wolverton BC, Wolverton JD (1993) Plants and soil microorganisms: removal of formaldehyde, xylene, and ammonia from the indoor environment. J Miss Acad Sci 38:11–15

Woudstra J (2000) The use of flowering plants in late seventeenth- and early eighteenth-century interiors. Garden Hist 28:194–208

Chapter 24
Biodiversity and Green Open Space

Ghillean T. Prance, Geoffrey R. Dixon and David E. Aldous

Abstract Biodiversity has been defined as the totality of genes, species, and ecosystems that inhabit the earth with the field contributing to many aspects of our lives and livelihoods by providing us with food, drink, medicines and shelter, as well as contributing to improving our surrounding environment. Benefits include providing life services through improved horticultural production, improving the business and service of horticulture as well as our environment, as well as improving our health and wellbeing, and our social and cultural relationships. Threats to biodiversity can include fragmentation, degradation and deforestation of habitat, introduction of invasive and exotic species, climate change and extreme weather events, over-exploitation of our natural resources, hybridisation, genetic pollution/erosion and food security issues and human overpopulation. This chapter examines a series of examples that provide the dual aims of biodiversity conservation and horticultural production and service; namely organic horticultural cropping, turf management, and nature-based tourism, and ways of valuing biological biodiversity such as the payment of environmental services and bio-prospecting. Horticulture plays a major role in the preserving of biodiversity.

Keywords Biodiversity · Ecosystems · Wildlife · Ecology · Horticulture · Natural value · Green space

---

Professor David E. Aldous – deceased 1st November 2013

G. T. Prance (✉)
The Old Vicarage, Silver Street, Lyme Regis DT7 3HS, UK
e-mail: siriain01@yahoo.co.uk

D. E. Aldous
School of Land, Crop and Food Science, The University of Queensland,
Gatton Campus, Lawes, Queensland 4343, Australia

G. R. Dixon
School of Agriculture, Earley Gate, University of Reading, RG6 6AR Reading, Berkshire, UK
e-mail: geoffrdixon@btinternet.com

GreenGene International, Hill Rising, Sherborne, Dorset, UK

G. R. Dixon, D. E. Aldous (eds.), *Horticulture: Plants for People and Places, Volume 2*,
DOI 10.1007/978-94-017-8581-5_24, © Springer Science+Business Media Dordrecht 2014

Introduction to Biodiversity

The term 'biological diversity' was first defined in 1980, in an Annual Report for the United States President's Council on Environmental Quality (Norse and McManus 1980). This described biological diversity as a two-level hierarchy, incorporating "genetic diversity and ecological diversity… the amount of genetic variability among individuals in a single species…(and) the number of species in a community of organisms". The term was further defined in the context of the multiple benefits that humans accrue from the natural world and as such proved a concept uniting environmental scientists and policy-makers alike. 'Biodiversity', a contraction of 'biological diversity', was first coined in 1985, in preparation for the National Forum on Biological Diversity, partially for conference publicity and promotion, and first appeared in print in the symposium's proceedings. By 1992 it was common practice to define 'biodiversity' in terms of genes, species and ecosystems. In 2001 Larsson and co-workers defined biodiversity as the "totality of genes, species, and ecosystems of a region" in the context of the traditional three levels at which biological variety has been identified; species, ecosystem and genetic diversity. In 2012 Morton and co-workers described biodiversity, that included *Homo sapiens*, as part of the approximate 9 million unique living organisms that inhabited the earth, although more recent research by Costello et al. (2013) has put the number of species on Earth today as some 5 ± 3 million species, of which 1.5 million have been named. By this definition the quality of an ecosystem has a value of 100 % in an undisturbed natural location and 0 % in a completely degraded area devoid of any wild species. Diversity embraces within and between species and of ecosystems, and extends from microbes to the largest mammals (Wilson 1992), to terrestrial and marine environments, to areas that may be the size of a garden to continents. Different classification systems exist to describe ecosystem diversity that range on a world scale from bio-geographic zones, biomes, eco-regions, and oceanic realms to smaller scale landscapes, ecosystems and communities (Fig. 24.1).

In 2007 Dunnett et al. and co-workers related biodiversity to the richness of life and richness in diversity to life forms in gardens to indicate the richness could be found in assets as small as the size of a small back yard through to a region, country or continent. The Scottish Biodiversity Strategy (2004) presented biodiversity as existing at all levels in the surrounding environment and having value to human beings. In a horticultural context biodiversity could be said to have both a production and amenity horticulture dimension, the former currently believing that we can well get along without biodiversity and the latter accepting the word "wildlife" as an alternative for biodiversity (Dunnett 2004; Dunnett et al. 2007). In this study wildlife actually engaged more people in discussion and was found to be more identifiable with the eco-friendly practices that are undertaken in their gardens. Under these circumstances wildlife richness was found to bring immense feelings of pleasure and well-being to the expert gardener rather than showing an indifference on principles of nature conservation.

The United Nations Food and Agriculture Organisation defined the term agro-biodiversity as the *variety and variability of animals, plants and micro-organisms*

Ecological diversity	Genetic diversity	Organismal diversity
Biomes		kingdoms
Bio-regions	populations	phyla
Landscapes	individuals	families
Ecosystems	chromosomes	genera
Habitats	genes	species
Niches	nucleotides	subspecies
Populations		populations
		individuals

Fig. 24.1 The composition and human interaction at all levels of biodiversity. (Source: Heywood and Baste 1995)

that are used directly or indirectly for food and agriculture, including crops, livestock, forestry and fisheries; perhaps it is time to introduce the term of horticultural biodiversity into this chapter which according to the authors would be *the process by which diverse organisms, ecosystems and ecological processes provide economic, environmental, and social benefit when interacting with managed open green space.* Regardless of the academic debate, there continues to remain strong moral and ethical reasons for protecting and conserving the many biological species and ecosystems (Attfield 1999), and now there is greater recognition that complex and diverse systems are often more functional and attractive than systems of lower diversity (Dunnett 2006; Dunnett et al. 2007).

Horticulture and Green Open Spaces

Green open space has been defined as "any piece of land covered with vegetation and usually refers to managed areas such as parks, golf courses, sports fields and other open land within the built-up area, whether publicly accessible or not" (State University of New York 2010) or "land that has been reserved for the purpose of formal and informal sport and recreation, preservation of natural environments, provision of green space and/or urban storm water management" (Anon 2013a). In almost all instances, the space referred to by the term open space is, in fact, green open space that is managed by humans in some way or that has been set aside for conservation purposes. Green spaces have both an urban (Aldous 2010, 2011) and rural dimension (Wilson 2006). The former often include assets such as parks and gardens, botanic gardens, outdoor sports facilities, community gardens, roof gardens, and natural/semi-natural urban green spaces where horticulture is usually involved in some way, the latter include the open spaces for cropping, forestry, orchards, plantations and wildlife habitat and are more the realm of agriculture, forestry and conservation.

In assessing the significance and value of biological diversity associated with green open space and environmental sustainability we need to have assess how much green open space we have on record. The more green space a municipality, region, or country has in the form of m^2 per capita of urban green space the greater the potential capacity to generate oxygen, absorb carbon dioxide, transpire water vapour, absorb heat and conserve biodiversity. The issue of required open green spaces per capita in urban systems has remained controversial; however the UN-Habitat (2009) has considered a minimum of 9 m^2 per capita (person) of green open space and upwards of the recommended international target of 60 m^2 per capita of green open space for sustainable development. Singh et al. (2010) noted that cities renowned for their green open spaces contain 20–30 % of their total geographical area under cover, or when the population of the city is taken into account, some 15–25 m^2 of urban green open space per capita. Ayata (2004) suggested that 40–60 m^2 of green space per person must be reserved in the development of open green space. Xie (1999) used a range of biological measurements to indicate that in general 60 m^2 per capita of urban green space is necessary to maintain optimum residential development and satisfactory environmental sustainability. It is important that a country acknowledges and knows the extent of its green open space assets as it is principally the environmental benefits that are of significance in providing the protection and conservation of biodiversity, nature and the natural processes (Bolund and Hunhammer 1999). Both Singh et al. (2010) and the Status of Urban Forestry in the Asia-Pacific Region (1998) reported on the percentage of green open space, as well as the m^2 per capita, for a number of countries in the Asia Pacific region (Table 24.1). In all cases the percentage green space and m^2 per capita can only be considered an estimate because of the differences in the range, type and distribution of the green open space assets in each country.

Biodiversity, Sustainability and Sustainable Development

Cameron (pers. comm.) outlines some of the processes highlighted by the unsustainable use of our natural resources and these are brought together in Fig. 24.2. This arranges the discriminating features of both sustainable and unsustainable land-management systems into contrasting causal pathways, highlights potential triggers and at each stage notes a few of the more important outcomes. What emerges from the flow chart is that sustainable management, in the form of traditional, locally

Table 24.1 Comparison of percent green space of total city and m² per capita (year) for capital cities in SE Asia[a]

Country/City	Green space % of total city[b]	m² per capita (year)[b]	Comments and reference source
Australia			Daytime population is 400,000 and residential 40,000 only; Figure refers to parkland, garden and recreational reserves (Stokie 1998 per. comm.) Estimated crown cover of about 24 million m² amounting to 80 m²/inhabitant (Brack 2002).
Melbourne (residents)	17.80 %	163.3 m² (97)	
Melbourne (daytime pop)	17.80 %	16.3 m² (97)	
China			Public parks and other green space increase to 40 % by year 2000 (or 8 m²/inhabitant (Shan 1994) (Jim 1998 per. comm.) (Jim 1998 per. comm.; Chan 1988 per. comm.) On an average China's cities have 32.54 % green cover. This varies greatly in Chinese cities like Nanjing and Wuhan, i.e., 44.3 m²/person and 10.3 m²/person respectively (Jim and Wendy 2009).
Average	23.8 %	5.7 m² (96)	
Beijing	28 (1994)	6.83 m²	
Hong Kong	39.2 %(97)	66.0 m²	
Hong Kong excl. country parks	1.5 % (97)	2.5 m²	
India			Unclear if these figures to public green space (Andresen and Plexman 1980; Pye-Smith 1996) [c]India, Delhi—Average tree and forest cover is about 20 % of geographical area and about 21 m²/inhabitant (FSI 2009, as per population data 2001). [c]India—Chandigarh—Average tree and forest cover is about 35.7 % of geographical area, i.e., about 55 m²/inhabitant (Action Plan 2009-2010, as per population data 2001).
Mean		0.003 (80s)	
Bombay		0.12 m²	
Indonesia			Parks/per capita (AIT 1998 per. comm.)
Jakarta		0.22 m² (86)	
Japan			5,000 ha (Yuji 1995 per. comm.) refers to parks, planned to increase to 6 m²
Takatsuki city urban forests	84% (1990)	4.52 m²	
Tokyo metropolitan area			
Korea South			Public green space (Park 1997 per. comm.)
Seoul	25.2%	14.57 m² (96)	
Malaysia			Public green space (Adnan 1998 per. comm.)
Kuala Lumpur	5 %	2.25 m² (97)	
New Zealand			Public green space, mainly sport fields with border trees (O'Reilly 1998 per. comm.)
Christchurch	12.2 %	0.018 m² (97)	
Singapore	17.8 %	7.5 m² (97)	Public parks and open space, increase to 8 by year 2000 and finally to 18 ha/capita (Yuen 1998). Singapore now has 46.5 % green cover to service a growing population of 4.6 million people (Conserving our Diversity 2009).

Table 24.1 (continued)

Country/City	Green space % of total city[b]	m^2 per capita (year)[b]	Comments and reference source
Sri Lanka			
Colombo	4.4 %		Green spaces 2.4 % private (golf course etc.) and 2.0 % public (municipal parks, etc.) (Wickramasinghe 1998 per. comm.)
Thailand			
Bangkok metropolitan area		1 m^2 (97)	Planned to increase to 4–5 m^2 by the year 2000 (Charmniern 1998 per. comm.)

[a] Modified from the Status of Urban Forestry in the Asia-Pacific Region (1998). FAO Corporate Document Repository. Retrieved from http://www.fao.org/docrep/003/x1577e/X1577E06.htm (Accessed 13th August, 2010)

[b] Estimate only[c] Modified from Singh, V.S., D.N.Pandey and P. Chaudhry (2010). Urban Forests and Open Green Spaces: Lessons for Jaipur, Rajasthan, India. Rajasthan State Pollution Control Board, Occasional Paper No. 1/2010:1-23. Retrieved from http://210.212.99.115/rpcb/RSPCB-OP-1-2010.pdf (Accessed October 20, 2010)

adapted production systems (column 1), has a number of clear features. It uses minimal external inputs (a) and a rate and scale of exploitation that is constrained by local circumstances (b). It also has a wide range of locally adapted plants, animals, landscapes, technologies, economies and cultures (types) (d), and involves the ongoing selection of 'types' that can more effectively exploit locally-distinctive circumstances (e). Traditional forms of exploitation, therefore, constitute a divergent process that not only depends upon inherited diversity (d) but also augments it (e). By contrast, unsustainable management, in the form of intensive, high-input land-management systems, enable production to override naturally occurring limits (h) and simplify the environment by favouring a narrow range of productive types (i). Although this process increases production, it causes widespread pollution, soil erosion and loss of biodiversity (h). It also displaces inherited diversity (j) in the form of locally adapted types and involves the convergent selection (k) of even more productive types, further simplifying the environment (the CAP legacy) and undermining its productive potential (l) and its ability to provide ecosystem services. The circumstances that trigger the movement from sustainable production (Column 1) to unsustainable production (column 2) appear to be management decisions that override local environmental constraints on large temporal and spatial scales (g).

Benefits of Horticultural Biodiversity

In recent years there has been increasingly strong recognition of how terrestrial ecosystems provide vital services for people and society whether they be rural or urban dwellers. The future capacity of Earth's ecosystems to provide these services is determined by the demands in socio-economic terms made from: agriculture and horticultural processes, the atmosphere, climate and soils. Biodiversity contributes to many aspects of our lives through providing us with food, drink, medicines and

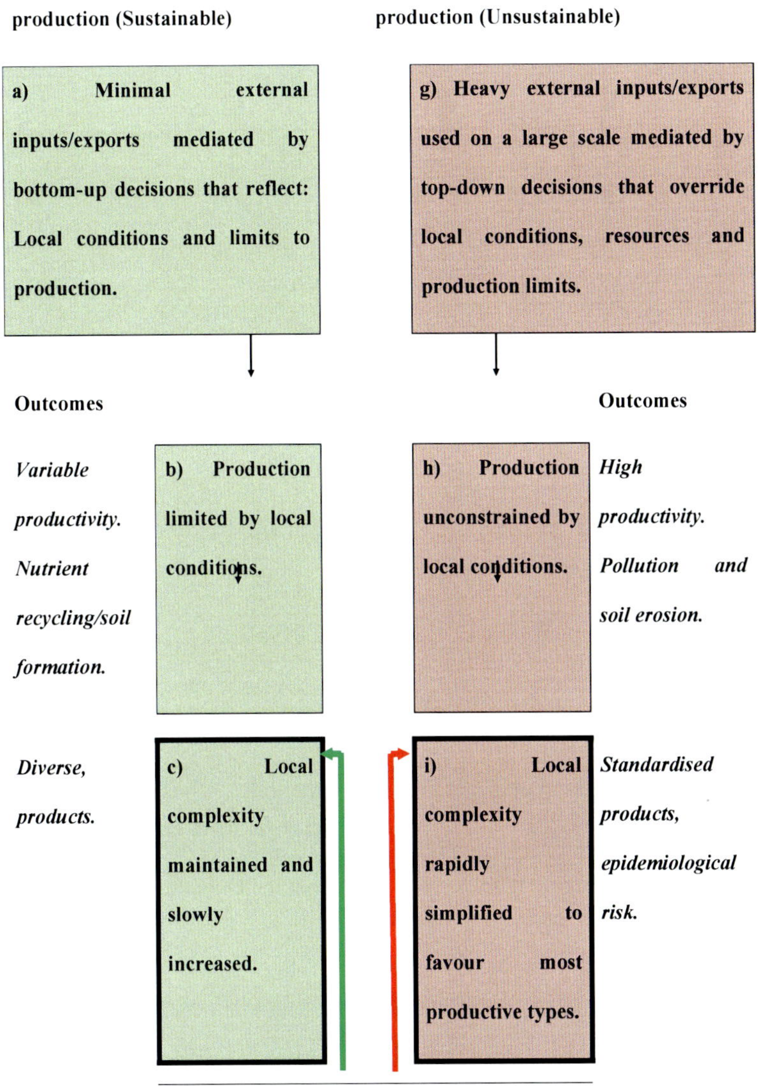

Fig. 24.2 Comparison of land management systems (sustainable *column 1* Traditional low-input) (unsustainable *column 2* Intensive high-input) in terms of their characteristics, environmental and economic effects and the conjectured causal processes involved in tipping land management systems from one state to another. *Green arrows* highlight a sustainable cycle that increases the diversity of both natural and economic components of the environment and increases system stability by creating new opportunities for exploitation. *Red arrows* highlight an unsustainable cycle. 'Types' refer to species, as well as crop cultivars and animal breeds, and perhaps even farming systems

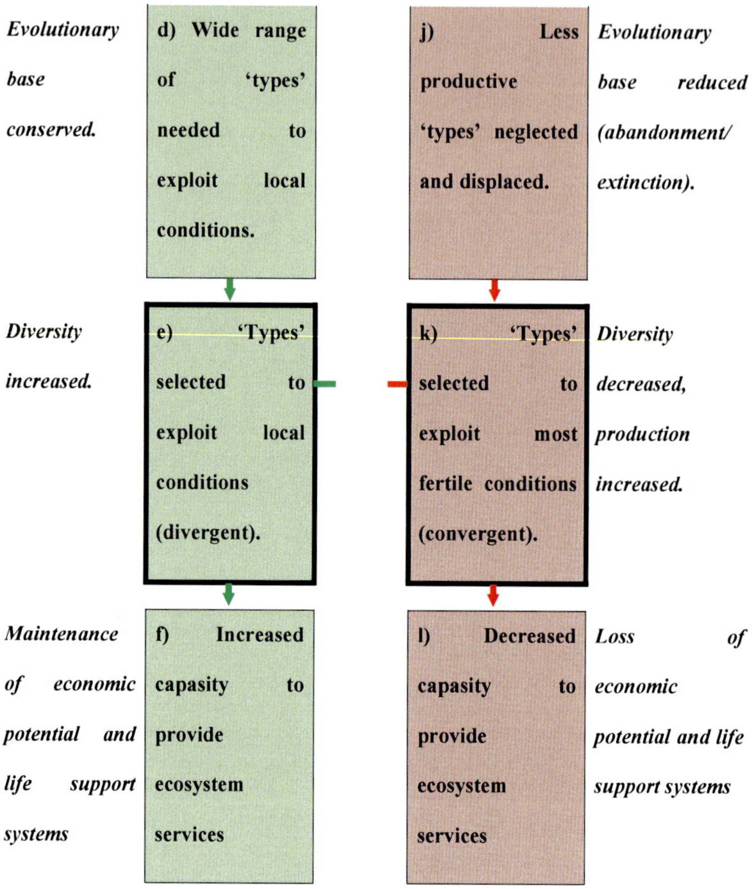

Fig. 24.2 (continued)

shelter in a horticultural production sense, as well as contributing to improving our surrounding environment, the economics of production and service, our health and wellbeing, and our social and cultural relationships (Anon 2013j).

Providing Life Services Through Improved Horticultural Production

Developing an understanding of the relationship between land-use and biodiversity has as much to do with economics and politics as with biological and the agricultural/horticultural sciences. Early economists of the eighteenth and nineteenth centuries such as Adam Smith, David Ricardo and Thomas Malthus were really agricultural economists since food production was then the primary rural activity and formed the bedrock for the urbanised Industrial Revolution. It was not until the Scott Report laid the principles for future use of rural land Lord Justice Scott commented, that the UK countryside was "ragged and unkempt" albeit with increased wildlife. Post-War rural support offered a dual approach, firstly achieving food and timber self-sufficiency and secondly, protecting and enhancing the beauty of our countryside. The English ecologist Arthur Tansley (1946) encapsulated this: "as we can plan for freedom as well as for order, efficiency, and material well-being in the political and social spheres, so we can plan for beauty and dignity as well as for convenience and comfort in our physical surroundings". Shortly afterwards the Huxley Report advocated establishing the first national nature reserves, the National Parks, the Nature Conservancy organisation and *inter alia* identifying Areas of Outstanding Natural Beauty (AONBs). These provisions were however, seen as secondary to the needs for national food security. In the United Kingdom environmental policies were developed that encouraged agricultural practices that would sustain and increase natural biodiversity in the context of farming activity. Almost concurrently with these changes came the implications for biodiversity arrived with the Inter-Governmental Environment Summit in Rio de Janeiro. In October 2006 Natural England, the government's advisor on the natural environment, came into existence by the Natural Environment and Rural Communities Act 2006 and was charged with overseeing farming's system of environmentally driven reimbursements and the protection of valued landscapes and biodiverse nature. This was the point at which biodiversity really emerged as an objective for policy makers. In the latter twentieth and early twenty-first centuries there has been an inexorable shift in the social and political role of the countryside in the United Kingdom which in turn affects the survival of biodiverse nature. The service sector now accounts for 70 % of economic activity and farming for 5 % in much of rural Great Britain. Across wide swathes of countryside land markets are now set by amenity values rather than by returns from agricultural enterprises.

The presence of diversity in horticultural crops assists in recovery when the dominant crop species is attacked by a pathogen or predator. Diseases in themselves are a form of biodiversity as was shown in the Irish potato blight of 1846, the grassy stunt virus of rice in India in the 1970s and the coffee rust attack in Sri Lanka, Brazil, and Central America in 1970 and most recently the loss of ash trees in England to the fungus *Chalara fraxinea* that was first reported in a nursery in Buckinghamshire and has spread rapidly throughout the UK. In addition despite the usefulness of artificial fertilizers, soil conditioners, wetting agents and pesticides, a large proportion of horticultural cropping would be impossible without essential ecosystem services. The processes of biodiversity are essential in the breakdown and recycling of nutrients within the topsoil as well as in pollination, where pollination is a necessity for their existence. As much as 50 % of pollination activity is carried out by native insects where the pollination of crops by insects and birds has been estimated to be worth upwards of AU$ 1.2 billion in Australia. Fitzpatrick (1994) reported that increased yields of 20–100 % were observed where crops of sheltered horticultural crops were compared with unsheltered crops. Effective pest control can also be threatened by loss of habitat, pollution, the introduction of non-native alien species, and compaction of soils from using inappropriate heavy machinery. Increased urbanization has threatened urban green spaced areas as a source of habitats for both species and production. Urban green space showed that urban green space, can provide substantial products such as fruits, vegetables, ornamentals and herbs, as well as act as a source of compost and energy as a result of retrofitted or purposely designed roof and community gardens (Mader 1996; Guoping 1999; Patel 1992).

Horticulture and Rare and Endangered Species

One way in which gardens and horticulture is helping with the conservation of endangered species is through encouraging both botanic gardens and individuals to propagate and cultivate some of the rarest plant species. The Center for Plant Conservation (CPC), based at the Missouri Botanical Garden coordinates the cultivation of some of the endangered species of the USA. It exists solely to prevent the extinction of some of America's most endangered species through cultivation ex situ in gardens and encourage restoration to the wild. Each participating garden is given the task of propagating and maintaining particular species appropriate for their climate and environment. This is maintaining and increasing the population of some of the most critically endangered species of the USA. Botanic Gardens Conservation International (BGCI) is a membership organization for botanic gardens and individuals and does much to promote conservation of plant species within botanic gardens and in the wild. A policy of BGCI is "to ensure that threatened plant species are secure in botanic garden collections as an insurance against loss in the wild." They support ex situ conservation of priority species with a focus on linking ex situ conservation with species conservation in natural habitats and they work with many botanic gardens on the development and implementation of habitat restoration. BGCI has a large and useful database of which plants are cultivated in which garden. This is a most useful tool for conservation planning.

Improving the Environmental and Ecological Services

Environmental and ecological services come mainly from large urban green spaces and as well as harboring much biodiversity they improve our air and water quality, absorb pollutants and release oxygen, reduce storm water runoff, cool the "heat island" in cities (Carne 1994; Zhang 1999), provide for cleaner water, air and soil (de Groot 1994), ameliorate the local climate in terms of temperature and humidity and protect water sheds from run-off (Biao et al. 2010). Plants modify temperatures by changing the rate at which energy is exchanged (Mastalerz and Oliver 1974). Finnigan et al. (1994) found that shade from trees could reduce local air temperatures by 4 °C, when the average surrounding temperatures were only 30 °C. Research has also shown that stable vegetative surfaces can control soil erosion and dust, improve the recharge and biodegradation of organic and inorganic compounds, reduce glare, noise and visual pollution, and improve the safety of vehicles on roadsides (Beard and Green 1994; Aldous 1996). Other environmental benefits include influencing wind flow, improving the sequestration of carbon (Brack 2002; Moore 2009), as well as impacting on chemical, pesticide and fertilizer loss (Balogh and Walker 1992). Both Mader (1996) and Xinian (1999) have shown that animal and bird diversity can be protected when urban and suburban natural habitats are stable and protected. Biodiversity supports many of these ecosystem services which are often taken for granted but continue to purify our water supply, and recycle nutrients providing ecosystem services that indirectly represent many millions of dollars per year to communities.

Improving Business and Service in Horticulture

Biodiversity is also important to the security of resources such as water, timber, paper, fibre, and food as well as for medicinal and pharmaceutical products. As a result, any loss of biodiversity can threaten business and longer term economic sustainability. Horticultural businesses and services provide employment for growers, managers and consumers along the production and amenity horticulture supply chain. Amenity horticulture services attract opportunities for business partnerships, and eco-tourism operations, as well as providing for event management for garden festivals, outdoor concerts and exhibitions (Aldous 2009). Municipalities and cities can often gain economically more from their urban green open space other than by producing food. For example, some 6–7 million tourists visit Singapore's parks and specialised nature-based attractions which contributed significantly to the economy (Arthur-Smith 1999) and the RHS Chelsea Flower Show attracting some 160,000 visitors per year and injecting some £ 14 million into the UK economy through receipts, direct revenue generation (leases, licenses, event hosting) and employment opportunities. The Eden Project in Cornwall, situated in a restored clay mine, attracts almost 1 million visitors a year and since opening in 2001 has added £ 1 billion to the local economy as more tourists now visit the region. Their policies of using local suppliers where possible has been one of the main factors. More broad economic benefits are found in the boosting of property values and bond ratings and energy savings made in terms of air conditioning costs, water savings from electricity generation, pollution and hydrological amelioration, and carbon sequestration. There are also indirect business and economic gains to be made with green open space. For example the City of San Antonio, Texas, in the United States, saved an estimated US$ 70 M a year in energy costs by maintaining an extensive shade tree cover (The Environmental Observatory 2002). Savings were also reported in storm water management, air quality and energy conservation with Kollin (2003) finding that trees provided direct energy savings (in air conditioning costs) of US$ 5.3 million per year to single family residences, as well as improve air quality by sequestering 10,000 t of carbon annually, and thus making savings of US$ 9.2 million. There are also growing business links between better individual and community health (Bird 2002) and between the cost of health services and the potential use of green space in reducing those health costs (McKenna 2003). In 2008 Chaudhry found, based on 2002–2003 figures, that the annual value of recreational use to the people of the Asian City of Chandigarh, India that their parks, gardens, boulevards, green avenues, reserve forests, wild life sanctuary and other landscape features, was in the order of ₹ 120.00 million, (AUD$ 279.0 million), ₹ 27.50 million (AUD$ 64.0 million) from residential business, and ₹ 92.40 million (AUD$ 214.0 million) from tourism. McKenna's (2003) study from Newcastle-on-Tyne in the United Kingdom showed that organised walks and physical activity programs, when referred by their doctors, can provide the necessary physical exercise to reduce health costs. Spottswoode (2001) found that regular walking would save on some 6.5 million working days and approximately € 6 million of production whereas Bird (2002) also concluded that a daily paced walk could reduce heart attack, stroke and diabetes by 50 %, and result in a 30 % drop in colon and breast cancer, as well as reducing obesity and save on direct health costs that were due to obesity, physical inactivity, and indirect health costs, such as absence from work and premature death.

Improving Human Health and Well-Being Through Nutrition, Recreation and Leisure

Apart from the human health aspects that emerge from a lack of nutrition and disease in many less developed countries (United Nations Environment Programme 1992), health in the developed world also depends greatly on biodiversity. Some of the health issues influenced by biodiversity include dietary health, infectious diseases, and physical, psychological, emotional, social, and spiritual health. Similarly the growing lack of drinkable water, pharmaceutical compounds, and medicines from nature that are used in primary healthcare require support from biodiversity sources.

These benefits extend to our well-being and quality of life. Not only are we attracted to natural landscapes that are largely the product of biodiversity, but most of us also value a healthy environment. Biodiversity changes in this environment can have a direct and positive impact on human well-being. Green space provides substantial intellectual, social, emotional and physical health benefits in which we see as therapy, rehabilitation, sport, recreation and as providing an overall improvement in our health and well being. There is a natural human yearning for open spaces. People like to have open views or seaside vistas from their homes. Orians and Heerwagen (1992) related this to the origins of humans in the open savannah habitats of Africa. People today still prefer to site their homes in habitats similar to those where our species evolved in Africa over several million years.

In 2005, Pretty and co-workers, demonstrated the positive contribution that natural green space can have on the health of people in the form of "green exercise". Engagement and involvement was considered at three levels—one of viewing nature through a window, two of being in the presence of a nearby natural event, such as walking or cycling to work, and three, being involved with a natural green space activity such as gardening, farming, and hiking. Involvement at levels two and three of "green exercise" have been shown to alleviate stress and reduce mental fatigue (Kaplan 1992; Bennett and Swasey 1996), increase wellbeing and self-esteem (Fjeld 2002; Pretty et al. 2005), reduce the potential for anger (Ulrich and Parsons 1992) and the risk of dementia, reduce the number of ailments and headaches (Kaplan et al. 1988), and aid in a more rapid recovery and less time spent in hospitals (Ulrich 1984, 2002). Maller et al. (2002, 2006) found that patients working in a natural environment experienced better health, visited their general practitioner less, took fewer prescription drugs, felt safer in their community, experienced less pain and discomfort, and had more opportunities to use their skills, when compared with their control. Franklin (2001) showed the importance of the natural environment for mental health. Participation in other "green exercise" activities has shown to improve work productivity (Honeyman 1992; Bergs 2002) as well as improve lifestyle. This relationship with plants and people has been comprehensively reviewed by Relf and Dorn (1995) and Relf (1998).

Horticultural therapy is a growing industry and many botanical gardens now offer training courses in this important discipline. It has evolved from various courses in gardens to a well controlled professional discipline. The American Horticultural Therapy Association defines this as: "The engagement of a person in gardening activities, facilitated by a trained therapist to achieve specific therapeutic goals." The Royal Hospital in Chelsea has installed a therapy garden for its pensioners. Horticultural therapy trusts are growing in both the UK and the USA as it is a proven help for many people (see for example the chapters in Haller and Kramer 2006).

Biodiversity within green spaces spawns and enriches the many leisure activities like gardening, hiking, bushwalking and bird watching through species richness as we look to these pursuits in order to improve our health. Dunnett and Qasim (2000) found that natural green space provided a soothing and restful environment and was non-judgemental in nature, with Wang (1999) finding that a "stroll about or rest" in green space could improve the health of people and enrich the culture of the nation. In 2003 Grahn and Ottosen (2003) stressed the importance of green areas as they contributed to ones peace and quiet, and the fascination for wildlife and the natural world. Recent studies have looked at the human psychological responses of plants in relation to psychological and physiological well-being and found that working with plants can influence a person's psychological response, as measured through blood pressure, heart rate and muscle tension (Townsend and Weerasuriya). Other researchers have found that relaxing in an environment such as a botanical garden can reduce mental fatigue, increase attentiveness, lower blood pressure and increase work productivity. In the Royal Botanic Gardens, Kew one frequently encounters individuals such as airline crews who have come to relax and escape from the stress of their jobs by walking or sitting on a bench surrounded by biodiversity of plants and animals.

Fig. 24.3 Green open space encourage human interaction, social cohesion as well as a safe play areas for children

Improving Our Social and Cultural Relationships

Societal change is also involved in working towards a greener sustainable space. Green space once was the centre of community development and community pride. The social benefits of green space can include those associated with recreation, tourism and environmental awareness education. Green space offers a myriad of ways to satisfy these elemental desires i.e. the opportunity to belong and be a part of a group is innate in all of us, whether the group consists of family members, relatives, or participants in a specific program. As a group, individuals come together to plan, organise, and implement programs and facilities and from these joint endeavours, develop bonds through networking that gives to social belonging. Natural green space can contribute significantly to the social capital of many countries by providing employment, education, and biodiversity and recreational benefit (Patel 1992; Lewis 1996; Cammack et al. 2002). From the social perceptive, natural green space has been shown to foster an active lifestyle (Scottish Executive 2003, 2004), provide safe play havens for children (Jacobs 1961), foster acquaintances with neighbours and develop closer friendships (Dunnett and Qasim 2000), and contribute positively to the social development in a community (Hart 1997; Bartlett 2005) (Fig. 24.3). Community and herb gardens have shown to reduce personal and neighborhood problems (Lewis 1996) with exposure to plants in offices shown to improve worker productivity and mood (Larsen et al. 1998). In an educational sense, green space can provide for environmental awareness, tuition of agriculture and horticulture practices, and academic achievement and development of life skills (Phibbs and Relf 2005). One of the more interesting areas of natural green space is the part it can play in reducing domestic violence, vandalism and crime by building interpersonal relationships between juvenile offenders and at risk juveniles and municipal officials (Bradley 1998). People working in, or coming in contact with natural green space, have been shown to be less aggressive and violent, with some studies showing a 50 % reduction in crime (Kuo 2010). Nature-based programmes have been of great assistance in providing employment opportunities and improving the self-esteem of the juvenile offender as well as providing safe places for young people at risk (Dawson and Zaicek 1998). Working with plants has also provided an opportunity for people with disabilities to become part of the community and to meet new friends, as well as providing a change from their institutional routine (Daubert and Rothert 1981).

Biodiversity in Decline

In a recent Future World Report the Commonwealth, Scientific and Industrial Research Organisation (CSIRO) reported that globally biodiversity continues to be in rapid decline (Hajkowicz 2012). Since life began on Earth, five major extinctions and several minor events have led to large and sudden drops in biodiversity, the most recent being the Holocene extinction, primarily caused by human impact that has resulted in both ongoing biodiversity reduction and accompanying loss of genetic diversity. The human effect has been so great that the term anthropocene has been coined for the present era where extinction is 100–1,000 times greater than natural (Crutzen and Stoermer 2000). However, Costello et al. (2013) argues that the increased conservation efforts, the numbers of taxonomists and species survival in secondary habitats have delayed the level of extinction.

Threats to Horticultural Biodiversity

Environmental threats now observed include an exploitation of productive land area or change of use, the degradation of natural systems, such as wetlands, grasslands, wild areas, coastal systems and tropical forests, a loss of function of these natural green spaces and ecosystems, and in other cases a loss of biodiversity and extinction of species. These threats to horticultural biodiversity may be specific to a country like Turkey whose threats include excessive use of agricultural inputs, misuse or excess irrigation, extensive erosion, turning the steppe areas to farm lands and settlement areas, economic policies and subsidies to define some crops, farmer's growing trends and growing urbanization or movement from rural to urban area (Uzun and Bayira 2009). The major threats to human health and wellbeing in less developed countries include poor nutrition and disease, and in developed countries, obesity, declining physical activity, growing rates of mental illness, high levels of family breakdown and declining community cohesion.

Horticultural biodiversity can be threatened by such things as fragmentation and degradation of habitat, invasive species, climate change, human overpopulation, hybridization, genetic pollution/erosion and food security and unsustainable use of natural resources and damaging events (FAO 1999). Almost all threats facing ecosystems today are the result of human and industrial activity whether direct or through climate change (Anon 2013f).

Fragmentation, Degradation and Deforestation of Habitat

Today human living is becoming more concentrated in the cities as they increasingly provide opportunities for business, employment, social, cultural and recreation pursuits. These opportunities led to concentrations in the use and activities associated with urban areas, particularly in the need for the amelioration of climate, transport and other light industry emissions, the various forms of pollution (air and water pollution and soil contamination), waste management and in the competition for use of the natural resources. In the rural landscape such degradation has often come about due to the impact of agriculture and horticulture, and many other land uses, and has been expressed through a loss of tree cover, native bushland, and grasslands. As habitats, natural green space plays an important part in the protection (Mader 1996; Wang 1999) and conservation (Aldous and Arthur 1995) of diverse populations of plants, animals and microflora, where otherwise they could well face extinction (Findlay and Bourdages 2000; Ceballos and Ehrlich 2002; Gaston et al. 2005, 2007). In urban areas as the human population increases, and we use up more and more land area for residences, industry, and commercial or recreational activity, habitat loss becomes a greater threat to biodiversity. Species are forced to live in higher concentrations, or move into habitats to which they are not adapted. Similarly the loss of native vegetation in the rural landscape has come about due to the impact of agriculture, emerging light industry, housing development and many other land uses that include tourism, has contributed to extensive land degradation often at the expense of tree cover, native bush, and/or grasslands (Cremer 1990). Clearing such vegetation can significantly add to the emissions of carbon dioxide, the main greenhouse gas, and as a result reduce the benefit of renewing oxygen and providing cleaner air. There is also clear scientific evidence that the retention of a good cover of native vegetation can prevent soil erosion and loss of water supply, reduce the effects of dryland salinity, provide a habitat for pest-eating birds as well as play a role in nutrient and water cycling.

Introduced and Invasive Species

Humans often also bring with them exotic or invasive species that are not native to a region or habitat. These invasive species sometimes carry with them diseases to which the local population is not adapted, and as such may cause a direct harmful effect. Without any form of natural barrier, such as with a river, ocean, mountain or desert, invasive species can occupy new territory, often either competing out native species by occupying their niches, competing with the resources that would normally sustain native species and as a result substantially reduce diversity. Human activities have often circumvented these natural barriers and allowed these invasive species to invade in much shorter time scales than historically have been required for a species to extend its range. For example introducing the oil palm to Indonesia and Malaysia, led to substantial economic benefits, but the benefits are accompanied by costly unintended consequences. Invasive species are not only plants but can also be introduced birds, fish, feral animals and plant and animal pathogens. Rotherham and Lambert (2011) provide detail on a wide range of case studies on invasive and introduced plants and animals. Invasive species have often been introduced by gardeners or botanic gardens. For example, *Phellodendron amurense* has escaped from the Botanical Garden in New York and has established itself as an invasive in the city. The current Flora of Hawaii consists of almost as many invasive species as native ones. Many of these were originally brought there as ornamentals. The forests are becoming overrun with species of the ornamental *Hedychium* especially the white ginger, *H. coronatrium,* and the kahili ginger, *H. gardnerianum*. The attractive African tulip tree (*Spathodea campanulata*) is invading many of the forests of Hawaii and is displacing the native tree species.

Climate Change

McPherson (1992) and others, state that the increase in greenhouse gases, the changes in precipitation patterns, the increase in carbon dioxide emissions, and rising sea levels will have major effects on the range and production levels of agricultural and horticultural food systems, the biodiversity of habitats and species, and would impact on the future sustainability of our natural green spaces. Research has shown that changes in long term environmental conditions can play a significant role in defining the function and distribution of plants and animals in a tropical forest landscape, not only in the past, but well into the present (Stork and Turton 2008). Natural climate change can result from internal forcing mechanisms, such as ocean variability, and external forcing mechanisms, such as orbital variation, solar output, magnetic field strength, plant tectonics, volcanism and the influences created by man (Anon 2013). If we consider biodiversity is in part a function of climate change the scientific evidence suggests that the global health implications of biodiversity loss are much more closely linked to the current human-caused climate change. A number of researchers have predicted that as temperatures exceed the 2 °C barrier that there will be a dramatic shift in species survival rates, particularly when working in association with other drivers such as pollution and habitat loss. If we assume 5.6 °C temperature increase over the course of the twenty-first century and a similar growth in emissions and greenhouse gases as at present then there will be important economic impacts on agriculture, horticulture, our natural resources and tourism (Dixon 2009). It is predicted that climate change will remain one of the major drivers of biodiversity patterns into the future. Elevated temperatures often observed in a city environment devoid of vegetation can contribute to water, air, and noise pollution. Urban landscapes that lack trees, turf and other vegetation, and that are overcrowded and isolated from the natural environment, are often given as causes for aggression, depression, and violence commonly found in modern society.

Many studies on the phenology of plants and the migration patterns of birds have shown that without any doubt climate change is having an effect of much of biodiversity. For example, Hepper (2003) studied the flowering times of selected garden plants in the Royal Botanic Gardens at Kew from 1961–2002 and found a distinct trend towards earlier flowering times. Many other studies have found similar results, for example, Fitter and Fitter (2002) and for bird migrations Jonzén et al. (2006), and Both et al. (2006). Climate change is causing plant species to migrate into new areas. Lowland species are migrating further up the mountains and this is threatening the alpine flora of Scottish mountains and plants of mountains in Japan (Tanaka et al. 1998). An increasing abundance of shrubs has been noted in the Arctic as the ice melts (Sturm et al. (2001).

Human Overpopulation

Last year more than half of the world's population, 3.17 billion out of 6.45 billion people, lived in cities, or in areas surrounding those cities, with future trends showing that this could exceed 5 billion people out of a total world population of 8.1 billion people by 2030 (State of the Worlds Cities, 2006–2007). About 75–90 % of people in North America, Europe and Australasia live in and around cities and this figure declines to about 40 % of people living in many less developed countries (Pinstrup-Anderson 1994). However it is in these less developed countries that some 1.4 billion people will be living in poor conditions by 2020. In 1900 there were 16 cities in the world whose population exceeded 1 million people. Leahy (2004) estimated that there are likely to be some 23 "mega-cities", each having more than 10 million people, 22 cities with 5–10 million people; 370 cities with 1–5 million people; and 433 cities with 0.5–1.0 million people in the world by 2015. Currently there are some 367 million people, or 30.4 % of the population of China, living in 668 cities (Callick 2006). The massive growth in the human population, and increased urbanization, through the twentieth century has had a significant impact on biodiversity when competing for natural resources (Satterthwaite 1999).

Hybridization, Genetic Pollution/Erosion and Food Security

With the development of the Green Revolution (Anon 2013i), which occurred principally between the 1940s and the late 1970s, there was an increase in research, development, and technology transfer that revolutionized agriculture and horticulture production and food security. Although conventional hybridization has brought with it increased yield, intensive hybridisation has resulted in genetic erosion and genetic pollution of the once large gene pools of the wild and indigenous breeds of the plant species that has resulted in a loss of genetic diversity and biodiversity as a whole as well as potentially destroying unique and rare genotypes on the planet making it difficult to further hybridize food crops and livestock against future resistant diseases and pests as well as adapting these plant and animal species to the elements of climatic change (Anon 2013d). It has been recognized that about 7,500 plant species are used to supply some 80 % of the world's food supply, but only about 95 % of these needs are obtained from some 30 plant species for human use (Uzun and Bayira 2009). Most people depend on these rather limited biological species for food, shelter, medicines and clothing. The real tragedy is the loss of diversity within our crop species. Traditional societies developed many landraces and cultivars of most crops. Today, as large scale farming replaces traditional methods, many of these varieties have been lost. This is a huge danger to the future of these crops as it is diversity that holds the genes for future developments in production, resistance to disease, adaptation to climate changes etc. In 1903 seed houses offered seeds of 497 varieties of lettuce, today there are only 38 on the market and 307 varieties of maize have been reduced to only 12 and this is the reality for all major crops.

Unsustainable Use of Natural Resources

Overexploitation occurs when any resource is consumed at an unsustainable rate. The unsustainable use of many of our natural resources can lead to loss of function of our natural cycles, such as in nitrogen fixation, which may lead on to a decline in soil fertility, or changes to the aquatic environment (Anon 2013c, k). This occurs in both terrestrial and marine environments through overgrazing, industrialization, over hunting and fishing, wild fire and flooding, deforestation, or the large-scale maintenance of agricultural and horticultural activities through artificially high levels of fertilizer, irrigation or pesticide. World Wildlife Fund (WWF) director-general James Leape said at a conference at Beijing's Tsinghua University. "If everyone around the world lived as those in America, we would need five planets to support us. For more than 20 years we have exceeded the earth's ability to support a consumptive lifestyle that is unsustainable and we cannot afford to continue down this path, "A natural resource that is increasingly being over used is water upon which our horticulture depends. Aquifers are being depleted and rivers are being increasingly diverted to irrigate crops. It takes 1,000 L of water to produce 1 kilo of wheat but 100,000 L to produce 1 kilo of meat (Godrej 2001).

Significance of Planning and Policy Support in Biodiversity

Many developed and developing countries now have governmental planning and policy authorities that are responsible for the conservation of biodiversity. In this planning there is a need to increase the awareness of the many different values of biodiversity to society to ensure that biodiversity is seen to contribute to the improvement of people's quality of life and lifestyle. Natural England is an agency working on this for the UK government (Anon 2013l, m). Their responsibilities cover much of England's green farming schemes, processes that influence the decline in biodiversity and manage their National Nature reserves and sites of specific scientific interest. Recently the England biodiversity group of Natural England has published a framework that will drive programmes on prioritising species and habitat by embedding an ecosystem approach into a UK Biodiversity Action Plan (BAP) (Anon 2013m), which is an internationally recognized programme that is designed to protect and restore biological systems that threaten habitats and species. Although the USA is an unratified signer of the accord that developed from the 1992 Convention on Biological Diversity (CBD) it has had in place a national program to protect threatened species in the form of the 1966 Endangered Species Act and is administered by the US Fish and Wildlife Service, the National Marine Fisheries Service. Some 7000 species are listed, both endangered or threatened and about half have approved Recovery Plans in place. Australia and New Zealand have developed detailed and rigorous BAPs. In Australia's case the total number of indigenous species may be 560,000, many of which are endemic, whereas New Zealand has implemented their BAP across ten separate themes (Anon 2013n).

Biodiversity and Sustainable Development Solutions

Sustainable horticulture builds on the long-standing desire of farmers, growers and managers to ensure that their land remains productive into the future. It also addresses the community's expectations and concerns for safe food and services and for environmental protection. Several organic organizations in Europe, North America and Australasia are already taking positive steps to incorporate biodiversity conservation planning into their sustainable growing and management practices. The sustainable utilization of biodiversity is already making significant contributions to the social and economic development of many countries that may include sustainable organic horticulture cropping, sustainable turf management, the use of perennial rather than annuals in agriculture, sustainable, nature-based tourism, the payment for environmental services and bio-prospecting (Anon 2013c, e).

Sustainable Organic Horticulture Cropping

Stolton and co-workers (2000, 2002) define organic farming as a "system of agriculture that relies largely on locally available resources and is dependent upon maintaining ecological balances and developing biological processes to their optimum" that emphasizes the use of management practices in preference to the use of off-farm inputs, that take into account that regional conditions require locally adapted systems. Threats have largely come from the loss of habitat, extensive use of conventional agricultural practices and climate change that result in the loss of natural biodiversity. Stolton et al. (2000) describe how organic farming contributes to biodiversity through organic practices such as crop rotation and associations, organic fertilizers, minimum tillage and increasing the density and diversity of indigenous invertebrates. These researchers also demonstrate the links between biodiversity and organic farming by maintaining vegetation that provides alternative food and refuge for many insect predators, reducing pesticide drift, and providing corridors for migrating species (FiBL 2000).

Sustainable Turf Management

Recent environmental issues and threats associated with sports turf operations, such as golf courses, have been with the alleged misuse of pesticides (Neylan 1997), nutrient leaching, or run off of chemicals, (Kaapro and Wyndham 1998; Lantzke 1999), public safety of golfers and staff, spray drift problems, wildlife conservation, native flora preservation (Kaapro and Wyndham 1998), and the resistance of pests to chemicals after prolonged use (Kadir and Barlow 1992). In order for sustainable turf management to be accepted into the broader industry the system must demonstrate enhanced production and service, provide for reduced pollution risk, protect the natural resource, prevent degradation of soil and water quality, as well as be economically viable and socially acceptable (Lefroy and Craswell 1997). The Chiba experiment, in Japan, is an example of a successful model in sustainable turf management. The Prefecture government proclaimed that only golf courses that avoided the use of agricultural chemicals would be permitted to open after April 1, 1990. Management strategies had to be developed that did not incorporate agricultural chemicals. Revised strategies adopted included the use of insect sex pheromones (Leal et al. 1993), the breeding of resistant turf grass cultivars, the installation of portable fans to reduce humidity, netted cages to trap adult moths, and the controlling of common diseases, such as dollar spot (*Sclerotina homoeocarpa*) with urea (Umemoto et al. 1995). Lake (1998) reported on the Petrik method of insecticide and fungicide usage where chemical usage was reduced by 60–80% by incorporating the practices of leaf analysis; spraying balancing biological regulators; soil microbes, enzymes and soil protection agents to develop a sustainable turf grass system. When integrated pest management (IPM) programs were introduced as part of biodiversity conservation, the golf course industry reported a 54% reduction in pesticide application without sacrificing turf quality (Foy 1988). Similarly the City of Waterloo, in the UK used to apply pesticides to 73% of its green space in the 1970's (Anon 2013o). Nowadays it has refined its plant health care program (PHCP) and now treats <0.1% of its green space as well reducing its maintenance costs per acre by 40% over the past 6 years.

Sustainable Nature-based Tourism

Tourism is the world's largest industry, with revenue that exceeds US$ 444 billion per year (Anon 2013q). Nature-based tourism comprises some 40-60% of this expenditure, and has been increasing at 10–30% annually (Wood 2002; Cereno 2009). It is evident that from an economic perspective, that the investment in biodiversity conservation has been the most productive with nature-based or ecotourism generating more income with significantly less environmental impact, than those enterprises like timber and cattle that directly exploit the natural resources of a country. Tourists seek enriching experiences and benefits, the diversity and extent of which can be influenced by the products, services and activities offered which in turn can be influenced by political, environmental and ecological, economical, lifestyle (aesthetic and visual, social, cultural, health) and demographic drivers (Jolliffe 1997; Pretty et al. 2005). In recent years these drivers have become more crucial on achieving sustainability of these green tourism assets with growing urban population pressures and carrying capacity and the need to adapt to the agents of climate change. In recent years tourism destinations promoting the theme "green and clean" have been so. In a simplistic sense green tourism could be considered a paradigm between the management of living green plant materials in the landscape, and the practice of traveling for pleasure. For nature-based tourism to be truly sustainable, the industry has to address two industry issues; the development of a satisfactory measurement of carrying capacity i.e. no destination can absorb thousands of visitors without doing itself serious environmental, as well as a satisfactory measurement in reducing the carbon footprint of an individual tourist (and tourism destination). Some examples of sustainable nature based tourism include the Green hotels in Hong Kong which make efforts to lower their energy and water usage, and reduce solid wastes (Anon 2013p), Singapore's Gardens by the Bay, a new generation of garden development and management practices, applying sustainable energy and water harvesting and usage solutions (Anon 2013r), Thailand's Tourism Authority of Thailand (TAT) (Anon 2013a, g) that supports both a Seven Greens programme, a conceptual framework and practical guidelines for carefully balancing tourism promotion and a healthy, sustainable environment (Anon 2013) as well as the promotion of Thailand Tourism Awards which emphasise the actions and programmes taken in response to global warming/climate change, participation, sufficiency economy, good governance, corporate social responsibility and overall integration (Anon 2013), and Malaysia's Forest Development and Management, National Parks which sustainably manage and develop their forest resources and optimize their contribution to the socio-economic development of the nation (Anon 2013b).

Many urban botanical gardens attract a large number of visitors to enjoy both the biodiversity and their calming atmosphere. Amongst others with a large number of visitors are the Royal Botanic Gardens, Kew in England, The New York Botanical Garden in the USA, the Singapore Botanic Garden, The Kirstenbosch Botanic Garden in South Africa and the Rio de Janeiro Botanic Garden in Brazil.

Payment of Environmental Services

Ecosystem services might broadly be called "the benefits of nature to households, communities, and economies" (Boyd and Banzhaf 2006) or, more simply, "the good things nature does" (Jenkins 2008). These environmental services are another form of economic valuation of biodiversity, where there is a payment for environmental services provided by ecosystems. Environmental benefits, such as valuation of water production, carbon dioxide fixation, biodiversity conservation and protection of scenic beauty by forests, and the corresponding payment of these ecosystem services, bring direct economic benefits to forest owners and contribute directly to the cost of conservation and protection of forests (Barrantes 2001) (Fig. 24.4).

Fig. 24.4 Preserving biodiversity by placing a value on "ecosystem" services like forests helps to conserve this hornbill in Singapore and the Orang-utan (*Pongo pygmaeus*) in Malaysia

Bio-Prospecting

Bio-prospecting is an umbrella term that describes the process of discovery and commercialization of new products based in biological resources, typically in less-developed countries (Anon 2013d; Tansley 1946). When implemented correctly, it can be considered as another form of sustainable utilization and economic valuation of biodiversity, as well as a means to supporting the conservation of biological diversity. Under this umbrella we usually think of the discovery of new medicines and agrochemicals, but new products for horticulture may also be involved. Scientists of the Eden Project in Cornwall, England collected a rare and endangered species of *Impatiens* in the Seychelles (*Impatiens gordonii*) with white flowers. This grew well in the tropical rainforest biome and one of their horticulturalists hybridized it with a red flowered *Impatiens*. The resultant pink-flowered hybrid has been introduced into cultivation and the proceeds from its sale are going to support plant conservation in the Seychelles. This work not only produced funds for conservation, but also boosted the number of individuals of the original rare species for its restoration in the wild.

Conclusions

- Biodiversity and its application in horticulture and horticultural science provide for a range of environmental, economic, social and health benefits
- Threats to biodiversity can include the exploitation of productive land area, the degradation of natural systems, and a loss of function of these natural green spaces and ecosystems. In horticultural cropping systems threats include exces-

sive use of agricultural inputs, misuse or excess irrigation, extensive erosion, growing urbanization and the threat of the increase of alien invasive species.
- Threats to human health and wellbeing in less developed countries include poor nutrition and disease and in developed countries, the overuse of chemical pesticides, obesity, declining physical activity, growing rates of mental illness, high levels of family breakdown and declining community cohesion.
- Successful sustainable development solutions can be undertaken at the local or national level and will continue to play an important role in facilitating the strategy of sustainable development and improving biodiversity of both urban and rural landscapes.

References

AIT (Asian Institute for Technology) (1998) Internet resource. http://www.hsd.ait.ac.th/cities/jakarta/envt.htm
Aldous DE (1996) Improving the physical urban environment with turf. Proceedings of the combined NZRA and IFPRA Asia/Pacific Regional Congress, November, Palmerston North, New Zealand, pp 8.1–8.6
Aldous DE (2009) Challenges on Southeast Asian Green Tourism. Proceedings of the urban forestry conference, Kuching, Sarawak, 17–19 November, pp 32–42
Aldous DE (2010) Green cities in Australia: adopting a national outlook to green open space planning. Australas Parks Recreation 13(4):10–14 (Summer Issue)
Aldous DE (2011) Green open space and environmental sustainability: Redland city council case study. Paper presented at the parks and leisure Australia national conference/international federation of parks and recreation administration Asia-pacific congress held in Fremantle, Western Australia over 18–21 September, 2011
Aldous DE, Arthur T (1995) Conservation and management of green open space. International federation of park and recreation administration. In: de Waal I (ed) Proceedings of the international federation of park and recreation administration. Belgium, Wellington, New Zealand, pp 1–11
Andresen JW, Plexman CA (1980) Proceedings of a symposium on urban forest management within the Commonwealth of nations, Agenda Subhead 5. Trees in rural and urban development. Eleventh Commonwealth Forestry Conference, Trinidad and Tabago, Trinidad
Anon (1996b) Fourth annual report on parks and maintenance practices, City of Gloucester, PO Box 8333, Gloucester ON KIG 3V5
Anon (2009) The Planning institute of Australia. http://www.planning.org.au/viccontent/2009-2. Accessed 13 March 2013
Anon (2013a) Thailand tourism rewards. http://www.thailand.com/travel/news/news-073112.html
Anon (2013b). Malaysia forest development. http://www.forestry.sarawak.gov.my/forweb/homepage.htm
Anon (2013c) Bio-prospecting. http://en.wikipedia.org/wiki/Bioprospecting
Anon (2013d) Biodiversity Wikipedia. http://en.wikipedia.org/wiki/Biodiversity. Accessed 13 March 2013
Anon (2013e) Guidelines for non-chemical pest control in golf courses. Agriculture and Forestry Department, Chiba Prefectural Government, Japan, 43 p
Anon (2013f) Garnaut climate change review. http://www.garnautreview.org.au. Accessed 12 March 2013
Anon (2013g) Thailands tourst authority. http://www.tatnews.org/ATF2009/4159.asp
Anon (2013h) Climate change. http://en.wikipedia.org/wiki/Climate_change. Accessed 12 March 2013
Anon (2013i) Green revolution. http://en.wikipedia.org/wiki/Green_Revolution

Anon (2013j) Black 2008. http://news.bbc.co.uk/2/hi/science/nature/7660011.stm
Anon (2013k) Overfishing. http://en.wikipedia.org/wiki/Overfishing
Anon (2013l) Natural England. http://www.naturalengland.org.uk/about_us/whatwedo/default.aspx
Anon (2013m) England's biodiversity framework-securing biodiversity. http://www.naturalengland.org.uk/ourwork/conservation/biodiversity/protectandmanage/framework.aspx
Anon (2013n) Biodiversity action plan. http://en.wikipedia.org/wiki/Biodiversity_action_plan#United_States
Anon (2013o) City of Waterloo Plant Health Care Program (PHCP), Service Center. http://www.city.waterloo.on.ca/pws/parks/operations/phcp.html
Anon (2013p) Hong Kong. http://www.rezhub.com/GreenTravel/GreenHotels/tabid/119/all/true/country/HK/Green/true/Default.aspx
Anon (2013q) World bank. http://web.worldbank.org/WBSITE/EXTERNAL/EXTDEC/EXTRESEARCH/EXTWDRS/0,contentMDK:22295143~pagePK:478093~piPK:477627~theSitePK:477624,00.html
Anon (2013r) Singapore's gardens by the Bay. http://www.sentosa.com.sg
Arthur-Smith R (1999) Singapore garden city. Proceedings of IFPRA-Asia/Pacific Congress. Hangzhou, September, China, pp 63–68
Attfield R (1999) The ethics of the global environment. Edinburgh University Press,Edinburgh
Ayata S (2004) Local government green space policy the case for Kartal and Sartyer in Istanbul. Dissertation, Master of Science Requirements, East Technical University, Istanbul, 177 pages
Balogh J, Walker W (1992) Golf course management and construction: environmental issues. Lewis Publishing, Boca Rato, 951 p
Bartlett S (2005) Urban children and the physical environment. http://www.araburban.org/childcity/Papers/English/Sheridan%20Barlett.pdf. Accessed 3 Sept 2010
Barrantes G (2001) Gasto y financiamiento ambiental en Costa Rica. 1992–2000. Informe a CE-PAL. Heredia, Instituto de Políticas para la Sostenibilidad. p 56
Beard JB, Green RJ (1994) The role of turf grass in environmental protection and their benefits to humans. J Environ Qual 223:542–460
Bennett E, Swasey J (1996) Perceived stress reduction in urban public parks. Hortic Technol 6(2):125–128
Bergs J (2002). The effect of healthy workplaces on the well-being and productivity of office workers. In Proceedings of plants for people international symposium, Floriade, Netherlands
Biao Z, Wenhua L, Gaodi X, Yu X (2010) Water conservation of forest ecosystem in Beijing and its value. Ecol Econ 69(7):1416–1426. 10.1016/j.ecolecon.2008:1009.1004
Bird W (2002) Green space and our health. Paper to London green space conference, countryside agency
Bolund P, Hunhammar S (1999) Ecosystem services in urban areas. Ecol Econ 29:293–301
Both C, Bouwhuis S, Lessells CM, Visser ME (2006) Climate change and population declines in a long-distance migratory bird. Nature 441:81–83
Boyd J, Banzhaf S (2006) What are ecosystem services—discussion paper. http://www.rff.org/Documents/RFF-DP-06-02.pdf. Accessed March 2013
Brack CL (2002) Pollution mitigation and carbon sequestration by an urban forest. Environ Pollut 116:S195–S200
Bradley R (1998) Public expectations and perceptions of policing. Home office, policing and reducing crime unit series, research, development and statistics directorate, policing and reducing crime unit, London
Callick R (2006) It's ecology v economics, China warns. The Australian, Tuesday. June 6:10
Cammack C, Waliczek TM, Zajicek JM (2002) The green brigade: the effects of a community based horticultural program on the self development characteristics of juvenile offenders. Hortic Technol 12:82–86
Carne J (1994) Urban vegetation: ecological and social value. Proceedings of the 1994 national greening Australia conference, October 4–6, Fremantle, Western Australia, pp 211–226

Ceballos G, Ehrlich PR (2002) Mammal population losses and the extinction crisis. Science 296(5569):904–907

Cereno RP (2009) Green Tourism Potential in the Country. UPLBRDE News Service Conserving our Diversity (2009) Singapore's National Biodiversity Strategy and Action Plan. National Biodiversity Centre, National Parks Board, Singapore Botanic Gardens, 1, Cluny Road, Singapore 259569, 21 p

Costello MJ, May RM, Stork NE (2013) Can we name earth's species before they go extinct? Science 339(6118):413–416

Cremer KW (ed) (1990) Trees for rural Australia. Inkata Press, Sydney

Crutzen PJ, Stoermer EF (2000) The 'Anthropocene'. Glob Change Newsl 41:17–18

Dawson C, Zajicek J (1998) The green brigade the effects of a community-based program on attitudes and behaviours of juvenile offenders. In Burchett M, Tarran J, Wood R (eds) Towards a new millennium in people-plant relationships: international people-plant symposium. University of Technology, Sydney, Printing Services, Sydney, pp 384–393

Daubert JR, Rothert E (1981) Horticultural therapy series (four manuals). Horticultural Therapy Services, Chicago Horticultural Society, Glencoe, IL 60022–0400

Dixon GR (2009) The Impact of Climate and Global Change on Crop Production. Chapter 17, pages 307-324. In: Climate Change: Observed Impacts on Planet Earth, T. M. Letcher (editor), publisher: Elsevier, Oxford.

Dunnett N (2004) The dynamic nature of plant communities. In Dunnett N, Hitchmough J (eds) The dynamic landscape. Spon Press, London

Dunnett N (2006) Green roofs for biodiversity: reconciling aesthetics with ecology. In 4th annual greening rooftops for sustainable communities conference, May 2005, Boston, MA

Dunnet N, Qasim M (2000) Perceived benefits to human well being of urban gardens. Hortic Technol 10(1):40–45

Dunnett N, Hitchmough J, Jenkins C, Tylecote M, Thompson K, Matthews Joyce R, Rae D (2007) Growing nature—the role of horticulture in supporting biodiversity (2007), Report No. 244 (ROAME No. FO6AB12). http://www.snh.org.uk/pdfs/publications/commissioned_reports/Report%20No244.pdf. Accessed March 2013

FAO (1999) The state of food insecurity in the world. Food insecurity: when people live with hunger and fear starvation. United Nations Food and Agriculture Organisation, Rome, Italy. http://www.fao.org/docrep/007/x3114e/x3114e00.htm. Accessed 14 March 2013

FiBL (2000) Organic farming enhances soil fertility and biodiversity. Results from a 21-year-old field trial. Research Institute of Organic Farming (FiBL), Frick. Dossier no. 1, August, Switzerland2000

Findlay CS, Bourdages J (2000) Response time of wetland biodiversity to road construction on adjacent lands. Conserv Biol 14(1):86–94

Finnigan J, Raupach M, Cleugh H (1994) The impact of physical environment of the impact of cities. In Proceedings of National Greening Australia Conference, October 4–6, Western Australia, pp 23–37

Fitter AH, Fitter RSR (2002) Rapid changes in flowering time in British plants. Science 296:1689–1691

Fitzpatrick D (1994) Money trees on your property: profit gained through trees and how to grow them. Inkata Press, Sydney

Fjeld T (2002) The effects of plants and artificial daylight on the well-being and health of office workers, school children and health care personnel. In Proceedings of plants for people international symposium Floriade, Netherlands 2002

Foy JH (1988) Integrated pest management: a different approach for the same old problems. USGA 26(5):9–11

Franklin H (2001). Beyond toxicity: human health and the natural environment. Am J Prev Med 20:234–240

Gámez R (n.d.) The link between biodiversity and sustainable development: lessons from INBio's bioprospecting program in Costa Rica. http://law.wustl.edu/centeris/Papers/Biodiversity/PDFWrdDoc/gamezfinal1.pdf. Accessed 15 March 2013

Gaston KJ, Warren P, Thompson K, Smith RM (2005) Urban domestic gardens (IV): the extent of the resource and its associated features. Biodivers Conserv 14:3327–3349

Gaston KJ, Warren PH, Devine-Wright P, Irvine KN, Fuller RA (2007) Psychological benefits of greenspace increase with biodiversity. Biol Lett 3(4):390–394

Godrej D (2001) The no-nosense guide to climate change. New International Publications, Oxford

Grahn P, Ottosen J (2003) The importance of nature in a life crisis—why are green areas so essential for people's health and quality of life? In Proceedings of the IFPRA Europe Conference, Stavanger, June 16-19, Norway, 10 p

Groot RS de (1994) Environmental functions and the economic value of natural ecosystems. In: Jansson AM, Hammer H, Folke C, Constanza R (eds) Investing in natural capital—the ecological economics approach to sustainability. Island press, Washington

Guoping X (1999). A preliminary study on the planning of the green land system. Proceedings of the IFPRA Asian-Pacific conference, Hangzhou, China, pp 105–109

Hajkowicz S (2012) Our future world: global megatrends that will change the way we live. CSIRO futures report. http://www.csiro.au/en/Portals/Partner/Futures/Our-Future-World-report.aspx. Accessed 1 March 2014

Haller RL, Kramer CL (eds) (2006) Horticultural therapy methods: making connections, healthcare, human service and community programs. Haworth Press, Philadelphia

Hart R (1997) Children's participation: the theory and practice of involving young citizens in community development and environmental care. In: Hagedorn R (ed) Therapeutic horticulture. Winslow Press, Florida

Hepper FN (2003) Phenological record of English garden plants in Leeds (Yorkshire) and Richmond (Surrey) from 1946 to 2002. An analysis relating to global warming. Biodivers Conserv 12:2503–2520

Heywood VH, Baste 1 (1995) Introduction. In: Heywood VH (ed) Global biodiversity assessment. Cambridge University Press, Cambridge, pp 1–19

Honeyman MK (1992) Vegetation and stress: a case study in the hotel industry. In Relf D (ed) The role of horticulture in human well being and social development: a national symposium. Timber Press, Portland, pp 220–222

Jacobs J (1961) The death and life of great American cities. Random House, New York

Jenkins M (2008) Mother nature's sum. http://www.psmag.com/business-economics/mother-nature-s-sum-4226. Accessed March 2013

Jim CY, Wendy YC (2009) Ecosystem services and valuation of urban forests in China. Cities 26(4):187–194

Jolliffe A (1997) Tourism and horticulture. NZ Gard J (Journal of the Royal New Zealand Institute of Horticulture) 2(2):15–21

Jonzén N et al (2006) Rapid advance of spring arrival dates in long-distance migratory birds. Science 312:1959–1960

Kaapro J, Wyndham S (1998) Environmental strategy for Australia golf courses. Bermuda Print, Burnley

Kadir AA, Barlow HS (1992) Pest management and the environment in 2000. CAB International, UK

Kaplan R (1992) The psychological benefits of nearby nature. In: Relf D (ed) The role of horticulture in human well-being and social development: a national symposium. Timber Press, Portland

Kaplan R, Kaplan S (1989) The experience of nature: a psychological perspective. Cambridge University Press, Cambridge

Kaplan S, Talbot JF, Kaplan R (1988) Coping with daily hassles: the impact of nearby nature on the work environment. Project report. USDA forest service, North Central forest experiment station, urban forestry unit cooperative agreement 23-85–08

Kollin C (2003) San Antonio: ripples of change, armed with new information on the benefits of their urban trees, official plans for a greener and more ecological based future. American Forests, Spring. http://www//findarticles.com/cf_0/m1016/mag.jhml. Accessed 12 March 2013

Kuo FE (2010) Parks and other green environments: essential components for a healthy human habitat. National recreational and park association, executive summary, research series 2010.

http://www.nrpa.org/uploadedFiles/nrpa.org/Publications_and_Research/Research/Papers/MingKuo-Summary.PDF. Accessed 12 March 2013

Lake B (1998) Club points the way to biological management. Turf Craft International Magazine 60: 58–59, Melbourne, Australia

Lantzke N (1999) Phosphorus and nitrate loss from horticulture on the swan coastal plain. Aust Turfgrass manag 2:10–13

Larsen LJ, Adams J, Deal B, Kweon B, Tyler E (1998) Plants in the workplace: the effects of plant density on productivity, attitudes, and perceptions. Environ Behav 30:61–281

Larsson T-B (2001) Biodiversity evaluation tools for European forests—Ecological Bulletin 50. Wiley-Blackwell, New Jersy

Leahy S (2004) An eco-town takes root. International development and environment article service. http://www.oneworld.ca/article/view/83997/1. Accessed 19 April 2013

Leal WS, Sawada M, Hasegawa M (1993) The Scarab *Beetle Anomala cuprea* utilizes the sex pheromone of *Popillia japonica* as a minor component. J Chem Ecol 19(17):1303–1313

Lefroy RDB, Raswell RT (1997) Soil as a filter for nutrients and chemicals: sustainability aspects. Food and Fertilizer Technology Center, 11 p. (Technical Bulletin 146, March)

Lewis C (1996) Green nature/human nature. Univ. Ill. Press, Chicago

Mader R (1996) Urban greening conference. International seminar on urban greening in latin America and the Caribbean, 2–5 December, Mexico City

Maller C, Townsend M, Pryor A, Brown P, St Leger L (2006) Healthy nature healthy people: 'contact with nature' as an upstream health promotion intervention for populations. Health Promot Int 21(1):45–54

Maller C, Townsend M, Brown P, St Leger L (2002) Healthy parks healthy people: the health benefits of contact with nature in a park context: a review of current literature. Social and Mental Health Priority Area Occasional Paper Series, Vol 1. Faculty of Health and Behavioural Sciences, Deakin University, Burwood

Mastalerz JW, Oliver CR (1974) Microclimate modification: development and application in the urban environment. Hortic Science 9(6):560–563

McKenna T (2003) Parks: for plants or for people?—How Newcastle develops its park policy with a health aspect. In Proceedings of the IFPRA Europe Conference, Stavanger, June 16–19, Norway, 8 p

McPherson EG (1992) Accounting for benefits and costs of urban green space. Landscape Urban plan 22:41–51

Moore GM (2009) People, trees, landscapes and climate change. In: Sykes H (ed) Climate change for young and old. Future Leaders. Brotherhood Books, Melbourne, Australia. http://www.futureleaders.com.au. Accessed 14 March 2014

Morton S, Sheppard A, Lonsdale M (2012). Explainer: what is biodiversity and why is it important. http://theconversation.edu.au/explainer-what-is-biodiversity-and-why-does-it-matter-9798. Accessed 1 March 2013

Neylan J (1997) Biological control agents in turfgrass management. Golf & Sports turf Australia, August/September, pp 15–19

Norse EA, McManus RE (1980) Environmental quality 1980: the eleventh annual report of the council on environmental quality. Executive Office of the President, Council of Environmental Quality. pp 31–80

Orians GH, Heerwagen JH (1992) Evolved responses to landscapes. In Barkow JH, Cosmides L, Tooby J (eds) The adapted mind: evolutionary psychology and the generation of culture. New York University Press, NY

Patel IC (1992) Socioeconomic impact of community gardening in an urban setting. In Relf D (ed) The role of horticulture in human well-being and social development: a national symposium. Timber Press, Portland

Phibbs EJ, Relf D (2005) Improving research on youth gardening. Hortic Technol 15:425–428

Pinstrup-Anderson P (1994) World food trends and future food security. International Food Policy Research Institute, Washington DC

Pretty J, Peacock J, Sellens M, Griffin M (2005) The mental and physical health outcomes of green exercise. Int J Environ Health Res 15(5):319–337

Pye-Smith CH 1996. Building green islands in Bombay. People & the Planet 1996, Volume 8.4, November 1996 (http://www.ourplanet.com/txtversn/84/smit.html).
Relf D (1998) Moving toward a new millennium in people-plant relations. In Burchett M, Tarran J, Wood R (eds) Towards a new millennium in people-plant relationships: international people-plant symposium. University of Technology, Sydney, Printing Services, Sydney, pp 1–7
Relf PD, Dorn ST (1995) Horticulture: meeting the needs of special populations. Hortic Technol 5(3):94–103
Rotherdam ID, Lambert RA (2011) Invasive and introduced plants and animals-human perceptions attitudes and approaches to management. Earthscan Publications, London, 352 pp
Satterthwaite D (1999) sustainable cities or cities that contribute to sustainable development? In: Satterthwaite D The earthscan reader in sustainable cities. Earthscan Publications, London
Scottish NH (2003) Scottish natural heritage's corporate strategy looking forward 10 years. (Perth: SNH Pubs.). http://www.snh.org.uk/58. Accessed 11 March 2006
Scottish E (2004) Scotland's Biodiversity—It's in your hands. A strategy for the conservation and enhancement of biodiversity in Scotland. Scottish Executive, Edinburgh
Singh VS, Pandey DN, Chaudhry P (2010) Urban forests and open green spaces: lessons for Jaipur, Rajasthan, India. Rajasthan State Pollution Control Board, Occasional Paper No. 1/2010:1-23. http://210.212.99.115/rpcb/RSPCB-OP-1-2010.pdf. Accessed 20 Oct 2010
Spottswoode J (2001) Making connections: a guide to accessible green space. Pace and Stoneham, the Sensory Trust 2001
State of the Worlds Cities (2006/2007) The millennium goals and urban sustainability. 2006/07. Report of the UN-HABITAT, 204 pp. ISBN: 1844073785
Status of Urban Forestry in the Asia-Pacific Region (1998). FAO corporate document repository. http://www.fao.org/docrep/003/x1577e/X1577E06.htm. Accessed 13 Aug 2010
State University of New York, New Paltz (2010) http://www.newpaltz.edu/green/definitions.html. Accessed 14 April 2010).
Stolton S, Geier B (2002) The relationship bewteen biodiversity and organic agriculture: defining appropriate policies and approaches for sustainable development. In: High Level Pan-European Conference on Agriculture and Biodiversity: Towards Integrating Biological and Landscape Diversity for Sustainable Agriculture in Europe, Strasbourg, 4 March 2002. STRA-CO/AGRI (2001) 3. UNEP, Council of Europe, Government of France. http://www.strategy\agriculture\conference\docs\agri03e.01
Stolton S, Geier B, McNeely JA (2000) The relationship between nature conservation, biodiversity and organic agriculture. IFOM Tholey-Theley, Germany
Stork NE, Turton NE (eds) (2008) Living in a Dynamic Tropical Forest Landscape: Lessons from Australia Blackwells 600 pp, 20, 2010.
Sturm M, Racine C, Tape K (2001) Increasing shrub abundance in the Arctic. Nature 411:546
Tanaka N, Taoda H, Omasa K (1998) Field studies on the effects of global warming on mountain vegetation in Japan. Global Environ Res 1:71–74
Tansley AG (1946) Our Heritage of Wild Nature. Cambridge University Press, Cambridge, England
Townsend M, Weerasuriya R (2010) Beyond blue to green: the benefits of contact with nature for mental health and well-being. Deakin University, April. http://www.beyondblue.org.au http://www.beyondblue.org.au. Accessed 21 July 2010).
Ulrich RS (1984) View through a window may influence recovery from surgery. Science 224:420–421
Ulrich RS (2002) Communicating with the healthcare community about plant benefits. In Shoemaker C (ed) Proceedings of the sixth international people plant symposium. Chicago Botanic Garden, Chicago
Ulrich RS, Parsons R (1992) Influences of passive experiences with plants on individual well-being and health. In: Relf D (ed) The role of horticulture in human well-being and social development. Timber Press, Portland, pp 93–105
Umemoto S, Yasuda M, Shingyoji T, Aoki K (1995) Control of Dollar Spot Diseases (*Sclerotina homoeocarpa*) of bent grass Green Turf with Urea. J Jpn Soc Turfgrass Sci 24(1), pp 13–17

UN-Habitat (2009) Global report on human settlements. Planning sustainable cities: policy direction. (United Nations Human Settlements Programme). Earthscan, London

United Nations Environment Programme (1992) Health. (Chap. 18) In: Tolba MK, El-Kholy OA, El-Hinnawi E, Holdgate MW, McMichael DF, Munn RE (eds). The world environment 1972–1992: two decades of challenge. Chapman & Hall, New York, pp 529–567

Uzun I, Bayir A (2009) Horticultural biodiversity in Turkey. Bulletin UASVM Horticulture, 66(2)/2009 Print ISSN 1843-5254; Electronic ISSN 1843–5394

Wang X (1999) Urban peripheral green space: type, quality and layout. In Proceedings of IFPRA-Asia/Pacific Congress, Hangzhou, September, China, pp 300–305

Wilson EO (1992) The Diversity of Life. Harvard University Press, Cambridge, 424 p

Wilson R (2006) Collaboration in context: rural change and community forestry in the four corners. Soc Nat Resour 19:53–70

Wood ME (2002) Ecotourism: principles, practices and policies for sustainability. http://www.pnuma.org/industria/documentos/Ecotourism1.pdf. Acccessed 12 March 2013

Xie Z (1999) Nanjing's gardens greening shall take the road of sustainable development. In Proceedings of IFPRA-Asia/Pacific Congress, Hangzhou, September, China, pp 98–102

Xinian Z (1999) Quantitative evaluation of environmental benefits of urban greenland in Beijing City. In Proceedings of IFPRA-Asia/Pacific Congress, Hangzhou, September, China, pp 315–323

Zhang X (1999) Quantitative evaluation of environmental benefits of urban Greenland in Beijing city. In Proceedings of IFPRA-Asia/Pacific Congress, Hangzhou, September, China, pp 315–323

Chapter 25
An Assessment of the Effects of Climate Change on Horticulture

Geoffrey R. Dixon, Rosemary H. Collier and Indrabrata Bhattacharya

Abstract Horticulture may be defined as the intensive cultivation and harvesting of plants for financial, environmental and social profit. Evidence for the occurrence of climate change more generally and reasons why this process is happening with such rapidity are discussed. These changes are then considered in terms of the effects which might alter the options for worldwide intensive horticultural cultivation of plants and its interactions with other organisms. Potentially changing climates will have considerable impact upon horticultural processes and productivity across the globe. Climate change will alter the growth patterns and capabilities for flowering and fruiting of many perennial and annual horticultural plants. In some regions perennial fruit crops are likely to experience substantial difficulties because of altered seasonal conditions affecting dormancy, acclimation and subsequent flowering and fruiting. Elsewhere these crops may benefit from the effects of climate change as a result of reduced cold damage and increased length of the growing season. There will be considerable effects for aerial and edaphic microbes, invertebrate and vertebrate animals which have benign and pathogenic interactions with horticultural plants. Microbial activity and as a consequence soil fertility may alter. New pests and pathogens may become prevalent and damaging in areas where the climate previously excluded their activity. Vital resources such as water and nutrients may become scarce in some regions reducing opportunities for growing

G. R. Dixon (✉)
School of Agriculture, Earley Gate, University of Reading, RG6 6AR Reading, Berkshire, UK
e-mail: geoffrdixon@btinternet.com

R. H. Collier
Warwick Crop Centre, The University of Warwick, Wellesbourne,
Warwick CV35 9EF, UK
e-mail: Rosemary.Collier@warwick.ac.uk

I. Bhattacharya
Department of Plant Pathology, Bidhan Chandra Krishi Viswavidyalaya, Mohanpur,
Nadia, West Bengal 741252, India
e-mail: indrabratabhattacharya@gmail.com

G. R. Dixon, D. E. Aldous (eds.), *Horticulture: Plants for People and Places, Volume 2,*
DOI 10.1007/978-94-017-8581-5_25,

horticultural crops. Wind and windiness are significant factors governing the success of horticultural plants and the scale of their impacts may change as climate alters. Damaging winds could limit crop growing in areas where previously it flourished. Forms of macro- and micro-landscaping will change as the spectrum of plants which can be cultivated alters and the availability of resources and their cost changes driven by scarcities brought about by climate change. The horticultural economy of India as it may be affected by climate change is described as an individual example in a detailed study.

Keywords Climate change · Temperature · Carbon dioxide · Water · Wind · Sustainability · Phenophase · Pest · Pathogen · Beneficial microbes · India case study · Potato blight study

Introduction

The world's climate is not static, there is ample evidence showing that it has been changing since the Earth was sufficiently solid that it possessed an atmosphere (Dixon 2009, 2012). Natural events are some of the best indicators of climatic change since the biology of plants and animals is regulated by their surrounding environment. Changes to the seasonal climatic rhythms are translated into the life and breeding cycles of both natural and cultivated plants and animals. Studies of changing biological rhythms were popular pastimes in the seventeenth and eighteenth centuries. This resulted in a legacy of long-term, phenological records which are valuable sources of primary data tracking the impact, scale and rate of altering climates. Great Britain owns some of the most extensive runs of this type of data. One of the longest archives from a single location was started by the landowner and naturalist Robert Marsham on his farm at Stratton Strawless, Norfolk from 1736 to 1797 (Spinks and Lines 2008; Fiske 2008). He traced 27 signs of spring during the period from the first snowdrop (*Galanthus* spp) flowering to the first cuckoo (*Cuculus canorus*) singing and noted how timing was affected by weather. He showed how a cold period delayed springtime and warm conditions speeded it up. Leafing by the mountain ash (*Sorbus intermedia*) for example, varied from March 5 to May 2 according to weather. In 1740 after a fearfully hard winter Marsham noted that spring was badly delayed and it was March before he heard a song thrush (*Turdus philomelos*) singing and hawthorn (*Crataguus* spp) bushes did not break into bloom until June, 2 months later than usual. Marsham's recording was continued by his descendants until 1958, providing a record of two centuries of continuous data. In Scotland there is a similarly substantial body of information starting in the early 1700s and illustrating the events on which changes in the environment may be identified (Last et al. 2003). These long-term records provide baselines from which to judge the pace of climatic change which has been recognised scientifically in the last couple of decades.

Climate change disrupts synchronies between temperature and photoperiod and subsequent biological behaviour. Darwin (1859) recognised this and formulated his relationship between genotype and the environment, which is then expressed as the phenotype of an organism. The application of this relationship was demonstrated for woody perennials such as *Rhododendron* species, varieties and cultivars by Dixon and Biggs (1996) and Cameron and Dixon (1997) who observed changes in dormancy and subsequent acclimation resulting from temperature changes. The life cycles of the fungi are especially influenced by environmental change. Fungal forays made around Salisbury, Wiltshire accumulated 64,000 records from 1950 onwards and demonstrated that the growing season for many fungi has increased as temperatures have risen (Gange et al. 2007). Over 50 years the reproductive season for some macro-fungi doubled in the length from 33 days to 74 days on average. As a result some of these organisms are now reproducing twice per year instead of once. It is suggested that changes in growing season for fungi brought about by climatic changes are the most extensive for any group of organisms on Earth. Over a 30-year period (1978 to 2007) observations of meteorological and botanical characters of flowering garden plants growing close to Edinburgh, Scotland were taken at weekly intervals (Last and Roberts 2013). During this time mean air temperatures significantly increased by 0.080, 0.044 and 0.026 °C yr^{-1} in the first, second and third quarters of the year and soil temperatures increased by 0.060 °C yr^{-1} in the first quarter. Over the 30-year period, measurements of trends in flowering showed that the early (February to March) species flowered circa 24 days sooner and the later flowering species (April to May) advanced by only circa 12 days.

The first detailed formal and statistically validated link between observed global changes in physical and biological systems and human induced climate change, predominantly from increasing concentrations of greenhouse gases, was demonstrated by Rosenzweig et al. (2008). These authors surveyed 29,500 data series of which 90% ($P << 0.001$) demonstrated that changes at the global scale are in the direction that would be expected as responses to the incidence of global warming. In biological systems 90% of the data sets showed that plants and animals are responding consistently to temperature change. This is mostly illustrated by phenological change with earlier blooming, leaf unfolding and spring arrivals. Events on the current scale have not happened on Earth in the past three quarters of a million years (King 2005). Previously, however, no one single species had full control of the Planet's entire store of resources and rapidly reproduced itself in unprecedented numbers. Man (*Homo sapiens*) now has these abilities and as a result of his activities the Earth's bank of resources is in imminent danger of being exhausted and its environment is changing in a manner that exacerbates that devastating erosion of resources (Solomon et al. 2007). The Earth's resource bank is being emptied at a rate far in excess of its capacities for natural replenishment.

Physical evidence for the effects of climate change began being presented with the Swedish physical chemist Arrhenius (1896). He estimated that doubling the carbon dioxide content of the atmosphere could increase the earth's temperature by 2.1 °C. These assessments were reinforced by the studies of Callendar (1938). He estimated that, in the preceding half century, by fuel consumption, man had added

150,000 million t of carbon dioxide into the atmosphere, of which 75% stayed there. Using the radiation absorption coefficients for carbon dioxide and water vapour he calculated that as a result temperature increased by 0.003 °C per annum. He studied observations from 200 meteorological stations around the globe and showed that in the period in question world temperature had in practice increased by 0.005 °C per annum.

Great Britain owns some of the longest physical meteorological records and these are now contained in The Central England Temperature Archive (Anon 2011a), dating back to 1659. These are the world's oldest continuous weather records describing an area within the triangle between Lancaster, Bristol and the outskirts of London and were initially compiled by Gordon Manley (1902–1980), a Durham University meteorologist. Manley meticulously worked through thousands of documents left by enthusiastic amateurs such as landed gentry and vicars who had made daily recordings between the seventeenth and twentieth centuries. He spent 30 years working through their journals, cross-checking and calibrating thermometer readings until he had achieved a standardised record. The Meteorological Office, Exeter now maintains this record. It demonstrates how climate has changed over this extended period. Based on predictions by the Intergovernmental Panel on Climate Change (IPCC) which utilise large scale computer models (Watson et al. 1998), the British Treasury economist Stern (2008) defined the implications of very rapid climatic change in the following terms:- "if no action is taken to reduce emissions, the concentration of greenhouse gases in the atmosphere could reach double its pre-industrial revolution levels as early as 2035, virtually committing us to a global average temperature rise of over 2 °C (as suggested by Arrhenius in 1896). In the longer term there would be more than a 50% chance that the temperature rise would exceed 5 °C."

Climate Change and Plant Phenophases

Phases in the life cycles of plants and animals are known as phenophases. Changes in these phenophase offer some of the clearest indications that the climate is changing and at a pace not seen previously in the Earth's history. It is the rapidity of change which is unique about the current changing climate and which has not been found in previous perturbations. In particular, changes in phenophase will have considerable impact on perennial and woody plants which have evolved with requirements for periods of dormancy, cold acclimation and vernalisation during their life cycles. Studies of climate change indicate that effects will vary with different species, so that there may be benefits for some while others may be damaged or extinguished. For example, studies worldwide demonstrate that some herbaceous perennial and woody plant phenophases are becoming earlier in spring and these changes closely correlate with rising temperatures. Initially these findings mainly came from studies in the Northern Hemisphere but now similar conclusions are emerging from studies of the flora in the Southern Hemisphere including Africa (Grab and Craparo 2011). Increasing temperatures affect flowering, pollination and harvesting dates. In arctic and alpine habitats some plants have been found blooming earlier and then flowering for longer periods, thereby increasing their ecological fitness (Gimenez-Benavides et al. 2011). By contrast, later-flowering species appeared to lose fitness under climate warming especially in extremely warm and dry years. Research by Haggerty and Galloway (2011) examined effects of changing climates on populations of the American native herbaceous plant *Campanulastrum americanum* (small American bell-flower). Plants were taken from either high or low altitudes and grown in environments that were at variance to their natural habitat. These demonstrated evolutionary malleability and an ability for modification of reproductive phenology in response to changed environments by this herbaceous perennial. Flexibility of this type is unusual.

Effects on Perennial Fruit

In many woody species there is a need for winter chilling. This characteristic defines the suitability of a geographical location for the production of many tree crops. Observed historic and projected future changes in winter chilling in California, USA for example were quantified with two chilling models, chilling hours and a dynamic model (Luedeling et al. 2009). Both models consistently predicted that climatic conditions by the middle to end of the twenty-first century will mean that California will no longer be capable of supporting some of the main temperate tree fruit species currently cropped there. The chilling hours model predicted bigger changes than the dynamic model. Similarly, the effects of reduced chilling conditions are demonstrated by Jay and Lichou (2010) with apricot (*Prunus armeniaca*) cultivars. Trees subjected to very cold winters produced abundant and high-quality flowers. By comparison, those subjected to very mild winter conditions developed a large numbers of malformed flowers. Big risks are apparent for those crop plants which require periods of both cold chilling and vernalisation before flowering and fruiting can take place. Transitions between the chilling and flower development phases and quantitative differences in the effectiveness of the winter chilling during dormancy will become apparent as climate change becomes greater. Temperate fruit and nut trees require adequate winter chilling before they can produce economically viable yields (Luedeling et al. 2001a, b). Global warming has the potential for reducing winter chilling and thereby greatly diminishing crop yields. Warmer regions of the world where growing temperate fruit is already difficult are most likely to experience severe reductions in crop yield and quality. A lack of winter chilling potentially threatens much of the temperate fruit production in these areas. By contrast, in some temperate regions, winter chilling is likely to remain relatively unchanged, and currently colder regions may even see an increase and hence fruit crop production will remain a viable enterprise or possibly expand. Mitigation allowing continued production in warm temperate regions and cool tropical regions will require the breeding of new tree cultivars which have lower chilling requirements or the development of husbandry routines for coping with insufficient winter chill. Each of these routes for mitigation demands much improved scientific understanding of the responses of tree crops to temperature.

Detailed evidence of climatic change affecting tree crops around the world is accumulating from the examination of medium- and long-term records. The reduced cropping resulting from warmer winters and advances caused by warm springs are apparent from a 20-year-long study of phenology in walnut (Luedeling and Gassner 2012) grown in California, USA. Here, the patterns of emerging first female and male blossoms and leaf development altered substantially. Similar results were obtained in France (Atauri et al. 2010). Marked increases in air temperature have been noted since the end of the 1980s in French fruit and vine growing regions. Here a long-term series of phenological data have been extracted from a national data base devoted to fruit trees and vine (PhenoClimReg.). Trends for advancing flowering dates in vines were apparent. In apple and pear there was an abrupt change in the mean dates of flowering stages, evident since the end of the 1980s. Chilling requirements took longer to be fulfilled in apple because of a lack of low winter temperatures. Heat requirements were more rapidly satisfied because of increasing temperatures from January to April. It is estimated that in warm temperate zones, growing stone- and pome-fruit cultivars requiring post-dormancy chilling will decline, with a parallel increase areas of cropping devoted to subtropical plants (Webb and Whetton 2010). This would have important implications in the very warmest fruit zones, e.g., the Southern Region of Brazil (Wrege et al. 2010) where industries serving worldwide supermarket retailers have only recently been established. Temperature increases by 1 °C lead to substantial decreases of accumulated chilling hours. The effect is more pronounced with an increase of 3 °C and should temperature increase 5.8 °C, opportunities for temperate fruit accumulating the required level of chilling hours will virtually disappear in parts of South America.

Examples of Regional Phenophase Effects

The effects of climate change vary greatly between different regions and there will be beneficiaries. For example, the main limitation to fruit production in the Okanagan Valley in Canada is currently winter injury (Quamme et al. 2010). Based on long-term records between 1916 and 2006 it is estimated that there will be a decrease in the frequency of injury and an increase in the minimum temperature of Arctic outflows. These changes appear to be associated with the warming trends of the region during winter and early spring. Also detected is a slight increase in frequency of Arctic outflows during late autumn and, as a result, production of pears, peaches and apricots might expand.

Meteorological data for the period 1958–2010 has been accumulated by the University of Bonn, Klein-Altendorf research centre (Blanke and Kunz 2011; Kunz and Blanke 2011). In this period it is apparent that there have been two distinct climatic phases:—1958 to 1987 with temperatures −0.4 °C below the long-term 50-year average of 9.4 °C followed by a ca. 20 year period of a +0.6 °C temperature rise (1988 to date). Comparisons of phenological data for phase II (1988 to date) with phase I (1958–1987) showed there was 4 days earlier full blooming, but only 2 days earlier harvest, without change in the date of leaf fall. These changes provided 2 days longer for fruit development in the cv 'Cox's Orange Pippin' and 5 days longer for cv 'Golden Delicious'. This resulted from temperatures exceeding the optimum for photosynthesis of pome fruit and leaves. Extending the period of leaf growth offers possibilities for growing more later-ripening cultivars.

There was no change in the amount of annual precipitation of 594 mm over the 50 years period but its distribution changed slightly with ca. 50 mm less precipitation in summer, and that could lead possible drought stress, especially in July, which has the effect of reducing fruit size and diminishing quality. Climate change will affect patterns of rainfall and evaporation reducing soil moisture content and run-off. Reduced winter and spring rainfall may indirectly cause increased frost risk. Physiological stress syndromes in temperate fruit, such as sunburn and reduced fruit colour development, are likely to increase in temperate regions.

Phenophases in Vegetable Crops

Less attention has been given to effects of climatic change on the life cycle of vegetable crops. It is evident however, that in crops which also require several seasonal cycles in which a vegetative phase leads on to a flowering and seeding reproductive phase and this change triggered by the accumulation of chilling units climate change will have implications resembling those identified for fruit crops. Commercially in field vegetable crops problems such as early bolting (premature flowering) could return. This will adversely affect the regional cropping of alliums, brassicas, beet and carrots for example.

Increasing night-time temperatures will raise the rate of respiration, altering the distribution of assimilates into reproductive sinks. As a result, fruit size and quality will be reduced making it more difficult to produce desirable high quality grades in crops such as cucurbits and tomatoes. That could result in re-evaluation of existing cultivars and breeding of new ones more closely suited to changing environments and market demands.

Climate Change and Benign Microbes

The soil food web (invertebrates and microbes) sequesters carbon, cycles nutrients and maintains ecosystem resilience and sustainability. Additionally, bacteria, fungi, nematodes, protozoa, and possibly other microorganisms, help suppress infection and colonisation of plant roots by pathogenic organisms (Chakraborty et al. 2012). Mutualism by mycorrhizae and endophytes confers benefits such as nutrient acquisition and stress tolerance to their plant hosts and, for example, mycorrhizae can help plants become more drought tolerant (Compant et al. 2010). Increasingly it is realised that understanding the ecology of soil microbes and their beneficial effects on crop growth will yield much improved means of horticultural husbandry. Manipulation of beneficial microbes can enhance soil fertility, help in the control of pests and pathogens and aid in the conservation of soil nutrients and water. Consequently knowledge of the impact of climatic change on beneficial microbes is an important means for the development of strategies for the mitigation of the consequences of climatic change.

Climate change will influence the relative importance, frequency and composition of the population groupings which constitute the soil food web, their trophic interactions and the processes controlling them. These changes will in turn have an impact on plants growing in the soil. As a result, there will be qualitative and quantitative alterations in plant physiology, tissue composition and signalling pathways associated with rising concentrations of carbon dioxide (CO_2) and increasing abundance and altered quality of plant litter. In turn, that will influence the processes of microbial decomposition and the release of nutrients (Chakraborty et al. 2012). Some studies of carbon dioxide-enrichment show that this leads to a stimulation of rhizosphere bacteria including nitrogen-fixing rhizobia and increased size and number of root nodules (Compant et al. 2010; Pritchard 2011) and increased bacterial biomass (Drigo et al. 2008). By contrast, other studies show no general trend (Zak et al. 2000). Much more information is required in order to resolve such apparently divergent sets of results.

It is known that elevated carbon dioxide concentrations can boost population size, colonisation, and/or diversity of mycorrhizal fungi (Compant et al. 2010), although the benefit to mycorrhizal associations from carbon dioxide enrichment will also be influenced by phosphorous and nitrogen availability in soil. Chakraborty et al. (2012) reviewed the impact of climate change on multitrophic interactions in soil and summarised a number of effects. Soil microbial activity and mineralization are increased in warmer soils, before acclimation reduces temperature sensitivity. This effect can be modified by changes in litter quality and physiology of soil organisms at high temperatures. Fungal populations and nematodes grazing on fungi and bacteria generally increase as a result of warming, while nematode diversity is reduced. Increasing soil temperature generally decreases soil microbial biomass in controlled experiments but in long-term field studies there is either no effect or a positive effect. Warming generally favours endophytic colonisation that protects the host plant from abiotic stress. Decreased precipitation and repeated wetting of soils reduce both bacterial and fungal biomass but fungal-dominated food webs respond more readily to changing precipitation. Altered precipitation regulates the population of microbial grazers and bacterial-feeding nematodes are more sensitive to drought than fungal feeders. Drier conditions can lead to changes at higher trophic levels, due to resource limitations starting at the base of the food chain, as soil organisms appear to be more sensitive to dry conditions than to increased precipitation. Increased precipitation generally benefits detritivores and herbivores at the second trophic level. Drought generally reduces mycorrhizal colonisation but the response varies with differing species. Pest insects are susceptible to infection by a range of bio-controlling organisms (bacteria, viruses, fungi, nematodes) which exist in the environment. Some of these organisms (e.g. the fungus, *Metarhizium anisopliae*) inhabit the rhizosphere. The ecology of these organisms has not been studied extensively (Bruck 2010), assessing the impact of climate change requires more research.

Climate Change and Pathogenic Microbes

Currently pre- and post-harvest damage caused by pests and pathogens destroys between 20 to 30 % of food before it reaches the consumer. Losses caused by pest and pathogens prior to harvest tend to cause greater devastation in under-developed and developing countries compared with developed nations. While it losses after harvest could be of greater significance in developed countries. This generalisation neglects consideration of the absence of processing and distribution infrastructure in developing countries and the effects of these deficiencies on the scale of losses. Globally, climate change may exacerbate the threat to food security posed by crop pathogens (Stukenbrock et al. 2006; Dixon 2012). In particular, weather conditions directly affect plant pathogens and the incidence of diseases. Pathogen epidemiology is influenced by prevailing weather conditions. This affects the spatial and temporal dispersal of propagules and their synchrony with sensitive crop growth stages and the frequency of suitable infection periods. In particular many fungal and bacterial plant pathogens require periods of wetness or high humidity for successful infection and colonisation. The speed of symptom development is affected by temperature and humidity through effects on pathogen growth and, for polycyclic types, the number of disease cycles. Pathogen survival is influenced by periods of low and freezing temperatures. Crop husbandry factors such as nutrition, irrigation and the use of rotations affects the potential for disease epidemic development following importation of propagules either from elsewhere or those which are endemic in the locality (Dixon 2012).

General Assessment

Studies are providing increasingly realistic suggestions for changes in the incidence of plant diseases affected by the magnitude and variability of temperature, precipitation and other climatic variables (Jeger and Pautasso 2008). Their recent examples include models predicting an increase following projected climate change in (i) severity of *Plasmopara viticola* epidemics on grapes near Turin in 2030, 2050 and 2080, (ii) the range and severity of epidemics of *Leptosphaeria maculans* on oilseed rape (*Brassica napus*) in the UK for the 2020s and 2050s, and (iii) the distribution and local impact of a range of forest pathogens (*Biscogniauxia mediterranea*, *Cryphonectria parasitica*, *Melampsora* spp., *Phytophthora cinnamomi* and *Sphaeropsis sapinea*) in France at the end of the twenty-first century. Short-term, local experiments have demonstrated the impacts of predicted global change on plant health. Recent examples include: (i) a study showing that elevated atmospheric carbon dioxide concentration increases the risk of infection with rice blast (*Magnaporthe oryzae*) and the percentage of rice affected by sheath blight; (ii) an experiment demonstrating species-specific responses to increased ozone concentrations of the susceptibility of young beech (*Fagus sylvatica*) and spruce (*Picea abies*) trees to *Phytophthora citricola*; and (iii) a 12-year warming experiment with heaters suspended over plots in a mountain meadow in Colorado, USA, in which there was a change in the prevalence of several different species of plant pathogens. A study by Shaw et al. (2008) provides an example of how the limitations of models and experiments can be overcome by making use of a long-term (1844–2003) UK data set on the occurrence of two key worldwide pathogens of wheat.

Since the action of some resistance genes is regulated by temperature, the efficacy of host resistance may also be affected by changing climatic conditions. It is suggested that increases in atmospheric carbon dioxide levels may also modify pathogen aggressiveness and/or host susceptibility and increase the fecundity and growth of some fungal pathogens (Luck et al. 2011).

Studies on the impact of climate change on plant pathogens have mostly concerned the impact of fungal, bacterial and virus pathogens of broad acre crops. The principles elucidated and effects identified are likely to apply with equal force to those affecting intensively grown horticultural plants. Microbes are generally reliant upon free water or high humidity in order to disperse, colonise, germinate, invade and propagate in both their sexual and asexual phases of the life cycles. Temperature, aerial humidity and soil moisture are key physical factors providing conditions conducive to these activities and, for that reason, form integral elements in most disease prediction and forecasting models (Dixon 1981). The impact of water on disease spread is graphically illustrated by Large (1950) who recounts the appearance of potato murrain (late blight, *Phytophthora infestans*) in Europe during the summer of 1845. Potato blight devastated crops across the Continent, but with gravest social consequences in Ireland where famines between 1845 and 1860 killed over 1 million people and lead to the emigration of a further 1.5 million because of the reliance on potato as a food source, particularly by poorer sections of the community.

Potato Blight as an Example

Today the potato is the world's fourth largest crop after maize, rice and wheat with an annual production of more than 325 million t (Anon FAO 2008). The crop provides an ideal food source in developing countries, where land is limited but labour is freely available and in advanced economies, where the crop is readily automated. Most notably there is rapidly increasing consumption in Asia where potato is forming an alternative to the traditional rice-based diet. The Indian sub-continent is now the world's third largest potato producer. Late blight (*P. infestans*) (Fig. 25.1) has become the most destructive disease of Indian potato crops with severe outbreaks destroying crops within 2–3 weeks. The disease occurs annually in the temperate hill States causing 40–85% yield losses. Elsewhere, in the subtropical Indo-Gangetic plains (80% acreage), late blight reaches epidemic proportions every 2 to 3 years causing 15–75% losses. In north-western Indian plains there is increasing occurrence of late blight epiphytotics increasing from severe outbreaks every 4 to 5 years to severe outbreaks now happening almost every alternate year.

Fig. 25.1 Severe infection by *Phytophthora infestans* causing potato blight on a potato crop (Ueli Merz)

The disease now appears earlier than was previously the case in the northern parts of India, in November. As climate is changing, temperatures have increased to a range of 14.0–27.5 °C from the 10–25 °C recorded previously. Genetic changes in the fungal population suggest that there is an appearance of heat tolerant A2 mating physiological races *P. infestans*. This increasing important economic problem has lead to studies being made of late blight in India for the last 15 years. Figures below show the seasonal severity of late blight in West Bengal from 1991–2011. These data suggest an that there is an increasing frequency of late blight epidemics in West Bengal in the last 5 years (Luck et al. 2010).

Studies of the impact of climate on late blight of potato are also reported from North America (Bourgeois et al. 2004; Baker et al. 2005). Currently, in the upper Great Lakes region of the United States, late blight of potato is a sporadic disease, occurring only when microclimatic conditions within the canopy are favourable and inoculum is present. Historical climatological trends in the upper Great Lakes region demonstrated increased occurrence of warmer and wetter growing season conditions, as well as local rises in total precipitation and in the frequency of days with precipitation. Consequently, the risks posed by potato late blight are becoming greater. These increasing disease risks are associated with upward trends in dry bulb and dew point temperature, especially during July and August. Increased risk of potato late blight has implications for extension agents and commercial horticulturists that demand increased emphasis on grower education and application of integrated disease management techniques. Climate change with warmer and more variable weather, may be similarly affecting tomato diseases in Egypt (Fahim et al. 2011). Increased night-time and winter temperatures, may be contributing to their greater prevalence. For example, tomato late blight (*P. infestans*) is the most destructive tomato disease in Egypt, causing fruit yield losses and has expanded in impact since the early 1990s. This has resulted in at least 2–3 additional agrochemical sprays being used on each crop.

Diseases of Brassicas

Oil seed crops in the family Brassicaceae are of major economic and social importance worldwide with huge acreages in grown Canada and China in particular. Studies of the impact of climatic change on disease patterns in oil seed rape (*Brassica napus*) are relevant to horticultural brassica crops. Diseases move from oil seeds into horticultural crops, since they provide green bridges and reservoirs for pathogen multiplication (Dixon 1975). Siebold and Tiedemann (2012) analysed the potential effects of higher mean temperatures on the life cycles of pathogens of oil seed rape in Northern Germany. Their study suggested that warming might lead to shifts in the future prevalence of these pathogens and that *Verticillium longisporum*, *Sclerotinia sclerotiorum* and *Alternaria brassicae* could be particularly favoured whilst *Pyrenopeziza brassicae* might become less significant. Their analysis focussed on temperature because it is more predictable than other climatic factors, such as precipitation, relative humidity, soil moisture or extreme weather events, but they did speculate on the effects of moisture. For example, since only short periods of leaf wetness are required for sporulation and infection of some foliar pathogens, they suggested that these processes are unlikely to be compromised by future precipitation patterns projected for the spring in northern Europe.

Traditionally, autumn-sown rape dominates the British oil seed crop with a very limited area of spring sown crops. Spring-sown rape crops grown in mainland Europe are very susceptible to devastation from the soil borne pathogen *Plasmodiophora brassicae,* causing clubroot disease (Fig. 25.2). The reason why British autumn sown crops escaped clubroot disease development was probably explained by the late Dr Cynthia Williamson (pers. comm) who showed that the soils in which the crop germinated and formed rosette growth stages were cooling and that lead to a cessation of activity by *P. brassicae*. Between February to early March when temperatures rose sufficiently for the microbe to resume activity, by then the host plants were already growing and had formed their components of yield. This resulted in production of profitable crops despite them being grown on infested land. This situation has now changed. Drilling of autumn-sown oil seed rape has now advanced such that it is really a late-summer (late July to early August) process. As a result the crop is sown into land where the soil temperatures are still high enough to allow *P. brassicae* activity. The British autumn season has extended at least to the end of November with the retention of higher soil temperatures accompanied by rain. This fulfils two predictions for climate change in Britain; that there will be longer both warmer and wetter autumns. As a result clubroot has now become a major and dominating pathogen of autumn-sown oil seed rape. Severe regular disease outbreaks are now recorded as far north as Aberdeenshire and Morayshire in North Eastern Scotland and here 75 % of land is infested with clubroot as a result of past generations of swede (*B. napus*) production for use as overwintering *in situ* as sheep forage. As result of changing agronomic and environmental conditions favourable to *P. brassicae,* substantial crop losses are happening and with consequential reductions in yields and quality in home produced rape oil (Dixon 2006).

Fig. 25.2 Clubroot galling symptoms on the root of Chinese cabbage, *Brassica rapa* var. *chinensis* caused by *Plasmodiophora brassicae*. (Geoff Dixon and Yara)

Tree Pathogens

White pine blister rust in North America, Europe and eastern Asia *Cronartium ribicola* in the white pines (five-needle pines in subgenus Strobus) cultivated or wild *Ribes* might serve as inoculum sources Hunt et al. (2010). But silviculture and horticulture can reduce the risk of serious impacts from blister rust using genetics for breeding and epidemiology for hazard assessment and disease control. Climate change threatens to cause major alterations in temperature and precipitation, resulting in maladapted conifers succumbing to various diseases and insect outbreaks. By contrast, many white pine species have broader ecological ranges and are tolerant of harsh environments these are traits that permit successful establishment and growth over wider geographic and altitudinal zones. Given appropriate management, white pines could thrive as valuable commercial crops and ecologically important keystone species.

Rhizosphere Effects

Climate change is likely to alter the distribution and severity of soilborne diseases as noted with *Plasmodiophora brassicae* causing clubroot in brassicas, and this will affect both intensive and low-input production systems (Mazzola 2010). Naturally occurring disease suppressive soils are well known. The introduction of alien biological agents into non-native soil ecosystems however, typically fails to yield commercially viable or consistent levels of artificially induced disease control. Greater emphasis is now placed on manipulation of cropping systems which manage resident beneficial rhizosphere microorganisms as a means for suppressing soilborne plant pathogens. One such strategy is the cropping of specific plant species or genotypes, or the application of soil amendments with the goal of selectively enhancing disease suppressive microbial communities (Dixon 2012).

Disease Prediction

Disease forecasting models can also be used to predict the impact of climate change on fungal pathogens. For example, an apple-scab forecasting model (Bourgeois et al. 2004) estimates an infection index for the ejection of ascospores calculated on an hourly basis from air temperature, relative-humidity and precipitation data. Using the model to investigate future climate scenarios for Quebec, Canada indicated that ascospore ejection would begin earlier, and that there would be a substantial increase in the number of infection periods between April and June. The model did not take the winter conditions into account however, although these may also be affected by climate change. Periods for high risk of European apple canker (*Neonectria galligena*) development in the United States, Chile, England, and Northern Ireland were determined from published data by Beresford and Kim Kwang Soo (2011). Greatest agreement occurred when it both rained on >30% of days per month and there was an average of >8 h/day with temperature of 11 to 16 °C. The highest risks of disease development were where the thresholds were exceeded more often and by greater amounts. Cropping areas at higher latitudes (>52 degrees) with frequent summer rainfall appeared to be most prone to European canker, including the fruit-rotting phase of the disease, probably because summer temperatures were more favourable compared with those at lower latitudes. There are a number of other models for predicting infection by the fungal pathogens of horticultural crops which could be used to investigate the impacts of climate change (e.g. Van Laer et al. 2005; Li et al. 2007; Wakeham and Kennedy 2010; Eastburn et al. 2010; Gilles et al. 2004; Sutton et al. 1986).

Direct, multiple effects on the epidemiology of plant pathogens including the survival of primary inoculum, the rate of disease progress during a growing season, and the duration of epidemics (Boland et al. 2004). Changes in the spectra of diseases are also anticipated. Similarly, abiotic stress conditions associated with environmental extremes are expected to increase as a result of climatic change. Interactions between biotic pathogens and abiotic conditions might represent an important effect of climate change on plant diseases and crop losses. The management of plant diseases will need to respond and plant breeding programmes will have important roles in the mitigation of the damaging effects of climatic change. There will be opportunities for the introduction of new crops and cultivars, but effective diagnostics must detect new pathogens and prevent them from entering previously unaffected regions with these new crops. Because of the long-lived nature of trees, forests are slow to adapt, and understanding the impact of climate change will have to become an integral part of forestry, woodland and parkland management plans. For example, adaptations in agriculture and forestry have been occurring in Ontario, Canada for over 100 years as commercial and social needs changed. But now these may need to occur at an accelerated rate because of more rapid changes in climate. In England and Wales the UKCIP98 Medium High scenario suggests a warmer climate (1.7–2.0 °C) and a higher annual rainfall (101–105%) producing wetter autumns and winters, but drier summers leading to a longer growing season and fewer frost days (Harris and Hossell 2002). These changes could mean changes in pest and pathogen patterns throughout the year and alterations in the profiles of species involved. Changes in opportunities for controlling pests and pathogens may themselves differ from what is currently feasible. The ECOMAC II model of available work days (AWD) indicated that in spring these may either remain at current levels or decline, whilst in early autumn, soil moisture deficits (SMD) will initially be higher, but the fields may become wetter at faster rates than at present.

The potential consequences of temperature changes on infectious plant, wildlife and marine diseases are reviewed by Harvell et al. (2002). Climate warming will affect host-pathogen interactions by (i) increasing pathogen development rates, transmission, and number of generations per year; (ii) relaxing overwintering restrictions on pathogen life cycles; and (iii) modifying host susceptibility to infection. For example, some plant diseases are more severe after mild winters or during warmer temperatures, which suggests that the direction of climate warming alters plant disease severity. Australian laboratory and field studies indicate that the fungus causing Mediterranean oak decline, *Phytophthora cinnamomi*, causes more severe root rot at higher temperatures than the current Mediterranean average. Similarly in a 14-year field study in England, the Dutch elm disease fungus, *Ophiostoma novo-ulmi*, caused greater defoliation in warmer years.

Bacterial Diseases

Bacteria are important pathogens of horticultural crops. *Xanthomonas campestris* pv. *campestris* is a Gram-negative bacterium which causes black rot, an important disease of vegetable brassica crops worldwide. This is primarily a seed-borne pathogen but the bacteria can disperse over short distances via wind, aerosols, irrigation water and rain. In general, *X. campestris* pv. *campestris* thrives as a severe disease causing agent in warm, humid climates and, consequently, is most serious in tropical, subtropical and humid continental regions. In view of the global distribution of *X. campestris* pv. *campestris*, black rot will become an increasingly important disease constraint favoured by climate change in more northern latitudes of vegetable production (Vicente and Holub 2013). This disease is now found commonly in English brassica crops whereas 20 years previously it was an infrequent curiosity. Hourly weather series were used for modelling the impact of climate change in Switzerland when applied to the number of predicted infection days of fire blight (*Erwinia amylovora*), a serious current and long-term threat to commercial production of apple and pear across Europe (Hirschi et al. 2012). The projected changes in infection days were affected by uncertainties in the leaf wetness approximation and the simulation of the blooming period. Two compensating effects were suggested however, warmer temperatures favouring infections were balanced by a temperature-induced advancement of the blooming period, leading to no significant change in the number of infection days.

Virus Diseases

Many important virus pathogens causing diseases of horticultural crops are transmitted by aphids. Generally, longer periods of migration and feeding activity of these insect vectors, caused by warmer conditions and longer growing seasons, will favour many virus diseases on a wider range of crops. An increased incidence of aphid-vectored viruses is predicted to occur as an effect of climatic change. This will be due to either increased winter survival of aphids or their earlier spring migration (Harrington and Stork 1995). Warmer soils will affect soil-borne viruses because vectors such as nematodes will be able to infect crops at earlier growth stages and the viruses will have greater impact on development and yield. Symptoms and yield-loss may also be exacerbated by heat and drought (West et al. 2012).

Alongside plant pathogens, microbes pathogenic to humans such as *Salmonella* are also cause for concern. Potentially this pathogen is particularly important when present on produce that is consumed raw. There are relatively few studies on the impact of climate change on this issue but in one study, the effects of extreme weather events such as drought and heavy rain on the global spread of *Salmonella typhimurium* were simulated using iceberg lettuce and green onions. These trials failed to produce clear cut results (Ge et al. 2012).

Climatic Change and Benign Invertebrates

Insects, particularly bees, are the main pollinators of many horticultural crops, including a large proportion of tree, cane and bush fruit crops. Both domesticated and wild pollinator populations might be affected by climate change, with unknown consequences for pollination. Most evidence for the effect of climate change on pollinators concerns butterflies. Climate change has already affected the distribution of butterflies (Hickling et al. 2006), (Fig. 25.3) and future changes are likely to have even more severe impacts (Settele 2008). Patterns for bees are consistent with the studies on butterflies; Williams et al. (2007) found a relationship between climatic niche and declines in British bumblebees, whereas Dormann et al. (2008) projected future general declines in bee species richness in Europe. Impacts of climate change include:- changes in the timing of activity, changes in phenology, declines due to narrower climatic niches or local or regional extinction of species, changing composition and functioning of pollinator communities (Potts et al. 2010). In particular, climate change-induced mismatches in temporal and spatial co-occurrence, and morphological and physiological interdependencies of differently responding plants and their pollinators can potentially disrupt these interaction between pollinators and host plants (Memmott et al. 2007).

Populations of phytophagous invertebrate pests are regulated naturally by a diversity of predators and parasitoids, although the level of control achieved is often deemed to be inadequate for commercial crops. Some of these species (such as coccinellids) are generalists and predate on a wide host range, whilst others are

Fig. 25.3 Peacock butterfly, *Inachis io*. (Geoff Dixon)

relatively specialised. As with pest invertebrates, the weather will have profound effects on survival, development, reproduction and dispersal. The impact of climate change is generally less-well researched however, than for pest invertebrates.

How environmental change affects species abundance will also depend on the food web within which particular the species interact. For example, Harmon et al. (2009) investigated the responses of pea aphid (*Acyrthosiphon pisum*) to increased frequency of episodic heat shocks, which affect the growth rates of pea aphid population by reducing fecundity or even sterilizing females, in the presence of one of two species on coccinellid predator. One species of predator ameliorated the decrease in aphid population growth with increasing heat shocks, because at low aphid densities aphid predation diminished, whereas a second predator remained equally effective at low prey densities. These results show that the impact of heat-shock disturbance on pea aphids depends not just on the presence of predators but also on their identity.

Climatic Change and Pest Invertebrates

The weather has profound impacts on populations of pest invertebrates, affecting their survival, development, reproduction and dispersal. Changes in environmental conditions will have direct effects on the pests themselves and also on the biotic factors that regulate their populations (such as pathogens and arthropod predators and parasitoids). As ectotherms, pest invertebrates have temperature-dependent rates of development for much of their life-cycle and have particular threshold temperatures above which development can occur and below which it ceases. These threshold temperatures are key factors in defining species distribution (Glauber et al. 2013). Such thresholds may not be static however, and may alter in response to acclimation (Alford et al. 2012) or to the presence of certain secondary (facultative) bacterial symbionts, as in the case of the pea aphid (*Acyrthosiphon pisum*) (Harmon et al. 2009).

In temperate climates pest species may be able to begin development earlier in the year and complete more generations per year as temperatures increase, provided these are below lethal temperatures. For example, emergence of adult cabbage root fly (*Delia radicum*) will start earlier in the spring and, if temperatures increase sufficiently, there may be at least one additional generation per year (Collier et al. 1991). Similarly, in northern Europe, spring migrations of pest aphids such as *Myzus persicae* have become increasingly early as temperatures have risen over recent decades (Hullé et al. 2010). Many species are able to enter periods of dormancy in order to survive adverse weather conditions during winter (diapause) and summer (aestivation), which means that they can survive a wider range of climatic conditions. In some species, diapause is obligate and in others facultative; species with an obligate diapause may not be able to increase the number of generations per year in response to increased spring and summer temperatures. Higher winter temperatures may increase survival of species that do not overwinter in diapause, such as anholocyclic *Myzus persicae* (Hullé et al. 2010).

Invertebrate pest populations can also be affected by the amount of precipitation. This may be through physical effects, leading to drowning or because raindrops dislodge the pest from its host plant (Mann et al. 1995) or because high humidity favours the development of entomopathogens (Lacey et al. 2001). Water intolerant pests include the cutworm (*Agrotis segetum*) and the amount of rainfall forms the basis of a forecasting model used in the UK (Bowden et al. 1983) for the behaviour of this pest. Conversely, slugs are susceptible to desiccation and the development of slug populations is favoured by relatively wet conditions (Willis et al. 2008). In addition to rainfall, other extreme weather events also affect survival for example, intense rainstorms or wind or high temperatures are related to the mid-season population crash in several species of pest aphid (Karley et al. 2004).

As an example of how climate affects insect behaviour is the work of Johnson et al. (2010) who studied how protected environments affected adult vine weevil (*Otiorhynchus sulcatus*) feeding and reproducing on red raspberry (*Rubus idaeus*). They focused on the period between adult emergence and the onset of oviposition (i.e. the pre-reproductive period), which represents the optimal period for control. The temperatures inside plastic covered tunnels were up to 4 °C warmer than in open-field plantations in 2008. Covered plants grew significantly faster (50 % increase in height and 16 % increase in leaf area) compared with those in open-field conditions. The carbon/nitrogen ratio in leaves was also higher in tunnels (12.07) than the field (10.89) as a result of a significant decrease in nitrogen concentrations (3.40 and 3.90 mg g^{-1}, respectively). Over 4 weeks, weevils consumed significantly more foliage in tunnels (370.89 mg) compared with weevils in the field (166.68 mg), suggesting compensatory feeding to counteract lower leaf nitrogen concentrations. Weevils in tunnels achieved sexual maturity 8 days earlier than those in the field and produced 20-fold more eggs by the time they were 5 weeks old. Applying a degree-day model showed good agreement between predicted and observed pre-reproductive periods for weevils in tunnels (36 and 30 days, respectively) and in field plots (41 and 38 days, respectively).

Changing climatic conditions and their span of variability could in future significantly affect the fitness of ectotherm species (Terblanche et al. 2010). Thus, there is an increased interest in understanding the effects of changes in means and their variances of temperature for example on traits such as climatic stress resistance. Current results suggest that large temperature fluctuations might reduce insect fitness. Increased mean temperatures in conjunction with its increased variability may therefore have stronger negative effects on an agricultural pest than just elevated temperatures alone.

Climatic Change and Benign Vertebrates

For vertebrates in general, the response to climate change will depend on their sensitivity to temperature change and by the extent of their dependence on a specific environment, particularly the dominant vegetation. Benign vertebrates include amphibia and reptiles and also certain species of bird. Amphibians and reptiles are ectotherms and environmental temperature exerts a great influence on their life and physiology (Pounds et al. 2006). Temperature changes can affect respiration, physiology, dispersal capacity and distributional ranges as well as physiological performance and fitness, which are highly temperature sensitive. Amphibians are the terrestrial vertebrates most at risk of extinction as a result of climate change (Alford 2011). Studies have indicated that climate change alone may not be the major cause of declines in amphibians and the pathogenic fungus *Batrachochytrium dendrobatidis* causes chytridiomycosis, a disease that infects amphibians across a wide range of taxa is an additionally very damaging factor. Climate change may however already be affecting the course of the chytridiomycosis pandemic itself, and its influence is likely to continue during this century, because vulnerability to the disease is strongly affected by weather and climate (Alford 2011).

Weather not only affects the metabolic rate of birds (e.g. cold weather requiring increased energy expenditure for body maintenance) but also exerts other indirect and direct effects on bird behaviour. For example, it can influence foraging conditions and the ability to carry out other essential behaviours, such as courtship. Weather also affects breeding success through, for example, chilling or starvation of young. Extreme weather events, such as prolonged low temperatures and droughts, can have catastrophic effects on bird populations, including long-term effects on whole cohorts (Stenseth et al. 2002). The implications for birds of climate change have only recently begun to be addressed. These effects include changes such as: earlier breeding; timing of migration; changes in breeding performance (egg size or nesting success for example); population sizes; population distributions; and in selection differentials between components of a population. There is growing evidence that some species may find it difficult to adapt to climate change because, for example, of the use of inappropriate environmental cues as phenological triggers, or because different parts of a food chain may respond differentially to climate change.

Climatic Change and Pest Vertebrates

Vertebrate pests of horticultural crops are primarily endothermic species (mammals and birds) and the key destructive species will vary depending on the geographical region. The European rabbit (*Oryctolagus cuniculus*) is a highly successful colonizing species across its worldwide introduced range, causing large-scale economic losses to field-grown crops, although, perhaps surprisingly, it is threatened within its native range (the Iberian Peninsula) (Fordham et al. 2012). Modelling studies by Fordham et al. (2012) indicated that rabbit survival is strongly influenced by disease outbreaks. Factors such as rainfall, pasture biomass and inter-specific competition for a food-limited resource are important additional regulatory forces, potentially having a preconditioning effect on disease severity. Further analysis is required to quantify potential synergies between climate, disease and biological factors. The woodpigeon (*Columba palumbus*) is another major pest of horticultural crops, particularly brassicas, in Europe. Its population is primarily determined by the level of the food supply in winter (Murton et al. 1964). Since the introduction of oil seed rape however, the number of fledged young has had a more significant effect on population size than mortality due to starvation (Inglis et al. 1997).

Climatic Change and Weeds

Weeds are non-crop plants which compete for resources with the crop and are a major constraint to horticultural production. Also included are ground keepers left over from previous crops such as potatoes or oil seed rape plants. Usually weeds are generalist plants capable of exploiting many environments and hence thrive as a result of utilising the resources such as water and nutrients provided for the crop. They have phenotypes which are probably capable of exploiting the effects of climate change such as rising temperature, increased rainfall and enhanced carbon dioxide concentration. In horticultural crops difficulties associated with controlling weeds are exacerbated because many of these crops are considered to be minor in terms of the land area which they occupy and there is therefore a limited incentive for agrochemical companies to develop and market selective herbicides because sales will be minimal. A number of studies indicate that weed competition and other problems associated with weeds could magnify as a result of increased variability in the climate (temperature and rainfall) and rising carbon dioxide concentrations (Ziska and Dukes 2011). In addition, weed genotypes are inherently more flexible (both within and between species) when compared with many crops with which they compete and may therefore be better fitted to withstand changes to the environment (Ziska and Dukes 2011). Weeds are also hosts of crop pests and pathogens providing green bridges between crops. As yet, it is difficult to quantify what the impact of climate change and increases in carbon dioxide will be on the role of weeds as reservoirs (Ziska and Dukes 2011)

Climatic Change and Invasive Species

The intentional introduction of non-native species, primarily for agricultural and horticultural uses, and the unintentional introductions as a result of globalisation have led to increasing numbers of invasive and alien species (Ziska et al. 2011). Some of these species have undoubtedly caused considerable damage in their new ranges and the arrival of new species is a continual threat. The development of populations of invasive species is currently facilitated by the absence of their natural enemies. They are able to thrive since competition is absent. Increasing globalisation, particularly international trade in plants and fresh produce, combined with climate change could be expected to extend the range and impact of invasive species with resultant adverse impact on commercial, environmental and social horticulture.

Predictions, Practice and Resource Use as Affected by Climate Change

Studies of the affects of climate change on horticultural husbandry and crop production indicate that a key issue will be the manner by which the potential for use of particular plants alters as a result of lost fitness for an environment. In parallel there will also be changes in husbandry practices influenced by the availability of weather windows for cultivations and harvesting operations. The basic principles of plant physiology which are likely to govern responses to climate change are broadly understood (for example see Morison and Morecroft 2006; Grace and Zhang 2006). While some elements in the changing environment may promote plant growth and reproduction others will be in short supply and cause physiological stresses. Here what matters is the magnitude of stresses which are likely to become increasingly more substantial. The growth of C_3 plants (temporal and boreal types) increases with rising carbon dioxide levels more than with C_4 plants (warm tropical types). Where the C_3 plants are in benign association with nitrogen-fixing microbes (legumes, for example) there appears to be added environmental benefit to be gained from some aspects of climate change. Benefits of additional carbon dioxide concentration are greater for annuals as compared with perennial plants. Leaf area increases as a result of rising photosynthesis with earlier and more complete light interception and resultant greater biomass. But maintenance costs increase with higher demands for energy and rising respiration. Leaf turnover rises partly due to shading effects and consequently photosynthesis per leaf falls. Stomatal opening is reduced with increased carbon dioxide. This is beneficial in limiting the impact of aerial pollutants like nitrogen oxides (NO_X), sulphur dioxide (SO_2) and ozone (O_3) but does inhibit water uptake. Stomatal conductance and transpiration rates drop as carbon dioxide concentrations rise. This effect is less marked when measured on a ground area (canopy evapo-transpiration) basis versus consumption measured against leaf–area. There is an increase in water-use efficiency in terms of dry matter formed relative to unit of water transpired. Consequently, leaf temperature increases raising the rate of plant development especially in early growth stages. Ultimately, however, reduced transpiration and resultant higher temperatures in the leaves leads to accelerated tissue senescence. Whether such an effect is beneficial or not depends on the extent to which temperatures rises and exceeds the optimum for efficient photosynthesis. Overall, the data suggest that elevated carbon dioxide may have positive benefits for C_3 plants including yield stimulation, improved resource–use efficiency, more successful competition with C_4 weeds, less damage from ozone toxicity and in some cases better pest and pathogen resistance. But such benefits may be counterbalanced by adverse effects of further rising temperatures. Although warming accelerates plant development it reduces processes such as grain filling, limits nutrient–use efficiency, increases water consumption and favours competition by C_4 weeds over C_3 crops plants. Changes in the water balance and amount of water available in the soil are crucial for efficient crop growth. In grasslands 90 % of the variance in primary production can be accounted for by the amount of annual precipitation received (Campbell et al. 1997). Calculations using the Penman–Monteith equation predict that potential evaporation increases by about 2–3 % for each 1 °C rise in temperature (Lockwood 1999). While biomass and yield increase with rising carbon dioxide concentrations dry matter allocation patterns to roots, shoots and leaves also change. Root to shoot ratios increase with elevated carbon dioxide favouring root and tuber crops.

There will be other atmospheric effects. The increasing concentrations of tropospheric ozone (O_3) a harmful air pollutant will damage plant growth and development (Leisner and Ainsworth 2012). Current ozone concentrations decrease forest productivity and crop yields and in future concentrations will increase if current emission rates continue. Data from 128 peer-reviewed articles published from 1968 to 2010 describing the effects of ozone on reproductive growth were studied. Current ambient ozone concentration significantly decreased seed number (−16 %), fruit number (−9 %) and fruit weight (−22 %) compared with charcoal-filtered air. In addition, pollen germination and tube growth were decreased by elevated ozone concentrations compared with charcoal-filtered air. Relative to ambient air, fumigation with ozone at between 70 and 100 ppb decreased yield by 27 % and individual seed weight by 18 %. Reproductive development of both C_3 and C_4 plants was sensitive to elevated ozone levels and changes in lifecycle, flowering class and reproductive growth habit did not significantly affect the response of plants to elevated ozone concentrations for many components of reproductive development. Elevated ozone concentrations decreased fruit weight and fruit number significantly in indeterminate plants, and had no effect on these parameters in determinate plants.

Combating the Destructive Effects of Wind

Predictions suggest that climate change will alter temperature, precipitation, air composition and wind frequency and speed. Changes in wind speed and velocity will affect many aspects of horticulture. Windbreaks are major components of successful cropping systems throughout the world Brandle et al. (2004). Windbreaks help control soil erosion and blowing snow, improve animal health and survival under winter conditions, reduce energy consumption of the production unit and enhance habitat diversity. They also provide refuges for predatory birds and insects. On the larger scale windbreaks offer habitats for wildlife and contribute positively to the carbon balance potentially easing some of the economic burdens associated with climate change.

Effective functioning of windbreaks requires the correct external structure in terms of width, height, shape and orientation and internal structure as in the correct volume and arrangement of branches, leaves and stems, these control the wind flow and amount of shelter which it can afford. Temperature is slightly raised and evaporation reduced near a windbreak. Some of the windiest locations are in the Southern Hemisphere as identified by Peri and Bllomberg (2002) in Patagonia. Here strong winds are a constraint to crop production and windbreaks are used to protect crops and animals. Porosity and distance from the windbreak have major effects on the degree of protection. Wind speed reduction is 85 % at one "H" (ie height of the windbreak) on the leeward side of the break. Benefits varied with crops, garlic for example showed no effect whereas the yield of tulip bulbs decreased by 25 % between 2H and 17H from the windbreak. Lucerne (alfalfa) yield was 40 % higher where protection was available compared with open field crops. Cherry and strawberry both suffered yield reductions caused by exposure to wind. It is likely therefore, that increased windiness will have considerable adverse effects on horticultural cropping and on the provision of environments designed and planted for rest and relaxation.

The use of trees in the landscape as mean of mitigating climate change is of major significance. In Australia tree loss is associated with land degradation, soil salinity, water and wind erosion, soil acidification, structural decline and nutrient degradation Bird et al. (1992). These authors estimate that 50 % of Victoria's crop and pasture land is at risk from degradation resulting from the loss of trees. In Western Australia 25 % of cleared agricultural land is badly wind eroded and increased salinity affecting 430,000 ha of land. At least 10 times that area (13 % of rangelands) are degraded by wind erosion caused by overgrazing and consequent loss of trees. While husbandry measures such as minimum tillage can help these should be accompanied by the establishment of windbreaks. The authors estimate that planting 10 % of land to windbreaks can achieve 50 % reduction in wind speed substantially improving the livestock and pasture production. Shelter improved wheat and oat yields at Rutherglen (Victoria) and lupin yield at Esperance (Western Australia) by 22, 47 and 30 % respectively. In semi-arid and dry temperate zones planting even only 5 % of land to windbreaks could reduce wind speed by 30–50 % and soil lost to erosion by 80 %. Windbreak trees eventually yield marketable timber and add value through previous carbon sequestration.

The Impact of Climate Change on Tree Species

Temperature and rainfall are considered to be the main factors determining the distribution of forest tree species. In Europe, climatic series show a global warming trend and an increase in frequency of summer droughts affecting growth and regeneration of tree populations (Silva et al. 2012). Beech (*Fagus sylvatica* L.) is particularly sensitive to drought and high temperatures. This species reaches the rear edge of its lowland European distribution in the south-west of France. At the moment, failure of beech establishment is not a significant problem at the regional scale in this rear edge of its distribution range. This study identified ecological factors related to beech recruitment by estimating seedling density on the forest floor respectively in 71 and 85 beech plots in the south-west lowlands of France. It also determined whether a relationship exists between seedling amount and mast-seeding. Local factors affecting the natural regeneration stages were established. The inherent capacity of the stand to produce fruits explained a greater part of variance in cupule density variance compared with the climatic factors. Meso-, micro- and pedo-climates were the main factors controlling seedling amount. Higher soil moisture, precipitation and temperature during the growing season increased seedling density, while late spring and early autumn frosts decreased it. Soil and stand conditions were also important. Fruit production increased in stands that showed tree crown degradation, while seedling amount decreased in this situation. The increase in the allocation of resources to reproduction could be a strategy of beech which currently copes with ecological constraints that tend to limit its establishment. Thus, seedling establishment is related to the factors controlling *F. sylvatica* presence at its southern distribution margins. This illustrates how natural regeneration is a key stage for beech success where the future of the species is jeopardized because of climate change. Similar effects are likely to be found with other major landscape tree species. These will suffer badly as a result of increasing inadequacy in their regeneration cycles. Here again the effects of such failures will be felt across all sectors of horticulture.

Using Trees as Means of Land Improvement

In mitigation schemes hundreds of thousands of rain-fed smallholder farmers in Zambia, Malawi, Niger, and Burkina Faso have been shifting their husbandry processes to farming systems which involve planting trees and this approach is restoring exhausted soils and increasing crop yields, household food security, and incomes (Garrity et al. 2010). (Evergreen Agriculture) is defined as the integration of particular tree species into annual food crop systems. The intercropped trees sustain a green cover on the land throughout the year which maintains vegetative soil cover, bolsters nutrient supply through nitrogen fixation and nutrient cycling, generates greater quantities of organic matter in soil surface residues, improves soil structure and water infiltration, increases greater direct production of food, fodder, fuel, fibre and incomes from products derived from the intercropped trees, enhances carbon storage both above-ground and below-ground, and induces more effective conservation of above- and below-ground biodiversity. Zambia, where conservation farming programmes include the cultivation of food crops within an agro-forest of the fertilizer tree *Faidherbia albida* is used. The Malawi Agroforestry Food

Security Programme is integrating fertilizer, fodder, fruit, fuel wood, and timber tree production with food crops on small farms on a national scale. A further example is the dramatic expansion of *Faidherbia albida* agroforests in millet and sorghum production systems throughout Niger via assisted natural regeneration. Additionally there is the development of a unique type of planting pit technology (zai) along with farmer-managed natural regeneration of trees on a substantial scale in Burkina Faso.

The future of horticulture in the arid and semi-arid zones of the world largely depends on the development of sustainable husbandry systems based on the choice of appropriate crops (Nefzaoui 2009). The suitable ones are those that successfully cope with water shortage, high temperature and poor soils, and easy management to provide food and forage for the subsistence life styles, in addition to valuable products and by-products. Desert cacti such as *Opuntia*, particularly *Opuntia ficus indica* (cactus pear), fit most of these requirements. They play an important ecological role in combating desertification as well as producing fruits and vegetables for human consumption, feed for livestock, biomass for energy purposes, cochineal for carmine production, and numerous by-products (beverages, vegetarian cheese, drugs, and cosmetics) in very arid environments. They also provide shelter and food for various wildlife species living in arid environments. Cacti can grow in severely degraded soils, which are not suitable for other crops. *Opuntia* spp. have great capacities for withstanding severe dry conditions and are ideal for favourably responding to global environmental changes. Their root characteristics avoid wind and rain erosion which encourages their growth in degraded areas.

Changes in Cropping Potential

Practical estimates suggest that the change in land suitability for cropping under progressive climate change will be region- and site-specific (Eitzinger et al. 2011). One possible solution is for the movement of crops to higher altitudes. This may be difficult where land at higher altitudes is lacking or unsuitable as in Uganda for example, where currently the tea crop is contributing substantially to the country's economic recovery following many years of decline. Those areas which currently have some cropping suitability will decrease by between 20–40 %. The optimum tea-producing zone is currently at an altitude between 1,450–1,650 m above sea level (masl) and by 2050, this will rise to an altitude between 1,550–1,650 masl. By 2050 areas of land at altitudes between 1,500–1,650 masl will suffer the biggest decrease in suitability for tea culture. Unfortunately, however, there is not much land available in higher altitudes and this leads to the conclusion that total land area available for tea cultivation will ultimately decrease. By 2050, the current tea growing area would only be suitable for one of the six potential alternative crops In the shorter term to 2020, however, some of these crops, especially maize, passion fruit, banana or citrus could be economic options for use where climatic suitability for tea production is declining but these may not produce wholly comparable incomes and may themselves be displaced as the effects of climate change become progressively more aggressive.

An alternative is moving cropping into higher latitudes as opposed to higher altitudes. Records of yearly and monthly average temperature, rainfall, first and last frost dates, frost-free period, sunshine hours have been collected in Xintai City, Hebei Province Northern China for the period 1957–2009 (Wang and Wang 2011). The rising winter temperature makes conditions favourable for planting yellow ginger (*Curcuma longa*, turmeric) and day lily (*Hemerocallus* spp). Hence here cropping can be enlarged as favourable conditions move northward.

Detailed horticultural crop studies are required which quantify the effects of climatic change on productivity and sustainability both for the grower and for the environment. Studies are now being reported for an increasing range of crops and from differing countries and regions. Reducing greenhouse gas emissions and optimizing energy consumption are important challenges for horticultural husbandry processes for mitigating climate change and improving resource-use efficiency. For example the protected strawberry (*Fragaria xananassa* Duch) crop has become a key component of the UK soft fruit sector and is a resource-intensive crop (Warner et al. 2010). Pre-harvest, the global warming potential varied between 1.5 and 10.3 t CO_2 equiv/ha/crop or 0.13 and 1.14 t CO_2 equiv/t of class 1 fruit depending on the intensity of the production system used. Understanding the impact of the effect of weather conditions and their variability on cropping is essential for developing husbandry practices compatible with climate change.

Field research towards these objectives has been reported from vegetable production areas in the Labe river basin (Polabi). This is one of the Czech Republic's traditional vegetable growing regions (Potop et al. 2011). Increasing number of tropical days caused reduced yield for the majority of vegetable crops, and negatively affected onion ($r=-0.69$), pea ($r=-0.44$), cabbage ($r=-0.31$), cauliflower ($r=-0.32$) and savoy cabbage ($r=-0.32$). There were negative correlations between longer heat waves and yields of pea and onion ($r=-0.50$), cauliflower ($r=-0.42$), carrot ($r=-0.30$) and celeriac ($r=-0.40$). Long-lasting extreme rainfall caused lower yields of garlic ($r=-0.36$), cucumbers ($r=-0.30$), and carrots ($r=-0.44$). Longer dry spells were associated with lower yields in the majority of root vegetables (from $r=-0.21$ to $r=-0.32$). Nearly all vegetable crops (root, bulb and leaf vegetables) are sensitive to dry and wet spells after sowing (planting) and in the last 3 weeks before harvest. Important for fruit producing vegetables is the finding that a wet period during June causes yield reductions. A tendency toward increased frequency of dry and heat episodes in Polabi is leading to decreased yields and increased yield variability. That, in turn, means rising costs for vegetable growing and economic losses for farmers.

Similarly, a study of combined greenhouse climate and a crop model which predicts hourly greenhouse climate with energy and mass fluxes and their consequences on crop production for future (2070–2099) and past (1960–1979) has started in the Avignon region of Southern France (Boulard et al. 2011). This uses a combined greenhouse climate and crop model which predicts hourly greenhouse climate and energy and mass fluxes and their consequences on crop production for future (2070–2099) and compares these with the past (1960–1979) climate in Avignon region (South-France). The future climate which is predicted corresponds to a moderate climate change scenario (+2.2 °C average temperature rise) and compares this with the past one using actual climate data. Future elevation of outside temperatures in winter will result in a 30 % energy saving for greenhouse heating. In summer time, air temperature rises which exceed 2.2 °C and reduces tomato growth. This affects leaf area index (LAI) (−10 %), fruit weight (−8 %) and yield (−7 %) linked with an increase of the plant stress duration. But when the effects of carbon dioxide concentration elevation from 380 presently to 700 ppm in the future are added then models predict a 20 % yield increase but still with a substantial increase in the duration of the plant stress during summer months. The effects of climate change on greenhouse production are both positive (energy saving, yield increase) and negative (greenhouse temperature rise, summer plant stress). It is evident that climate change will lead to excessive changes in the environments within greenhouses and glasshouses, particularly during the summer (Dannehl et al. 2012). Therefore, a new environmental strategy for protected cropping is required. Interactions between changing microclimatic conditions resulting from several climatic regimes, plant growth, fruit yield and the formation of secondary plant compounds were studied in 2008 to 2009. A combined application of a high pressure fog system and carbon dioxide enrichment can be applied to decrease the internal temperature and to increase the levels of relative humidity and carbon dioxide concentrations at a high ambient temperature, accompanied by raising mean temperature. This accelerated plant growth, increased dry matter in leaves, and promoted the formation of more tomato fruit set per truss in comparison with those grown under current conventional climate conditions. This led to a yield increase by 20 %, to a reduction of blossom-end rot in tomatoes and to a pronounced increase in fruit size during the spring experiments. Also secondary metabolism was promoted resulting in increased contents of lycopene (49 %), beta-carotene (35 %), and phenolic compounds (16 %) as well as associated antioxidant activity in the water-insoluble (18.5 %) and water-soluble (35.4 %) fractions compared with the conventionally treated plants. The new climate strategy increases yields and improves fruit quality and health-promoting properties of tomatoes.

Case Study—The Impact of Climate Change for Indian Horticulture (Based on the Findings Published by Singh et al. 2010 and supplied by Indrabratta Bhattacharya)

India has the second largest human population following that of China and is the world's largest democracy. Horticulture has become a major part of the Indian agricultural economy. Production in 2007 was 185 million t compared with 25 million t in 1950–1951 (Singh et al. 2010). The current target is 300 million t. Currently, horticulture contributes 30 % of agricultural GDP from 8 % of land area. Crops range across: fruit, nuts, vegetables, tubers, mushroom, spices, floriculture, medicinal, aromatic and plantation cropping. Scientific progress has been achieved with developing new and more productive cultivars, increased efficiency in water and nutrient management, effective plant health policies and strategies for the reduction of post-harvest losses.

India is considered as one of 27 countries likely to be most adversely affected by climate change. The effects of climate change potentially will increase wind damage through the added intensity of tropical storms. There is a predicted 15 % increase in the intensity of tropical cyclones. Shallow rooted fruit and plantation crops will be especially vulnerable to this type of damage.

India is the second largest producer of fruit and vegetables following China. Its land area spans temperate, sub-tropical and tropical zones. Evidence of global warming is evident from the receding snowfall in the Himalayas, retreating glaciers, movement of the temperate fruit belt northward and shortening of the rabi (winter) season accompanied by erratic and irregular rainfall patterns. This resembles the impact of climate change reported for the production of apple and satsuma mandarins in Japan (Suguira and Yokozawa 2004). There cultivation of these crops is also migrating northwards.

Current climatic conditions make India highly suitable for cultivation of medicinal plants, spices and plantation crops such as: rubber, tea, coffee, banana and coconut and but they also demonstrate the vulnerability of fruit cropping. At present, Indian tropical subtropical and temperate fruit cultivars are well adapted to particular soil and climatic conditions. This is an important economic factor since India is the second largest world fruit producer with 10.9 % of global production compared with China's 16.7 %. There have been some long-term studies of climate, initiated before Independence in 1947. For example, mean temperatures in coastal region of Andra Pradesh (East Coast) and in Kerala (West Coast) from 1901 to 2006 showed increases of 0.8 °C and 1.2 °C, respectively. Rain fall increased during the same period by 3.47 % in the East and 2.58 % in the West.

Banana for example, ranks first among India's fruit crops covering large areas in southern, western and eastern regional states of India. Temperature is the main factor controlling growth and development of this crop requiring optimum monthly average temperatures around 26.5 °C. The low temperature stress effect "choke throat" is encouraged by temperatures < 15 °C among the Cavendish cultivars grown in southern states of Kerala and Karnataka. Low temperatures delay harvesting, shooting and finger filling mainly due to a reduced rate of leaf production. Currently, low temperature stress can be avoided by changing the planting time and subsequent cultural practices. All types of citrus (Mandarin, sweet orange, lime, acid lime, grapefruit and lemon) are grown widely in India from the tropical to subtropical to temperate hilly regions. Satisfactory growth and development and quality fruit are temperature regulated. High soil temperatures and moisture stress reduce fruit acidity and improve the quality of Mandarin oranges in Eastern Himalayan regions. Grapes in India are cultivated in tropical subtropical and small areas in the temperate north western Himalayan region. Total production and productivity are highest in the arid tropical region. High humidity and rain adversely affect the quality and productivity of grapes in subtropical climates. Pineapple requires humid tropical and subtropical environments. While papaya (*Carica papaya*) is highly sensitive to sudden increases in temperature which stimulate the development of male flowers compared with female. The optimum temperature for cropping is 21–33 °C. By comparison guava (*Psidium guajava*) is highly tolerant of environmental induced stresses and well suited in the agro-climatic regions of India.

Apple crops cover 46% of the total cropped area under fruit and form 76% of total fruit production of Himachal Pradesh State in the western Himalayan region of India. Recently, while the apple growing area has gradually increased its productivity has decreased at the rate of 0.016 t/ha annually between 1985 and 2009 (Vijyashri Sen 2010). An extensive study of climate change on apple productivity in three different districts of Himachal Pradesh at different elevations. The effects of climate change were evident as increasingly erratic rainfall and rises in temperatures during the snowfall period. Normally, apple and stone fruit buds remain dormant until they have accumulated sufficient chilling units which overcome dormancy and that then permits normal flower and fruit setting (Byrne and Bacon 1992). Where sufficient chilling is not achieved then physiological disorders develop such as delayed flowering, reduced fruit set and increased buttoning and reduced fruit quality. The results of the survey showed that apple cultivation is moving towards higher altitudes where sufficient chilling units can still be Achieved. Indian meteorological data obtained from 1984/1985 to 2005/2006 taken by 22 observatories at different elevations from 2000–4000 msl showed decreases of 36.8 mm in annual snowfall. There is a trend for reduced snowfall during January (27 mm per year) and February (23 mm). Lower snowfall is one of the reasons of the loss of chilling units in apple growing regions of Himachal Pradesh. During the last two decades there was a decrease in snowfall correlated with rising temperatures evident from the information obtained in the apple growing region. Delays in snowfall occurrence and its earlier disappearance resulted in fewer chill units and lower productivity. Winter temperatures and precipitation especially as snow are thought to be crucial for reduction of dormancy and subsequent bud breaking (Jindal et al. 2001). At least 1,200 chilling units are required for effective bud breaking in Himachal Pradesh (Jindal and Mankotia 2004). Subsequently apple fruit size and quality depend upon suitable summer temperatures since they influence fruit development.

Climate change leading to variable weather patterns will increase temperature and water stress. In general predictions suggest an increase of 3–6 °C in northern areas compared with southern India. But rises of only 1 °C are sufficient to influence degree-day accumulations for dormancy breaking. Crops vary in their response to temperature triggers. Mango for example has a temperature bias and become more vegetative at higher temperatures. This affects the size and number of leaves initiated in the vegetative buds potentially reducing yield and quality. Sharp reductions in temperature will cause cold stress damage in mango. Increases in pathogen and pest damage are predicted with pathogens of mango such as *Oidium magifera* (powdery mildew), *Colletotrichum gleosporoides* (anthracnose), *Bacterocera dorsalis* (fruit fly) and various Lepidoptera species being encouraged. Changes which increase rainfall at fruiting will be deleterious. Lowering of light intensity at fruiting reduces the ascorbic acid and sugar content of mango crops. Rising temperatures will affect the grape crops and are likely to delay fruit maturation and reduce quality. Coconut production in Western India will benefit from climate change provided there is sufficient water available for its production. But in areas such as Tamil Nadu, Karnataka, Andra Pradesh, Orissa and Gujarat cropping will be adversely affected. Oil palm also requires high temperatures for growth and yield, but temperature increases will also encourage pest and pathogen populations. This crop has only been introduced to India in the last two decades in response to rising world demand for oil crops. Mean temperatures of 24–18 °C are optimal for growth and yield. Higher temperatures, heavy rainfall and increased rainy days are associated with diseases such as bunch end rot, spear and bud rot.

Potato crops have increased in India since 1950 with rising area (6 fold), production (15 fold) and productivity (3 fold). Crops are mainly concentrated in the Indo-Gangetic plains with mild cool winters. The winter and post-monsoon seasons are most likely to be affected by climate change. Indirect effects include reduced precipitation, increased salinity and more unpredictable events such as erratic rainfall, flooding and frosting. Increased carbon dioxide concentrations raise yield as mean tuber weight since tuber number is not affected. Higher temperatures lead to etiolated growth with smaller sized compound leaves and leaflets reducing Leaf Area Index (LAI). Increased carbon dioxide does not appear to compensate for the detrimental effects of elevated temperature while quality is severely affected with reduced marketable grade tubers and increased physiological internal disorders. Because of the shallow root system potato is severely affected by drought. Consequently, almost the entire Indian potato crop is irrigated. Water shortage alters dry mater partitioning to root, shoot, stem and leaf as related to developmental growth stage and root:shoot ratio. Seed potato production can be adversely affected through rises in virus vectors (aphids) and increased viral diseases. Under climate change there is likely to be increased aphid populations and reduced aphid-free seasons. Global warming diminishes the "time window" for the availability of potatoes for processing where they are kept in reduced temperature stores of limited periods before being sent to the factories. Excessive rain delays planting and emergence reducing tuber yield, produces flooding of fields, at harvest encourages tuber rotting pathogens and increases late blight incidence causing severe crop losses.

In 2005–2006 India produced 109.05 million t of vegetables from 7.16 million ha of land. There is rising demand for vegetables in urban areas. Leafy and succulent vegetables are sensitive to climatic conditions. Temperature rises will affect growth and yield but may also compromise vernalisation. It is suggested that indeterminate vegetable crops are less affected by heat stress because flowering time is extended compared with determinate types. Elevated carbon dioxide has positive effects (24–51 %) on vegetables such as cucumber, bell pepper, tomato, eggplant and onion. But rising temperature can adversely affect crop duration, flowering, fruiting, fruit size and ripening resulting in lost productivity and economic yield. The result will be changing sowing and planting dates and modified nutritional strategies conserving soil organic matter, increased use of fertigation and drip irrigation and reductions in sprinkler applications and more mulching with crop residues and plastic sheets, use of raised beds to avoid flooding problems.

Medicinal and aromatic plants have gained interest recently in India. Climate change will affect them by causing lost protein quality and increased antioxidant content. Where these plants are native to India they may respond by moving to higher altitudes when higher temperatures have adverse effects, some species may become extinct. Foodwebs and co-evolved relationships as between pollinators and pollen producers will be disrupted whereby certain combinations no longer exist at similar times leading to the extinction of both animal and plant. Pathogens, pests and invasive aliens from both flora and fauna may exploit changed conditions colonising vacant spaces caused by the extinction of natives. Medicinal and aromatic plants are used in traditional and modern medicine providing a valuable source of income for rural and tribal harvesters. This could be impaired leading to lower revenue for local economies.

The implications of climate change for crop pathogens in Indian horticulture are considerable since in general they are encouraged by both increased moisture and temperature. This applies across the board for tropical and temperate fruit, plantation and vegetable crops. Indirect effects include changed movement of pathogens, altered host resistance and new requirements for disease management strategies. Virus vectors and virus diseases are most likely to be promoted. In Northern India spread of vectors, pests and pathogens will be accelerated since former climate barriers will lose their effectiveness. Vectors can survive from one vegetative phase to another leading to earlier and faster development of the transmitted diseases.

India in the main has a tropical climate which is summarised as wet and warm. But there is a great diversity in local weather and climates. But overall Indian agriculture and horticulture are largely dependent on the south-west monsoon system. Monsoons have seasonal trends with typical winds and rainfall patterns occurring in June to September. Unstable and erratic monsoon periods are evident from changing patterns of rainfall and increased frequency of extreme weather such as flooding and droughts. This will influence cropping and produce consequent economic and social problems.

An example of biological changes is seen with the B-biotype of whitefly (*Bemisia tabaci*) which were first recorded in the Kolar District of Karnataka, south India during the summer growing season (March to June) of 1999. In India the average population model for this pest was 11 whitefly per 100 leaves in 1984 and that increased to 24 by 2004. The numbers of whiteflies increase 1,000-fold on tomato, cabbage and other vegetables which were previously seldom infested by local whiteflies. An epidemic of tomato leaf curl virus disease followed which has since continued to affect tomato production seriously in India. Since 1999 the B-biotype has spread 600 km (Mysore to Raichur) in north-south and about 250 km (Shmoga to Chittor) in an east-westerly direction. This was associated with the appearance of the whitefly spread virus potato apical leaf curl which is now causing damage to seed stocks. New husbandry strategies have been needed in order to contain this infection. Since these insects can adapt rapidly to changing environments they may have increased flexibility in their geographical range. The spectrum of species causing damage could also increase as in the recent upsurge of *Bemisia afer* and the appearance of new and more aggressive biotypes such as the B-biotype of *B. tabaci*. Populations of the leaf hopper (*Empoasca fabae*) increased from 16 per 100 leaves in 1984 to 24 in 2004 and this was followed by a rise in leaf hopper burn from 48 to 68 %. Leaf hoppers have become significant pests of early planted potato crops in the sub-tropical plains of India. The appearance of aphid (*Myzus persicae*) has advanced by 5 days in the last 20 years and populations have risen from 567 per

100 leaves in 1984–1985 to 653 in 2003–3004. This has the effect of reducing the seed production period and adversely affecting yield. Aphids have become more significant pests of the autumn planted crops. Overall, there has been a perceptible increase in pest damage over the past 20 years which is detrimental to crop yield.

Climate change and resultant rising vector and virus incidence is predicated to have a substantial adverse effects on India horticulture. Increasing temperatures will favour many insect vectors of virus pathogens also it is suggested that the rate of virus multiplication in the host plants will increase leading to earlier symptom expression and heavier crop losses. Pest problems will be aggravated by climate change. Pest population composition will change in different areas. The length of the growing season for tropical and sub-tropical pests will increase, resulting in more generations and greater damage. Over-wintering pests will increase producing larger populations as a base for build-up in the following seasons, In some cases pests may not need hibernation periods thereby producing more generations than currently. Many pests may have diapause strategies disrupted as linkages between temperature and moisture regimes and day lengths are altered. The reproductive ability of pests might be affected by changes to the nutritional value of plants on which they feed due to elevated carbon dioxide levels under climate change. Some pests will become more serious while others may decline in importance. The economics of cost:benefit relationships for control will also alter and biological control is expected to become a more important part of mitigation strategies.

It is suggested that late blight of potato (*Phytophthora infestans*) disease outbreaks will become more severe associated with increased ambient temperatures and higher RH. This has been seen already in India in the periods 1997–1998 and 2006–2007 when crop losses in the State of the Punjab exceeded 40% resulting in an acute shortage of potatoes. There also appeared to be a shift in disease epidemiology since lesions were more frequent on stems compared with leaves causing "stem blight" as opposed to "leaf blight". By comparison early blight which is encouraged by high temperatures is less likely to become more severe. Soil borne pathogens requiring cool growing conditions, such as *Synchitrium endobioticum* (wart disease), *Rhizoctonia solani* (black scurf) and *Streptomyces scabies* (common scab), will be discouraged by increasing temperatures. Whereas *Sclerotium rolfsii* (sclerotinia wilt), *Macrophomina phaseolina* (charcoal rot) are predicted to increase in severity and spread into previously unaffected regions. Increasing temperatures will favour many insect vectors of virus pathogens also it is suggested that the rate of virus multiplication in the host plants will increase leading to earlier symptom expression and heavier crop losses.

Social science studies of the perception and response of populations in the Western Himalayas demonstrate that growers and their families accept that climate change is taking place by reference to indigenous local knowledge of how crops normally respond to weather conditions and the effects of altered environments (Vedwan and Rhoades 2001).

The risks associated with climate change have become an insurable commodity. Weather insurance providing cover against the occurrence of extreme conditions has been developed based on a weather index. The weather indices may be defi-

Table 25.1 An example of the weather insurance schedule for apple crops

Time Period	December to March	March to May	April to August
Stage	Dormancy (Rest)	Flowering	Fruit set and development
Risk	Non availability of sufficient chilling and moisture	Extreme temperature fluctuations	Non availability of sufficient water
Weather Index	Chilling units as per Utah model Aggregrate precipitation	GDD, a Tmin and Tmax based index that capture deviations on higher and lowerside	Aggregate rainfall during subphase of the crop phase

cit or excess rainfall, extreme fluctuations of temperature, relative humidity and/or combinations of all weather factors. The historic data is used in order to derive the index. For countries such as India the World Bank is providing support in order to cope with the effects of climate change (Table 25.1).

The effects of climate change on environmental and social horticulture

Climate change will affect environmental and social horticulture at least as importantly as the effects on commercial production industries. For leisure and hobby gardeners there will be a necessity for coping with drier soils in summer and wetter soils in winter while maintaining soil fertility. Intensification of pest, pathogen and weed competition will make it more difficult to achieve reasonable yields and quality of outcome. Lawn maintenance will be more difficult due to water logging and competition from mosses, liverworts and other similar plants during autumn, winter and spring periods. In the summer months there will be added costs and complications determined by the rising cost of water used for irrigation and limitations to supply especially during drought periods. There are pundits who advocate the use of drought tolerant perennials and bulbous species but these will be difficult to manage during wet winter conditions. It is likely that shoot extension growth in trees and shrubs will continue through most of the year if not for full 12 month periods making the care of species which need winter dormancy difficult. Summer time planting of container-grown and containerised will be more difficult with greater risks of failure. This will pose significant challenges for the garden centre industry particularly in formulating marketing programmes which provide retail customers with plants which have the capacity for successful growth and flowering. On the positive side it is possible that a wider range of plants may be cultivated in the open. Plants previously considered as too tender for northerly regions will become available for use. Warmer and drier summers and autumns will increase the enjoyment of gardens and hobby production.

The management of historic sites and tourist attractions will be more difficult. Historic gardens, their associated archaeology, associated wildlife, and specimen plant collections will be affected by changing climatic conditions. Additional pressures will come from increasing length of opening periods and numbers of visitors making garden management more demanding especially with less predictable summer weather conditions. Storms and floods will increase as weather patterns become more extreme, violent and unpredictable. Maintenance costs will increase particularly for lawns and sports turf which contains fine grass species.

But if the visitor season is extended then income should rise but that may be offset by increased costs. There will be opportunities for developing educational roles as centres of excellence in environmentally sustainable gardening techniques and in the science and practice of climate change.

Drought resistance becomes important for both commercial fruit, flower and vegetable crops and for ornamentals used in parks and gardens. Some effects of climate change appear conflicting, as with elevated carbon dioxide enhancing photosynthesis and changes to water-use efficiency. Increasing rainfall in some regions will contrast considerably with shortages elsewhere, affecting the activities of benign and pathogenic soil borne microbes. In the marketing chain there will be changes in the availability, cost and quality of produce. Towards the end of the twenty-first century these changes will become marked and associated with substantial price rises, reflecting the increasing scarcity of what are currently common commodities of food, flowers and plants. That could have substantially adverse effects for health and welfare if the amount of fruit and vegetables in diets is reduced because of increasing prices and scarcity.

The commercial industry which supplies hobby gardeners will experience difficulties maintaining container-grown stock through summers where water is scarce and expensive. There will be problems in coping with the effects of extreme weather on business activities such as resupplying outlets. Marketing wider ranges of plants will provide challenges in terms of the knowledge and expertise required from staff. But there will be extensions of peak times of gardening activity and therefore increased business opportunities. More equipment will be required by the hobby gardeners adding to marketing potentials (Bisgrove and Hadley 2002).

Potential risks related to the use of chemical pesticides have encouraged research of low impact alternatives as biocontrol agents (BCAs) and the implementation of a more environmentally sound horticultural husbandry as the integrated pest management (IPM) and organic agriculture were reviewed by Moser et al. (2012). Climate change is becoming an important issue for the general public and mitigation practices are under discussion. The level of consumers' awareness of mitigation practices and their willingness to pay for products produced with low carbon emission is unknown in any detail. But there are a few indications, in one survey consumers' attitudes and preferences to fruit (apples) produced by using BCAs instead of chemical pesticides and/or climate change mitigation practices were evaluated in a choice experiment. The survey administered to a very small sample of 96 consumers in different supermarkets in Northern Italy during autumn 2009 showed factors that govern the purchasing decision, origin and price are the major determinants, followed by organic production and good appearance. Using BCAs and climate change mitigation practices increased the probability of purchasing apples. The only coefficient associated to climate change however, that is statistically significant was that respondents were willing to pay a premium price of about 0.50 €/kg when their awareness was raised. Moreover, results indicate that when buying apples most people have specific requirements in mind regarding method of production, origin, appearance and price, but comparing choices made by the respondents with their individually-stated minimum requirements, the majority violated them, overall there is low consumer loyalty to ethical ideas. This indicates that a great deal more has to be achieved in communicating the effects and implications of climate change to consumers.

Horticulture worldwide is already being affected by climatic changes. There will be some apparent beneficiaries but the overall outcomes are likely to be disadvantageous. Horticultural science can provide means by which the worst aspects can be mitigated. For this to happen requires the provision of well educated and expert scientists and research facilities. These must then feed new processes and procedures through to able and skilled practitioners. The time available in which this can be achieved is reducing rapidly (Dixon 2012).

Acknowledgement The authors acknowledge with gratitude the use of information contained in Singh et al. (2010) especially that of: Singh (2010), Kumar et al (2010), Laxman et al (2010), Bhiguvanshi (2010), Suresh and Babu (2010), More and Bhargava(2010), Singh et al (2010), Rao et al (2010), Das (2010), Chowdappa (2010), Reddy (2010) Sridhar et al (2010), Singh and Bhat (2010) and Ambast and Srivastava (2010).

References

Alford RA (2011) Ecology: bleak future for amphibians. Nature 480:461–462

Alford L, Blackburn TM, Bale JS (2012) Effect of latitude and acclimation on the lethal temperatures of the peach-potato aphid *Myzus persicae*. Agric For Entomol 14:69–79

Ambast SK, Srivastava RC (2010). Impact assessment and adaptation strategies to mitigate the effects of climate change on island ecosystem. In Challenges of climate change – Indian Horticulture (Singh HP, Singh JP, Lal SS (eds)). Westville Publishing House, New Dehli, pp. 191–195

Anon (2008) Climate change and food security: a framework document. United Nations Food and Agriculture Organisation, Rome. http://www.fao.org/docrep/010/k2595e/k2595e00.htm Accessed 20 March 2014

Anon (2011a) Hot news from long ago. The Times, Monday 2 May 2011, p. 15

Arrhenius S (1896) On the influence of carbonic acid in the air upon the temperature of the ground. Philos Mag 41:237–276

Atauri IGC, Brisson N, Baculat B, Seguin B, Legave JM, Calleja M, Farrera I, Guedon Y (2010) Analysis of the flowering time in apple and pear and bud break in vine, in relation to global warming in France. Acta Hortic 872:61–68

Baker KM, Kirk WW, Stein JM, Andresen JA (2005) Climatic trends and potato late blight risk in the upper Great Lakes Region. HortTechnology 15(3):510–518

Beresford RM, Kim KS (2011) Identification of regional climatic conditions favorable for development of European canker of apple. Phytopathology 101(1):135–146

Bhiguvanshi SR (2010) Impact of climate change on mango and tropical fruit crops. In Challenges of climate change – Indian Horticulture (Singh HP, Singh JP, Lal SS (eds)). Westville Publishing House, New Dehli, pp. 31–35

Bisgrove R, Hadley P (2002) Gardening in the global greenhouse: the impacts of climate change on gardens in the UK. UK Climate Impacts Programme, UKCIP, Union House, 12-16 St Michael's Street, Oxford OX1 2DU

Bird PR, Bicknell D, Bulman PA, Burke SJA, Leys JF, Parker JN, Vandersommen FJ, Voller P (1992) The role of shelter in Australia for protecting soils, plants and livestock. National Australian Conference on the role of trees in sustainable agriculture. October 1991, Albury, Australia. Agroforestry Syst 20(1–2):59–86

Blanke MM, Kunz A (2011) Effects of climate change on pome fruit phenology and precipitation. Acta Hortic 922:381–386

Boland GJ, Melzer MS, Hopkin A, Higgins V, Nassuth A (2004) Climate change and plant diseases in Ontario. Can J Plant Pathol 26(3):335–350

Boulard T, Fatnassi H, Tchamitchian M (2011) Simulating the consequences of global climate change on greenhouse tomato production in South-France: preliminary results. Acta Hortic 919:71–80

Bourgeois G, Bourque A, Deaudelin G (2004) Modelling the impact of climate change on disease incidence: a bioclimatic challenge. Can J Plant Pathol 26:284–290

Bowden J, Cochrane J, Emmett BJ, Minall TE, Sherlock PL (1983) A survey of cutworm attacks in England and Wales, and a descriptive population model for Agrotis segetum (Lepidoptera: Noctuidae). Ann Appl Biol 102:29–47

Brandle JR, Hodges L, Zhou XH (2004) Windbreaks in North American agricultural systems. 1st World Congress of Agroforestry, June 27th-July 2nd, 2004, University of Florida, Orlando, FL., USA. Agroforest Syst 61–62(1):65–78

Bruck DJ (2010) Fungal entomopathogens in the rhizosphere. BioControl 55:103–112

Byrne DH, Bacon TA (1992). Chilling estimation: its importance and estimation. The Texas Horticulturist 18:5–9

Callendar GS (1938) The artificial production of carbon dioxide and its influence on temperature. Quart J Roy Meteorol Soc 64(275):223–240

Cameron RWF, Dixon GR (1997) Air temperature, humidity and rooting volume affecting freezing injury to Rhododendron and other perennials. J Hortic Sci 72(4):553–562

Campbell BD, Stafford Smith MD, McKedon GM (1977) Elevated CO_2 and water supply interactions in grasslands: a pasture and rangelands management perspective. Global Change Biol 3:177–187

Chakraborty S, Pangga IB, Roper MM (2012) Climate change and multitrophic interactions in soil: the primacy of plants and functional domains. Global Change Biol 18:2111–2125

Chowdappa P (2010) Impact of climate change on fungal diseases of horticultural crops. In Challenges of climate change – Indian Horticulture (Singh HP, Singh JP, Lal SS (eds)). Westville Publishing House, New Dehli, pp. 144–151

Collier RH, Finch S, Phelps K, Thompson AR (1991) Possible impact of global warming on cabbage root fly (Delia radicum) activity in the UK. Ann Appl Biol 118:261–271

Compant S, Van Der Heijden MGA, Sessitsch A (2010) Climate change effects on beneficial plant-microorganism interactions. FEMS Microbiol Ecol 73:197–214

Dannehl D, Huber C, Rocksch T, Huyskens-Keil S, Schmidt U (2012) Interactions between changing climate conditions in a semi-closed greenhouse and plant development, fruit yield, and health-promoting plant compounds of tomatoes. Sci Hortic 138:235–243

Darwin C (1859) On the origin of species by means of natural selection, or the preservation of favoured races in the struggle for life. John Murray, London

Das M (2010) Climate change and its impact on medicinal and aromatic plants. In Challenges of climate change – Indian Horticulture (Singh HP, Singh JP, Lal SS (eds)). Westville Publishing House, New Dehli, pp. 137–143

Dixon GR (1975) The reaction of some oil rape cultivars to some fungal pathogens. Proceedings 8th British insecticide and fungicide conference 2, 503–506

Dixon GR (1981) Vegetable crop diseases. Messrs Macmillan Co. Ltd., London, pp. 404 (also translated into Arabic)

Dixon GR (2006) The biology of Plasmodiophora brassicae Wor.–A review of recent advances. Acta Hort, 706:271–282

Dixon GR (2009) Chapter 17, pp 307–324: The impact of climate and global change on crop production. In: Letcher TM (ed) Climate change: observed impacts on planet earth. ISBN: 978-0-444-53301-2. Elsevier, Oxford, p 444

Dixon GR (2012) Climate change, plant pathogens and food production. Can J Plant Pathol 34(3):362–379

Dixon GR, Biggs MP (1996) The impact of environmental change on acclimation and dormancy in Rhododendron. In: Froud-Williams RJ, Harrington R, Hoking TJ, Thomas TH (eds) Implications of global environment change for crops in Europe. Aspects Appl Biol 45:93–100

Dormann CF, Schweiger O, Arens P, Augenstein I, Aviron S, Bailey D, Baudry J, Billeter R, Bugter R, Bukácek R, Burel F, Cerny M, Cock RD, De Blust G, DeFilippi R, Diekötter T, Dirksen J, Durka W, Edwards PJ, Frenzel M, Hamersky R, Hendrickx F, Herzog F, Klotz S, Koolstra B, Lausch A, Le Coeur D, Liira J, Maelfait JP, Opdam P, Roubalova M, Schermann-Legionnet A, Schermann N, Schmidt T, Smulders MJ, Speelmans M, Simova P, Verboom J, van Wingerden W, Zobel M (2008) Prediction uncertainty of environmental change effects on temperate European biodiversity. Ecol Lett 11:235–244

Drigo B, Kowalchuk GA, van Veen JA (2008) Climate change goes underground: effects of elevated atmospheric CO_2 on microbial community structure and activities in the rhizosphere. Biol Fertil Soils 44:667–679

Eastburn DM, Degennaro MM, DeLucia EH, Dermody O, McElrone AJ (2010) Elevated atmospheric carbon dioxide and ozone alter soybean diseases at SoyFACE. Global Change Biol 16(1):320–330

Eitzinger A, Laderach P, Quiroga A, Pantoja A, Gordon J (2011) Future climate scenarios for Uganda's tea growing areas. Final report July 2011, p 23. International Center for Tropical Agriculture (CIAT), A.A. 6713, Cali, Colombia

Fahim MA, Hassanein MK, Hadid A, F A, Kadah MS (2011) Impacts of climate change on the widespread and epidemics of some tomato diseases during the last decade in Egypt. Acta Hortic 914:317–320

Fiske R (2008) Robert Marsham 'The Man of Feeling' with brief bibliographical notes relating to his Indications of Spring, p. 6. Parochial Church Council of St. Margaret's Church, Stratton Strawless, Norfolk, Great Britain

Fordham DA, Sinclair RG, Peacock DE, Mutze GJ, Kovaliski J, Cassey P, Capucci L, Brook BW (2012) European rabbit survival and recruitment are linked to epidemiological and environmental conditions in their exotic range. Austral Ecol 37:945–957

Gange AC, Gange EG, Sparks TH, Boddy L (2007) Rapid and recent changes in fungal fruiting patterns. Science, 316(5821):71

Garrity DP, Akinnifesi FK, Ajayi OC, Weldesemayat SG, Mowo JG, Kalinganire A, Larwanou M, Bayala J (2010) Evergreen agriculture: a robust approach to sustainable food security in Africa. Food Secur 2(3):197–214

Ge C, Lee C, Lee J (2012) The impact of extreme weather events on Salmonella internalization in lettuce and green onion. Food Res Int 45:1118–1122

Gilles T, Phelps K, Clarkson JP, Kennedy R (2004) Development of MILIONCAST, an improved model for predicting downy mildew sporulation on onions. Plant Dis 88:695–702

Gimenez-Benavides L, Garcia-Camacho R, Iriondo JM, Escudero A (2011) Selection on flowering time in Mediterranean high-mountain plants under global warming. Evol Ecol 25(4):777–794

Glauber SF, daSilva AN, Mogens L, Glass B, Luiz GS, Branco A (2013) Temperature and respiratory function in ectothermic vertebrates. J Therm Biol 38:55–63

Grab S, Craparo A (2011) Advance of apple and pear tree full bloom dates in response to climate change in the southwestern Cape, South Africa: 1973–2009. Agric For Meteorol 151(3):406–413

Grace J, Zhang R (2006) Predicting the effect of climate change on global plant productivity and the carbon cycle. In: Morison JIL, Morecroft MD (eds) Plant Growth and Climate Change (pp. 187–208). Blackwell Publishing Ltd., Oxford

Haggerty BP, Galloway LF (2011) Response of individual components of reproductive phenology to growing season length in a monocarpic herb. J Ecol (Oxford) 99(1):242–253

Harmon JP, Moran NA, Ives AR (2009) Species response to environmental change: impacts of food web interactions and evolution. Science 323:1347–1350

Harrington R, Stork NE (1995) Insects in a changing environment: 17th Symposium of the Royal Entomological Society, 7-10 September 1993 at Rothamsted experimental station, Harpenden, p 535

Harris D, Hossell JE (2002) Pest and disease management constraints under climate change. The BCPC conference: pests and diseases, Vol 1 and 2. Proceedings of an international conference held at the Brighton Hilton Metropole Hotel, Brighton, UK, 18–21 November 2002. pp 635-640. British Crop Protection Council Farnham UK

Harvell CD, Mitchell CE, Ward JR, Altizer S, Dobson AP, Ostfeld RS, Samuel MD (2002) Climate warming and disease risks for terrestrial and marine biota. Science (Washington) 296(5576):2158–2162

Hickling R, Roy DB, Hill JK, Fox R, Thomas CD (2006) The distributions of a wide range of taxonomic groups are expanding polewards. Global Change Biol 12:450–455

Hirschi M, Stoeckli S, Dubrovsky M, Spirig C, Calanca P, Rotach MW, Fischer AM, Duffy B, Samietz J (2012) Downscaling climate change scenarios for apple pest and disease modeling in Switzerland. Earth Syst Dyn 3:33–47

Hullé M, Coeur d'Acier A, Bankhead-Dronnet S, Harrington R (2010) Aphids in the face of global changes (Review). Comptes Rendus Biol 333:497–503. http://0-www.scopus.com.pugwash.lib.warwick.ac.uk/record/display.url?eid=2-s2.0-77956626798&origin=resultslist&sort=plf-f&src=s&st1=aphid+and+harrington&sid=6F430162AA3CFDD-8B71F9B29C5D72AA6.ZmAySxCHIBxxTXbnsoe5w%3a90&sot=q&sdt=b&sl=40&s=TITLE-ABS-KEY-AUTH%28aphid+and+harrington%29&relpos=6&relpos=6&searchTerm=TITLE-ABS-KEY-AUTH%28aphid+and+harrington%29-corrAuthorFooter

Hunt RS, Geils BW, Hummer KE (2010) White pines, Ribes, and blister rust: integration and action. (Special Issue: White pines, Ribes, and blister rust.). Forest Pathol 40(3/4):402–417

Inglis IR, Isaacson AJ, Smith GC, Haynes PJ, Thearle RJP (1997) The effect on the woodpigeon (Columba palumbus) of the introduction of oilseed rape into Britain. Agric Ecosyst Environ 61:113–121

Jay M, Lichou J (2010) Climate change and flowering of apricot. [French] Changement climatique et floraison de l'abricotier: incidence des temperatures hivernales sur la qualite de la floraison. Infos-Ctifl 260:44–47

Jeger MJ, Pautasso M (2008) Plant disease and global change-the importance of long-term data sets. New Phytol 177(1):8–11

Jindal KK, Chauhan PS, Mankotia MS (2001) Apple productivity in relation to environmental components. In Productivity of Temperate Fruits, (Jindal KK, Gautam DR (eds)), pp. 12–20

Jindal KK, Mankotia MS (2004) Impact of changing climatic conditions on chilling units, physiological attributes and productivity of apple in Western Himalayas. Acta Hort. 662: 111–117

Johnson SN, Petitjean S, Clark KE, Mitchell C (2010) Protected raspberry production accelerates onset of oviposition by vine weevils (Otiorhynchus sulcatus). Agric For Entomol 12(3):277–283

Karley AJ, Parker WE, J.W. P, Douglas AE (2004) The mid-season crash in aphid populations: why and how does it occur? Ecol Entomol 29:383–388

King D (2005) Climate change: the science and the policy. J App Ecol 42:779–783

Kumar SN, Bai KVK, Thomas GV (2010) Climate change and plantation crops: impact, adaptation and mitigation with special reference to coconut. In Challenges of climate change – Indian Horticulture (Singh HP, Singh JP, Lal SS (eds)). Westville Publishing House, New Dehli, pp. 9–22

Kunz A, Blanke MM (2011) Effects of global climate change on apple 'Golden Delicious' phenology-based on 50 years of meteorological and phenological data in Klein-Altendorf. Acta Hortic 903:1121–1126

Lacey LA, Frutos R, Kaya HK, Vail P (2001) Insect pathogens as biological control agents: do they have a future? Biol Control 21:230–248

Large EC (1950) The Advance of the Fungi. Jonathon Cape, London

Last F, Roberts A, Patterson D (2003) Climate change? A statistical account of flowering in East Lothian 1978–2001. In Baker S. (ed) East Lothian 1945-2000: Fourth Statistical Account Vol. 1: The County (pp. 22-29). East Lothian Library Service for the East Lothian Fourth Statistical Account Society, UK

Last FT, Roberts AMI (2013) Onset of flowering in biennial and perennial garden plants: association with variable weather and changing climate between 1978 and 2007. Sibbaldia 10:85–132

Laxman RH, Shivashankara KS, Rao NKS (2010) An assessment of potential impacts of climate change on fruit crops. In Challenges of climate change – Indian Horticulture (Singh HP, Singh JP, Lal SS (eds)). Westville Publishing House, New Dehli, pp. 23–30

Leisner CP, Ainsworth EA (2012) Quantifying the effects of ozone on plant reproductive growth and development. Global Change Biol 18(2):606–616

Li BH, Yang JR, Dong XL, Li BD, Xu XM (2007) A dynamic model forecasting infection of pear leaves by conidia of Venturia nashicola and its evaluation in unsprayed orchards. Eur J Plant Pathol 118:227–238

Lockwood JG (1999) Is potential evapotranspiration and its relationship with actual evapotranspiration sensitive to elevated atmospheric CO_2 levels? Climatic Change 41:193–212

Luck J, Asaduzzaman M, Banerjee S, Bhatacharya I, Couglan K, Detnath GC, De Boer D, Dutta S, Forbes G, Griffiths W, Hosain D, Huda S, Jazannathan R, Khan S, O'leary G, Miah G, Saha A, Spooner-Hart R (2010) (compliers) The effects of climate change on pests and diseases of major food crops in the Asia Pacific Region. Asia Pacific network for global change research, Final report ARCP 2010-05CMY-Luck. APN Secretariat, East Building 4F 1-5-2 Wakinohama Kaigen Dori, Chuo-ku Kobe 651-0073 Japan

Luck J, Spackman M, Freeman A, Trebicki P, Griffiths W, Finlay K, Chakraborty S (2011) Climate change and diseases of food crops. Plant Pathol 60:113–121

Luedeling E, Gassner A (2012) Partial least squares regression for analyzing walnut phenology in California. Agric For Meteorol 158/159:43–52

Luedeling E, Zhang MH, Girvetz EH (2009) Climatic changes lead to declining winter chill for fruit and nut trees in California during 1950–2099. PLoS ONE 2009 July:e6166

Luedeling E, Girvetz EH, Semenov MA, Brown PH (2001a) Climate change affects winter chill for temperate fruit and nut trees. PLoS ONE 2011 May:e20155

Luedeling E, Kunz A, Blanke M (2001b) More winter chill for fruit trees in warmer winters? Modelling winter chill, as affected by global warming, in the Meckenheim fruit growing region. [German] Mehr Chilling fur Obstbaume in warmeren Wintern? Modellierung des Kaltereizes-unter dem Einfluss des rezenten Klimawandels-fur das Meckenheimer Obstbaugebiet. Erwerbsobstbau 53(4):145–155

Mann JA, Tatchell GM, Dupuch MJ, Harrington R, Clark SJ, McCartney HA (1995) Movement of apterous Sitobion avenae (Homoptera, Aphididae) in response to leaf disturbances caused by wind and rain. Ann Appl Biol 126:417–427

Mazzola M (2010) Management of resident soil microbial community structure and function to suppress soilborne disease development. In: Reynolds MP (ed) Climate change and crop production 200–218. CABI Wallingford, UK

Memmott J, Craze PG, Waser NM, Price MV (2007) Global warming and the disruption of plant-pollinator interactions. Ecol Lett 10:710–717

More TA, Bhargava R. (2010) Impact of climate change on productivity of fruit crops in arid regions. In Challenges of climate change – Indian Horticulture (Singh HP, Singh JP, Lal SS (eds)). Westville Publishing House, New Dehli, pp. 76–84

Morison JIL, Moorecroft MD (2006) Plant Growth and Climate Change. Blackwell Publishing Ltd., Oxford

Moser R, Pertot I, Raffaelli R (2012) Consumers' attitude to fruit produced by using biocontrol agents and climate change mitigation practices. IOBC/WPRS Bulletin 2012. 78:299-303. In: Pertot I, Elad Y, Gessler C, Cini A (eds) Proceedings of the IOBC/WPRS working group "Biological control of fungal and bacterial plant pathogens", Graz, Austria, 7–10 June 2010. International Organization for Biological and Integrated Control of Noxious Animals and Plants (OIBC/OILB), West Palaearctic Regional Section (WPRS/SROP), Dijon

Murton RK, Westwood NJ, Isaacson AJ (1964) A preliminary investigation of the factors regulating population size in the woodpigeon Columba palumbus. Internatl J Avian Sci IBIS 106:482–507

Nefzaoui A (2009) Cactus: a crop to meet the challenges of climate change in dry areas. Ann Arid Zone 48(1):1–18

Peri PL, Bloomberg M (2002) Windbreaks in southern Patagonia, Argentina: a review of research on growth models, windspeed reduction, and effects on crops. Agroforest Syst 56(2):129–144

Potop V, Koudela M, Mozny M (2011) The impact of dry, wet and heat episodes on the production of vegetable crops in Polabi (river basin). Sci Agric Bohem 42(3):93–101

Potts SG, Biesmeijer JC, Kremen C, Neumann P, Schweiger O, Kunin WE (2010) Global pollinator declines: trends, impacts and drivers. Trends Ecol Evol 25:345–353

Pounds JA, Bustamante MR, Coloma LA, Consuegra JA, Fogden MPL, Foster PN, La Marca E, Masters KL, Merino-Viteri A, Puschendorf R, Ron SR, Sánchez-Azofeifa, GA, Still CJ, Young BE (2006) Widespread amphibian extinctions from epidemic disease driven by global warming. Nature 439:161–167

Pritchard SG (2011) Soil organisms and global climate change. Plant Pathol 60:82–99

Quamme HA, Cannon AJ, Neilsen D, Caprio JM, Taylor WG (2010) The potential impact of climate change on the occurrence of winter freeze events in six fruit crops grown in the Okanagan Valley. Can J Plant Sci 90(1):85–93

Rao NKS, Laxman RH, Bhatt RM (2010) Impact of climamte change on vegetable crops. In Challenges of climate change – Indian Horticulture (Singh HP, Singh JP, Lal SS (eds)). Westville Publishing House, New Dehli, pp. 113–123

Reddy MK (2010) Climate change and virus diseases of horticultural crops. In Challenges of climate change – Indian Horticulture (Singh HP, Singh JP, Lal SS (eds)). Westville Publishing House, New Dehli, pp. 152–165

Rosenweig C, Karoly D, Vicarelli M, Neofotis P, Casassa G, Menzel A (2008) Attributing physical and biological impacts to anthropogenic climate change. Nature 453:353–357

Sánchez-Azofeifa GA, Still CJ, Young BE (2006) Widespread amphibian extinctions from epidemic disease driven by global warming. Nature 439:161–167

Settele J, Kudrna O, Harpke A, Kuehn I, van Swaay C, Verovnik R, Warren M, Wiemers M, Hanspach J, Hickler T, Kühn E, van Halder I, Veling K, Vliegenthart A, Wynhoff I, Schweiger O (2008) Climatic risk atlas of European butterflies. BioRisk 1:1–710

Shaw MW, Bearchell SJ, Fitt BDL, Fraaije BA (2008) Long-term relationships between environment and abundance in wheat of Phaeosphaeria nodorum and Mycosphaerella graminicola. New Phytologist 177:229–238

Silva DE, Mazzella PR, Legay M, Corcket E, Dupouey JL (2012) Does natural regeneration determine the limit of European beech distribution under climatic stress? For Ecol Manage 266:263–272

Singh BP, Bhat NM (2010) Implications of climate change on potato diseases and insect pests. In Challenges of climate change – Indian Horticulture (Singh HP, Singh JP, Lal SS (eds)). Westville Publishing House, New Dehli, pp. 178–184

Singh HP (2010) Impact of climate change on horticultural crops. In Challenges of climate change – Indian Horticulture (Singh HP, Singh JP, Lal SS (eds)). Westville Publishing House, New Dehli, pp. 1–8

Singh HP, Singh JP, Lal SS (2010) Challenges of climate change-Indian horticulture. Central Potato Research Institute, Shimla; Westville Publishing House, New Delhi

Singh JP, Lal SS, Govindakriishnan PM, Dua VK, Pandey SK (2010) Impact of climate change on potato in India. In Challenges of climate change – Indian Horticulture (Singh HP, Singh JP, Lal SS (eds)). Westville Publishing House, New Dehli, pp. 90–99

Siebold M, Tiedemann von A (2012) Potential effects of global warming on oil seed rape pathogens in Northern Germany. Fungal Ecology 5:62–72

Solomon S, Qin D, Manning M (eds) (2007) The physical science basis. Contribution of working group 1 to the fourth assessment report of the intergovernmental panel on climate change (IPCC). Summary for policy makers. Cambridge University, Cambridge, pp. 1–18

Spinks T, Lines J (2008) *Chapters in the life of Robert Marsham Esq., FRS of Stratton Strawless, Norfolk.* Parochial Church Council of St. Margaret's Church, Stratton Strawless, Norfolk, Great Britain

Sridhar V, Kumar N K K, Verghese A. (2010) Implications of climate change on horticultural crop pests – Indian Scenario. In Challenges of climate change – Indian Horticulture (Singh HP, Singh JP, Lal SS (eds)). Westville Publishing House, New Dehli, pp. 166–177

Stenseth NC, Mysterud A, Ottersen G, Hurrell JW, Chan K-S, Lima M (2002) Ecological effects of climate fluctuations. Science 297:1292–1296

Stern, N (2008) The Economics of Climate Change: The Stern Review. Cambridge University Press, Cambridge

Stukenbrock EH, Banke S, McDonald BA (2006) Gobal migration patterns in the fungal wheat pathogen Phaeosphaeria nodorum. Mol Ecol 15:2895–2904

Sugiura T, Yokozawa M (2004) Impact of global warming on environments for apple and satsuma manadarin estimated from changes of annual mean temperature. J Japan Soc Hort Sci 59:471–477

Suresh K, Babu MK (2010) Impact of climate change on oil palm in India. In Challenges of climate change – Indian Horticulture (Singh HP, Singh JP, Lal SS (eds)). Westville Publishing House, New Dehli, pp. 36–41

Sutton JC, James TDW, Rowell PM (1986) Botcast-a forecasting system to time the initial fungicide spray for managing botrytis leaf-blight of onions. Agric Ecosyst Environ 18:123–143

Terblanche JS, Nyamukondiwa C, Kleynhans E (2010) Thermal variability alters climatic stress resistance and plastic responses in a globally invasive pest, the Mediterranean fruit fly (Ceratitis capitata). Entomol Exp Appl 137(3):304–315

Van Laer S, Hauke K, Meesters P, Creemers P (2005) Botrytis infection warnings in strawberry: Reduced enhanced chemical control. Commun Agric Appl Biol Sci 70:61–71

Vedwan N, Rhoades RE (2001) Climate change in the western Himalayas of India: a study of local percdeption and response. Climate Research 19:109–117

Vicente JG, Holub EB (2013) Xanthomonas campestris pv. Campestris (cause of black rot of crucifers) in the genomic era is still a worldwide threat to brassica crops. Mol Plant Pathol 14:2–18

Wakeham AJ, Kennedy R (2010) Risk assessment methods for the ringspot pathogen Mycosphaerella brassicicola in vegetable brassica crops. Plant Dis 94:851–859

Wang YuHe Wang ChengJun (2011) Climate change and its influence on crop farming in Xintai City. Meteorol Environ Res 2(5):24–27, 30

Warner DJ, Davies M, Hipps N, Osborne N, Tzilivakis J, Lewis KA (2010) Greenhouse gas emissions and energy use in UK-grown short-day strawberry (Fragaria xananassa Duch) crops. (Special Section: Climate change and agriculture). J Agric Sci 148(6):667–681

Watson RT, Zinyowera MC, Moss RH, Dokken DJ (eds) (1998) The Regional Impacts of Climate Change–An assessment of vulnerability. Cambridge University Press, Cambridge

Webb L, Whetton PH (2010) Adapting agriculture to climate change: preparing Australian agriculture, forestry and fisheries for the future 119-136 Centre for Australian Weather and Climate Research, CSIRO and the Bureau of Meteorology, 107-121 Station St, Aspendale, Vic 3195, Australia. Editors Stokes C & Howden M. CSIRO Publishing Collingwood Australia

West JS, Townsend JA, Stevens M, Fitt BDL (2012) Comparative biology of different plant pathogens to estimate effects of climate change on crop diseases in Europe. Euro J Plant Path 133:315–331

Williams PH, Araújo MB, Rasmont P (2007) Can vulnerability among British bumblebee (Bombus) species be explained by niche position and breadth? Biol Conserv 138:493–505

Willis JC, Bohan DA, Powers SJ, Choi YH, Park J, Gussin E (2008) The importance of temperature and moisture to the egg-laying behaviour of a pest slug, Deroceras reticulatum. Annals of Applied Biology 153:105–115

Wrege MS, Caramori PH, Herter FG, Steinmetz S, Reisser Junior C, Matzenauer R, Braga HJ (2010) Impact of global warming on the accumulated chilling hours in the Southern Region of Brazil. Acta Hortic 872:31–40

Zak D, Pregitzer K, King J, Holmes W (2000) Elevated atmospheric CO_2, fine roots and the response of soil microorganisms: a review and hypothesis. New Phytologist 147:201–222

Ziska LH, Dukes JS (2011) Weed biology and climate change. Blackwell Publishing Ltd.

Ziska LH, Blumenthal DM, Runion GB, Hunt ER, Diaz-Soltero H (2011) Invasive species and climate change: an agronomic perspective. Climatic Change 105:13–42

Chapter 26
Concepts and Philosophy Underpinning Organic Horticulture

David Pearson and Pia Rowe

Abstract Many argue that food products from certified organic production systems are a vital component of meeting global food security challenges into the middle of the twenty-first Century. Whilst the concept of organic in this context, with its emphasis on minimizing the use of artificial chemicals and other external inputs, is not new—as it exists in all systems operating without human contributions, its philosophical position emerged as a reaction against the increased 'industrialization' of food production that occurred in developed countries around the 1940's. Increasing formalization saw the emergence of global coordination through the International Federation of Organic Agricultural Movements and its four principles of health, ecology, fairness and care that are now embodied in formal independent certification systems in most countries around the globe. Sales of certified organic horticultural products are a major component, at around 30%, of what is now a US$ 60 billion global industry. Continued expansion of sales (and private production) of organic horticultural products is likely to continue due to their natural affinity with local sourcing of healthy fresh products.

Keywords Organic food movement · History · Organic principles · Organic certification

Introduction

As the world population grows, the issue of food security is once again becoming topical all around the world (FAO 2012). In light of the far reaching ramifications of conventional agriculture and the associated chemical use in the 1900s, organic agriculture is now established as one of the crucial components of the global food system. The world is also facing unprecedented issues related to human diets and sustainability of the food system. The population in the industrialized countries is fatter

D. Pearson (✉) · P. Rowe
Faculty of Arts and Design, University of Canberra,
Australian Capital Territory 2601, Australia
e-mail: david.pearson@canberra.edu.au

P. Rowe
e-mail: pia.rowe@canberra.edu.au

G. R. Dixon, D. E. Aldous (eds.), *Horticulture: Plants for People and Places, Volume 2*,
DOI 10.1007/978-94-017-8581-5_26,

than ever before, consuming too much animal protein and not enough horticultural products. Further the consumption of beef and dairy in particular have been linked to negative ecological impacts on the planet (McMichael et al. 2007). For these reasons, it will be important to promote and increase the overall consumption of fruit and vegetables, and keep increasing the market share of organic agriculture around the world.

While the increased output from organic production systems has brought the concept of 'organic' into mainstream society, the idea itself is not new. It could be argued that humans have practiced 'organic' farming since the origins of agriculture. This continued until the Industrial and Scientific Revolution in the 1800s enabled radical changes to the farming practices and started moving the production towards what these days is called conventional agriculture (Treadwell et al. 2003).

Despite growth, overall land used for organic agriculture in the whole world is still quite low, with only 0.9 % of agricultural land being organic at the end of 2009. Again, there are large differences between countries, with some having far higher shares (For example: Falkland Islands 36 %, Liechtenstein 27 % and Austria 18 %). Of the 37 million ha of organic agricultural land most is used for grazing, approximately 5.5 million ha (15 %) is arable land of which 0.22 million ha is used for vegetables. Permanent crops, of which most important are coffee, followed by olives, cocoa, nuts and grapes, amount for 2.4 million ha (6 %) (Willer and Kilcher 2011). In Europe in 2009, the market shares for organic food and drink were still fairly low, with Denmark having the highest shares of sales at 7 %, followed by Austria at 6 % and Switzerland at 5 % (Willer 2011).

Organic horticulture refers to growing fruits, vegetables, flowers, ornamental plants, nuts, olives, medicinal and aromatic plants, root crops, as well as beverage crops such as coffee and tea, using the organic principles. Fresh produce such as fruit and vegetables have so far had the largest market shares in the organic market, and it has been considered the 'gateway' into the world of organic food (Dettman and Dimitri 2009). The growing consumer demand for a wide variety of fresh vegetables has contributed to the fact that many retailers offer organic produce all year round, and the sales of convenient items such as baby carrots and bagged salads has increased rapidly in recent years (ibid). In addition to being of comparable quality with conventional products in terms of freshness, taste and convenience, consumers are attracted to organic horticultural products due to the perception that they are better for their health (Pearson and Henryks 2008) although the scientific evidence for this is not conclusive (Smith-Spangler et al. 2012).

The International Federation of Agricultural Movements has developed the following definition for organic agriculture. It is important to note that its scope is beyond a widely held misconception that organic agriculture simply relates to the use, or lack thereof, of chemical inputs such as fertilizers and pesticides.

> Organic agriculture is a production system that sustains the health of soils, ecosystems and people. It relies on ecological processes, biodiversity and cycles adapted to local conditions, rather than the use of inputs with adverse effects. Organic agriculture combines tradition, innovation and science to benefit the shared environment and promotes fair relationships and a good quality of life for all involved. (IFOAM 2011)

What makes this definition particularly useful is that it takes into account the whole supply chain—from farmer to consumer. Thus the focus is not solely on production factors, but also on the social impacts of the process of producing, selling and consuming food. It is inclusive of all products, people and places. Therefore the principles of organic agriculture as specified in the previous definition are adapted to suit a wide variety of different situations.

To fully understand how organic agriculture has developed into the industry it is today, we will begin this chapter by reviewing the history of organic farming. By doing this, we will be able to examine both the environmental as well as the social and economic implications of the organic movement, and how these have formed over the years. We will then look at the requirements for organic certification. Finally, we will investigate the current market for organic products around the world.

History of Organic Farming

In many ways, the relationship most people have with food has always been coloured with anxiety. The main issue throughout our history has been the supply, or lack thereof, of food. This has been exacerbated by an uneven distribution of the available food products. However, as Reed and Holt (2006) point out, for the affluent ones around the world, this concern has been largely replaced by the concern for the quality attributes of food. In many of the developed countries the food supply far exceeds demand, and how much an individual can consume depends solely on their wealth. This may be one of the reasons why the question of food safety has gained a foothold in many households, and issues such as pesticide use in food production and their impact on human health have become topical. Such concerns have in turn paved the way for organic agriculture, and the whole organic movement has been enjoying unprecedented growth over recent decades.

The organic movement "…has a history of almost 100 years, with over 50 years of continuous production on some farms." (Kristiansen and Merfield 2006). However, the concept of organic farming is not new. Organic farming has been practiced since humans started moving away from hunter-gatherer origins and first started cultivating land, as all farming was initially organic by default. In other words, people weren't following organic principles because they were exercising a choice or because they made a conscious decision to do so, but rather because of "absence of a choice". Simply put, farmers did not have access to synthetic fertilizers or mechanical apparatuses that have been the cornerstone of industrial agriculture. However, as El-Hage Scialabba (2007) points out, today "…true organic agriculture is practiced by intent, not default. You do not automatically become organic simply because you never used prohibited chemicals anyway."

Organic farming is not just about the farming practices itself, but also its wider reaching socio-cultural aspects that continue to shape, and be influenced by, our society. Therefore, we will have to look beyond the location of production to gain an understanding of the origins of organic farming.

The first inorganic fertilizers that marked the start of the 'agricultural revolution' and allowed farming to move towards being 'industrialized', or what is now referred to as 'conventional agriculture', were produced in the 1840s (Kristiansen and Merfield 2006). These new agricultural methods were thought to increase the yields and improve food security. Instead, by the early 1900s, agriculture and agricultural science was in crisis, not just because of ecological, but also economic and social problems (Vogt 2007). The consequences of the increased use of inorganic fertilizers started to become obvious as the soil quality began degrading with widespread erosion occurring in many countries including the USA, South Africa and Australia (Holt and Reed 2006) and despite the use of the chemicals that were designed to improve the crops, yields started dropping, and even the nutritional content of the products was proven to have suffered. Consumers started questioning the safety of food, and at the same time, rural lifestyles and traditions changed as more and more people moved into cities.

As a result of these issues with conventional agriculture, many scientists and people in the agricultural area started to look for alternatives (Francis and Van Waart 2009). While Walter Nortborne may have been the first to use the term 'organic farming' in his book 'Look to the Land' in 1940, the origins of the organic movement had already been ignited in the early 1900s. What makes the history of this movement particularly fascinating is the fact that it was not limited to just one or two areas, but was global from the very beginning.

In Germany anthroposophist Rudolf Steiner was approached by two groups of people in the early 1920s. The first group consisted of farmers who were concerned about the rapid degeneration of seed-strains. The second group brought to his attention increases in animal diseases (Pfeiffer 1958). Prompted by these concerns, Steiner conducted his famous lectures on 'biodynamic agriculture' in 1924, which provided one of the starting points for the organic movement. Steiner's approach was not limited to the health of the soil from a scientific perspective, as the agricultural methods promoted by him included a spiritual element and were linked to cosmic forces. Because of this spiritual aspect, his methods have attracted a lot of criticism and ridicule, but as Vandermeer (2011) points out, biodynamic, as this approach is now called, was part of a bigger movement, the Antroposophism Movement, which was especially important in the Western Europe and continues to attract thousands of followers.

Underpinned by the spiritual imperatives of Antroposopism, biodynamic agriculture values the interrelationships between all biological elements of the farming system (Kristiansen 2006). Biodynamic farming methods consequently focus on developing agricultural systems which nurture the health of the whole ecosystem and enable it to be self-sustaining. The movement is not only characterized by its specific methods, such as the integration of farm animals, the use of local crop varieties, and planting based on astronomical cycles, but their belief that a farm must be true to its 'essential nature'(Paull 2011). In this sense, the farm is perceived as an individual entity, and thus everything the farm should be available within the boundaries of the system.

The application of Steiner's philosophy to farming was first publically articulated in 'The Agriculture Course' which he ran at Koberwitz in Silesia in 1924 (Paull

2011). The ideas presented in the course were then further developed and compiled in Ehrenfried Pfeiffer's 1938 publication *Bio-Dynamic Farming and Gardening.*

However, the methods used in biodynamic farming are not necessarily connected with their spiritually grounded Anthroposophic origins. As far back as the Marienstein Farmers' Conference in 1928 (Paull 2011) farmers external to the Antroposophic tradition had observed that these methods improved the quality of foodstuffs (Von Pilasch 1928).

Despite the gradual decoupling of biodynamic methods from their metaphysical origins, the Antroposophic conception of the farm as an 'organism' has still been greatly influential in the organic movement more broadly. This is largely because the idea was adopted as a foundational concept by pioneers of what became known as the organic food movement, such as Lord Northbourne in his book *Look to the Land* first published in 1940 (Paull 2011).

Steiner was not the only one associated with the origins of the organic movement whose interests and beliefs may have caused concern amongst those who adhered to the strictly scientific principles of agriculture. Vandermeer (2011) has noted two curious trends that occurred simultaneously and that may have created subtle barriers to the "scientific approach to the alternative agricultural movement". According to him, the magazine 'New English Weekly', which was founded in 1932 and commonly associated with the organic movement, "... was much more focused on 'Christian Sociology'". Furthermore, a contributor to this magazine, the agricultural scientist and farmer, Jorian Jenks, was also associated with the origins of the organic movement, and was a member of the British Union of Fascists (ibid).

In the English speaking world, a British scientist, Albert Howard, who worked both in India and the UK is seen by many as one of the founders of the organic movement (Pretty 2005; Francis and Van Waart 2009). Howard saw food as being integral to good health and drew direct links between the health of soil and the quality of food eaten to the health of people:

> Real security against want and ill health can only be assured by an abundant supply of fresh food properly grown in soil in good health. The first place in post-war plans of reconstruction must be given to soil fertility in every part of the world. (Howard 1945)

As the first person to start a comparison of organic and conventional farming, in 1939 in the UK, Lady Eve Balfour was also a pioneer of organic farming. She called this the Haughley Experiment and based on her research published a book 'The Living Soil' in 1943. She then cofounded The Soil Association in 1946, which over the years has developed from a small organization based on a farm and mainly focused on research, into the leading charity campaigning for sustainable agriculture in the UK.

In the US, F. H. King of the United States Department of Agriculture traveled in Japan, Korea and China in 1905 and documented the organic systems used in their local agriculture. Another key figure in the origins of the organic movement in the US was the University of Missouri soil chemist William Albrecht, who much in the same vein as Howard, linked soil and food quality with human health. One of Steiner's old colleagues, Ehrenfried Pfeiffer, who had become a vocal proponent of biodynamic agriculture also brought the movement to the US in the late 1930s (Vandermeer 2011). In 1942 J. I. Rodale created a magazine called 'Organic Farming and Gardening' in which he chronicled the experiences of innovative farmers and taught people how to grow food without chemicals.

Despite the dispersed global interest in organic farming, it remained a minority business in most parts of the world in the years following the Second World War. As the world was trying to recover from the physical and economic consequences of the war, the pressure was on the farmers to produce more food than ever before, and the use of pesticides increased rapidly in many countries. Advances in technology also contributed to this, as for example, aircrafts that were able to spread fertilizers and pesticides quickly and efficiently over the large areas were developed post Second World War (Woods 2003). What's more, the post-war years saw the conception and rapid uptake of the 'Green Revolution' initiatives that were developed to save people from starvation around the world (Hazell 2009). These initiatives included the use of pesticides, herbicides and fertilizers to increase food supply.

The term 'Green Revolution' is used to refer to a range of research, development, and technology initiatives that ran from the 1940s to the 1970s (Hazell 2009). In particular, it involved the development of high-yielding varieties of grains, the modernization of farm management techniques and the expansion of large-scale irrigation infrastructure. The distribution of hybridized seeds and synthetic fertilizers and pesticides to farmers were also key elements of the movement. The 'Revolution' was facilitated by research bodies, such as The International Maize and Wheat Improvement Center, which were created specifically for the task of increasing yields and promoting new 'improved' crop varieties around the world (Hazell 2009).

Initially, the 'Green Revolution' was heralded as a great success and was credited with significantly reducing famine in many countries. However, the limitations of the new methods and crop varieties became increasingly apparent from the 1960s onward.

As pesticide use increased, the scientific evidence of the negative effects of synthetic chemicals started to become more and more apparent. Rachel Carson, an American marine biologist, started to investigate the environmental problems they caused in the 1950s. Her book 'Silent Spring' (1962) brought consumers' attention to these issues and provided a step towards environmentalism amongst the general public. The book also played a significant part in the final ban of the use of DDT (dichlorodiphenyltrichlorethane) as a pesticide in 1972 the USA.

In terms of the history of the organic movement, 1972 also signaled the foundation of the International Federation of Organic Agriculture Movements (IFOAM) in Versailles, France. The initiative for the movement had come from the president of the French farmer organization, Nature et Progrès, and had initially five founding members;

> The Soil Association from Great Britain represented by Lady Eve Balfour, the Swedish Biodynamic Association with Kjell Arman, the Soil Association of South Africa in the person of Pauline Raphaely, Rodale Press from the United States of America whose representative was Jerome Goldstein and of course, Nature et Progrès with Roland Chevriot.

However, what makes the history and rising popularity of organic agriculture particularly fascinating is that it started gaining attention amongst the general population even before the majority of the advocacy organizations were established. Lockeretz (2007) lists the following as factors that contributed to the growth of what has become known as the organic food movement:

- Great social and political upheaval worldwide, combined with heightened public awareness of environmental threats, such as the pesticide use and 'blue baby' syndrome from increased nitrate levels in drinking water from the use of fertilizers;
- Greater suspicion of synthetic chemicals in all aspects of food; the farmers' growing concerns of their own health;
- The general strong anti-establishment activism that paved way for both environmental and antiwar activists to join their forces; and finally,
- The countercultural hippie revolution which also promoted 'back-to-the-land' principles.

Given its strong roots in countercultural and anti-establishment movements, it is somewhat surprising that the organic movement did not diminish after the years of upheaval were over and the economic climate improved around the world. Instead it grew to find supporters from all walks of life, ranging from the stereotypical anti-establishment hippies to conservative policy makers (Pearson et al. 2011a). Mitchel et al. (1992) examined the rise of environmental organizations between 1970–1990 in the US, and noted that by the end of 1960s, education campaigns alone were found to be insufficient to inspire and mobilize people. At the same time, direct action was considered as too aggressive a method by many mainstream organizations. Instead, policy reform became the new preferred choice for environmental organizations. Prior to the 1960s, only government agencies were granted standing in the administrative proceedings in the implementation of environmental laws and the Reagan government in general had been very dismissive of any environmental issues, but the introduction of interest representation in the administrative law in the US enabled the environmental lobby to engage in the political activity and influence the policy at a totally new level (ibid).

The organic food movement both benefitted and contributed to the increased presence of collective action in the political arena. This no doubt increased its support from a basis in the countercultural 'alternative' lifestyle beginnings to having a more mainstream appeal associated with organic principles in general, and the health aspects of organic diet in particular.

By the late 1980s, IFOAM had expanded rapidly from its humble beginnings of five founding members to an organization that networked more than 100 member organizations in 50 countries, representing a total of 100,000 individuals (Geier 1998). A part of this development was, according to Geier, due to the growing interest in organic agricultural methods in the developing worlds, and the countries behind the 'iron curtain'. IFOAM also started active lobbying to influence the planned regulation for organic agriculture in the European Union, and organized conferences, workshops and scientific meetings to further develop the movement (ibid).

Table 26.1 Price premium of organic above conventional (Source: Derived by the authors from Lin et al. 2008)

	Premium (%)
Fruits	
Apple	34
Banana	36
Grape	22
Orange	19
Strawberry	*34*
Average for fruits	*29*
Vegetables	
Carrot	15
Onion	18
Pepper	35
Potato	82
Tomato	*19*
Average for vegetables	*34*

From the consumers' perspective, increased production and therefore availability of organic products also contributed to rapid growth (Pearson et al. 2011a). As the availability organic products spread from health food shops and into the more mainstream shopping outlets, such as supermarkets, this ease of access enabled many of those who would not have made the effort to seek out organic products and start purchasing them, and to purchase them more often.

However, in many cases the prices of organic products were, and still are, a lot higher than those produced using conventional methods, and this in turn acts as a prohibiting factor even amongst those who would like to have a completely organic diet (Pearson et al. 2011a).

Retail price fluctuations for horticultural products are much larger than those associated with most other grocery items with the reasons associated for this including variations in product quality and seasonal fluctuations in supply and demand. Further the role that price plays, and the amount of attention that a consumer gives to it is complex, and will depend on the buyer, product and situation (Pearson 2001, 2005). Hence caution needs to be applied to any attempts at making generalizations about the price premiums associated with purchases of organic products. Table 26.1 above provides an indication of the price premiums.

Whilst on average organic products were more expensive, the variation in actual prices paid for a specific product ranged widely—such as the difference between the cheapest and most expensive organic apples being over 10 times (Lin et al. 2008).

In many countries, governments are increasingly recognizing the benefits of organic farming, particularly for the natural environment, and are therefore supporting its expansion.

It is also interesting to note that while the issue of increasing food production is seen by many as being the main driving factor behind increasing industrialization of the food system, some advocates of sustainable agriculture also place blame on the consumers for these developments. For example it may be that 'careless eating' has driven the industrialization and created the demand for unsustainable practices such as all year round availability of fruit and vegetables (Kleiman 2009). The famous quote from Wendell Berry (1990), "how we eat determines, to a considerable extent, how the world is used" exemplifies this way of thinking well. There is little doubt that consumers are a significant stakeholder in the industrialization of food system, and, perhaps to an even greater extent, they have led and sustained the expansion of the organic food movement.

Requirements for Organic Certification

Organic Production Principles

As previously mentioned, organic production has been defined, by the International Federation of Organic Agricultural Movements (IFOAM), as "a production system that sustains the health of soils, ecosystems and people (IFOAM 2011).

IFOAM provides four principles for organic production, namely, health, ecology, fairness, and care. According to the first principle of health, IFOAM goes on to state that organic agriculture "should sustain and enhance the health of soil, plant, animal, human and planet as one and indivisible." In this context health is considered to be the wholeness and integrity of living systems. Hence it is not simply the absence of illness, but the maintenance of physical, mental, social and ecological well-being. Thus organic agriculture aims to produce high quality, nutritious food that contributes to wellbeing as a form of preventive healthcare. This is achieved by minimizing the use of fertilizers, pesticides, animal drugs and food additives that may have adverse health effects (IFOAM 2011).

The principle of ecology states that organic agriculture "should be based on living ecological systems and cycles." Thus organic farming, pastoral and wild harvest systems should fit the cycles and ecological balances in relation to the culture and scale in the natural local environment. Inputs should be reduced by reuse, recycling, and efficient management of materials and energy. And finally, organic agriculture should aim to protect and benefit the common environment including biodiversity as well as variety of habitats and landscapes (IFOAM 2011).

The principle of fairness states that organic agriculture should "build on relationships that ensure fairness with regard to the common environment and life opportunities." This is amongst people, and with other living beings. Individuals should have a good quality of life, hence organic agriculture aims to contribute to food self-sufficiency and the reduction of poverty. Further, animals should be provided with the conditions that accord with their natural behavior (IFOAM 2011).

The fourth and final principle, that of care, states that organic agriculture should "be managed in a precautionary and responsible manner to protect the health and well-being of current and future generations and the environment." This incorporates a blending of scientific developments with traditional wisdom. Thus organic agriculture aims to prevent significant risks by adopting appropriate technologies and rejecting unpredictable ones, such as genetic engineering (IFOAM 2011).

Based on these principles IFOAM has developed standards and associated certification systems for organic products (IFOAM 2012a). Organic standards are used to create an agreement about what an "organic" claim on a product means, and in many cases it is communicated to consumers through a certification symbol or logo. Groups of organic farmers and their supporters, often in one geographic region, began developing organic standards back in the 1940's. Today there are hundreds of these private organic standards around the world. In addition there are those developed by more than 60 national governments which involve certification by an independent third-party and aim to regulate any kind of an "organic" claim on a product label.

IFOAM provides a minimum set of requirements for organic production and processing (IFOAM 2012b). This substantial document (126 pages) provides details for production (crops, animals, aquaculture), processing, handling, labeling and social justice. Both private standards and government regulations are eligible for consideration for official endorsement through a process referred to as equivalence with this IFOAM baseline. For example the UK based Soil Association has built upon the IFOAM baseline to develop its own standards for certification of organic products (SA 2012).

Organic Certification of Product and Grower

The existence of credible certification process for organic products is extremely important for both the producers and consumers. Certification aims to ensure consistency of the quality attributes embodied in organic products. Thus it reduces opportunities for fraud and increases consumers' confidence in the product they buy. This helps to maintain and increase sales. As consumer trends throughout the world are towards 'one-stop-shopping' for food products, such as weekly purchases from supermarkets, where there is no direct connection with producers, organic certification identifies the production method used. This helps to establish a relationship between consumer and producer whilst also adding to consumer confidence in these products (Zagata and Lostak 2012).

In 2011, 84 countries had organic standards, and 24 were in the process of drafting legislation. The total number of certification bodies in 2011 was 549, with the most located in the European Union, Japan, the United States, South Korea, China, Canada, India, and Brazil (Willer and Kilcher 2011). By way of example, there are seven organic certification bodies Australia (DAFF 2011) each with a different logo.

The United States is a large and well developed marketing for organic products. As such it provides an example of how the organic market may develop in other countries. Their Department of Agriculture (USDA) has a National Organic Program (NOP) that regulates the "standards for any farm, wild crop harvesting, or handling operation that wants to sell an agricultural product as organically produced." Labeling requirements apply to "raw, fresh product and products that contain organic agricultural ingredients" and are based on the percentage of ingredients, with products labeled "100 % organic" only containing only organic ingredients, products labeled 'organic' must consist of at least 95 % of organic ingredients, and products with at least 70 % of organic ingredients can use the term "made with organic ingredients". However, only the first two can use the USDA seal on their package (USDA 2008).

In addition to organic certification of organizations, whether they be producer, processor, or retailer, for specific products, there are locally focused approaches that provide access to organic certification for small holders. These Participatory Guarantee Systems (PGS) are supervised by IFOAM and it is estimated that there are about 40 PGS initiatives established worldwide, with Latin America and India having the most farmers certified through the system (ibid).

Market for Organic Products

Organic agriculture, including horticultural products, is continuing to experience rapid growth globally. A lot of this expansion is being driven by consumers who are reconnecting with the food production chain and placing value on organic farming (Pearson et al. 2011a). While the global economic crisis with its far reaching ramifications on both funding of organic agriculture as well as the spending power of consumers slowed down the growth in 2009, growth rates are projected to increase rapidly again as countries around the world start coming out of the recession (Willer and Kilcher 2011). Global sales of organic food and drink have grown over three-fold over the last ten years, $US 18 billion in 2000 to US$ 60 billion 2010. There has been significant variation between countries though, with highest growth rates reported in France (+ 19 %) and Sweden (+ 16 %) (Willer 2011).

While the market for organic food has continued to expand, the concept of 'local food' is seen by many to be offer a suite of benefits to both producers and consumers that may, over time, take sales away from certified organic products (Pearson et al. 2011b). The main reason for this lies behind the fact some organically grown food is produced in "industrial-scaled monocultures far from the places it is consumed" (Kleiman 2009). This is particularly relevant to organic horticulture, as consumers are far more likely to value locally grown fruit and vegetables instead of those that are air freighted across the world.

Conclusions

The concepts and philosophies underpinning organic production have evolved since the 1940's as an alternative to the increasing industrialization of food production. Today a wide range of organic products are grown and made available to consumers throughout the world through distribution channels ranging from subsistence farming though to supermarkets with multinational supply chains.

Through global leadership provided by the International Federation of Organic Agricultural Movements its principles of health, ecology, fairness and care are now manifest in certification systems and regulations for organic horticultural products throughout the world. As such they provide an example of a food system, and associated products for consumers, that gives explicit emphasis to human health and environmental sustainability.

The continuation of global sales growth for organic horticultural product is anticipated to continue as consumers increasingly reconnect with the source of their food and, as part of this, place value on organic certification. Further, sales of organic horticultural products benefit from their natural synergy with the trend towards local sourcing of healthy fresh products.

References

Berry W (1990) The pleasures of eating. http://www.ecoliteracy.org/essays/pleasures-eating. Accessed 23 March 2012
Carson R (1962) Silent spring. Houghton Mifflin, New York
DAFF (2011) AQIS organic approved certifying organizations. Department of Agriculture Fisheries and Forestry (Australian Government). http://www.daff.gov.au/aqis/about/contact/aco. Accessed 1 Nov 2012
Dettman R, Dimitri C (2009) Who's buying organic vegetables? Demographic characteristics of U.S. consumers. J Food Prod Mark 16(1):79–91
El-Hage Scialabba N (2007) Forward. In: Lockeretz W (ed) Organic farming: an international history. Oxfordshire: CAB International
FAO (2012) Committee on world food security. Food and Agriculture Organization of the United Nations 2012. http://www.fao.org/cfs/en/ Accessed 1 Nov 2012
Francis C, Van Waart J (2009) History of organic farming and certification. In: Francis C (ed) Organic farming: the ecological system. American Society of Agronomy, Maddison, Wisconsin, USA
Geier B (1998) A look at the development of IFOAM in its first 25 years. http://www.ifoam.org/about_ifoam/inside_ifoam/pdfs/First_25_Years.pdf. Accessed 10 March 2012
Hazell P (2009) The Asian green revolution. International Food Policy Research Institute
Holt G, Reed M (2006) Sociological perspectives on organic agriculture. CABI International, Wallingford
Howard A (1945) The post-war task. In: Pretty J (ed) The earthscan reader in sustainable agriculture. Earthscan, London
IFOAM (2011) Principles of organic agriculture. International Federation of Organic Agriculture Movements. http://www.ifoam.org/. Accessed 13 Jan 2012
IFOAM (2012a) Organic standards and certification. International Federation of Organic Agriculture Movements. http://ifoam.org/about_ifoam/standards/index.html. Accessed 8 Nov 2012
IFOAM (2012b) The IFOAM norms for organic production and processing- Version 2012. Bonn, Germany, International Federation of Organic Agricultural Movements, p 126
Kleiman J (2009) Local food and the problem of public authority. Tech Cult 50(2):399–417
Kristiansen P, Merfield C (2006) Overview of organic agriculture. In: Kristiansen P, Taji A, Reganold J (eds) Organic agriculture: a global perspective. CSIRO Publishing, VIC, pp 1 23
Lin B-H, Smith TA, Huang CL (2008) Organic premiums of US fresh produce. Renew Agr Food Syst 23(3):208–216
Lockeretz W (2007) What explains the rise of organic farming? In: Lockeretz W (ed) Organic farming: an international history. CAB International, Oxfordshire

McMichael A, Powles J, Butler C et al (2007) Food, livestock production, energy, climate change, and health. Lancet 370:1253–1263
Mitchell R, Mertig A, Dunlap R (1992) Twenty years of environmental mobilization: trends among national environmental organizations. In: Dunlap R, Mertig A (eds) American environmentalism: the U.S. environmental movement 1970–1990. Taylor and Francis New York, Inc, Washington DC
Paull J (2011) Biodynamic agriculture: the journey from Koberwitz to the world, 1924–1938. J Org Syst 6(1):27–41
Pearson D (2001) How to increase organic food sales: results from research based on market segmentation and product attributes. Australas Agribus Rev 9(8):1–8
Pearson D (2005) Marketing fresh fruits and vegetables: exploration of individual product characteristics and their relationship to buyer's attention to price. Australas Agribus Rev 13(9):1–13
Pearson D, Henryks J (2008) Marketing organic products: exploring some of the pervasive issues. J Food Prod Mark 14(4):95–108
Pearson D, Henryks J, Jones H (2011a) Organic food: what we know (and don't know) about consumers. Renew Agric Food Syst 26(2):171–177
Pearson D, Henryks J, Trott A, Jones P, Parker G, Dumaresq D, Dyball R (2011b) Local food: understanding consumer motivations in innovative retail formats. Brit Food J 113(7):886–899
Pfeiffer E (1938) Bio-dynamic farming and gardening: soil fertility renewal and preservation (Trans: F. Heckel). Anthroposophic, New York
Pfeiffer E (1958) Preface. In: Steiner R (ed) Agriculture course: the birth of the biodynamic method. Rudolf Steiner Press, UK
Pretty J (2005) Agrarian and rural perspectives. In: Pretty J (ed) The earthscan reader in sustainable agriculture. Earthscan, London
Reed M, Holt G (2006) Sociological perspectives of organic agriculture: an introduction. In: Holt G, Reed M (eds) Sociological perspectives of organic agriculture: from pioneer to policy. CAB International, Oxfordshire
SA (2012) Standards soil association. http://www.soilassociation.org/organicstandards. Accessed 7 Nov 2012
Smith-Spangler C, Brandeau M, Hunter G et al (2012) Are organic foods safer or healthier than conventional alternatives? A systematic review. Ann Intern Med 157(5):348–366
Treadwell D, McKinney D, Creamer N (2003) From philosophy to science: a brief history of organic horticulture in the United States. HortScience 38(5):1009–1014
USDA (2008) United States department of agriculture, agricultural marketing service. Organic labeling and marketing information. http://www.ams.usda.gov/AMSv1.0/getfile?dDocName=STELDEV3004446&acct=nopgeninfo. Accessed 13 March 2012
Vandermeer J (2011) The ecology of agroecosystems. Jones and Bartlett, Sudbury
Vogt G (2007) The origins of organic farming. In: Lockeretz W (ed) Organic farming: an international history. CAB International, Oxfordshire
Von Pilasch S (1928) Farmer's conference. Anthroposophical Mov 5(34):267–268
Willer H (2011) Organic agriculture in Europe 2009: production and market. Research Institute of Organic Agriculture, FiBL, Switzerland. BioFach Congress Nürnberg February 18 2011. http://orgprints.org/18365/2/willer-2011-european-market.pdf. Accessed 12 March 2012
Willer H, Kilcher L (eds) (2011) The world of organic agriculture-statistics and emerging trends. IFOAM and FiBL, Bonn
Woods N (2003) The aerial application of pesticides. In: Wilson M (ed) Optimising pesticide use. Wiley, West Sussex
Zagata L, Lostak M (2012) In goodness we trust. The role of trust and institutions underpinning trust in the organic food market. Sociol Ruralis 52(4):470–487

Index

G. R. Dixon, D. E. Aldous (eds.), *Horticulture: Plants for People and Places, Volume 2*,
DOI 10.1007/978-94-017-8581-5,

Active cooling, 361
Active ingredient, 506, 610
Active lifestyle, 17
A. cunninghamii, 457
Adam Smith, 794
Adam's pome, 1273
Adansonia digitata, 723
Added value, 10
Addis Ababa, 718
Adelaide, South Australia, 680
Adenosine signaling cascade, 980
Adhatoda vasica, 721
Adipocytes, 986
  differentiation of, 986
Adiposity levels, 986
Adrenal glands, 984
Adult phase, 104
Adventitious embryony, 113
Advertising, 641
Advisory service model, 1129
Aechmea fasciata, 768
Aegiphila, 720
Aeneas, 1261
Aeneid, 1261
Aerating the soil, 1033
Aeration, 376, 386
Aerial photography, 703
Aerial pollutants, 773, 778
Aerodynamic resistance, 366
Aeroponics, 378
Aesthetic, 650, 698, 732, 767, 773, 1272
  awareness of, 1005
Aesthetic benefits, 694
Aestivation, 834
Afghan farmers, 723
Afghanistan, 1243
AFLP markers, 304
Africa, 17, 84, 124, 264, 267–270, 274, 282–285, 287, 289, 419, 438, 440, 455, 471, 606, 608, 662, 708, 715–718, 721, 723, 733, 798, 820, 1173, 1175, 1176, 1181, 1182
African cassava mosaic virus, 1182
African cocoa, 285
African forest products, 273
African locust bean, 717
African Plant Protection Organization, 1181
African tulip tree, 803
African Union, 1181
After-ripening, 623
Age of Enlightenment, 679, 1278
Age-related neuronal declines, 988
Aggression, 6, 803
Aggressiveness, 17
Aging, 1047, 1049, 1053
Aglycone, 976
Agnes Arber, 1220
Agostino Chigi, 1212
Agostino Gallo, 1257, 1270, 1272, 1276
Agrarian reform, 1147
Agribusiness, 1124, 1146, 1151, 1153, 1155, 1157
  sectors of, 1153
  supply chains of, 1155
  system, 1140, 1150–1152, 1154, 1155, 1163
Agricultural development, 1126
Agricultural economies, 727
Agricultural education, 1129
Agricultural environments, 1066
Agricultural extension, 1118
Agricultural Extension Service, 1125
Agricultural knowledge, 1158
Agricultural land, 713
Agricultural lands and soils, 513
Agricultural movement, 863
Agricultural producers, 1147
Agricultural productivity, 1140–1143, 1148
Agricultural runoffs, 724
Agricultural science and horticultural science, 1124
Agricultural services, 1149
Agriflor, 416
Agrifood industry, 1143
Agrobacterium tumefaciens, 232, 636
Agroecology, 613
Agroforestry, 127, 713
Agroforestry Food Security Programme, 841
Agroforests, 724, 841
Agronomic risk, 753
Agrostis, 737
Agrostis capillaris, 734, 739
Agrostis stolonifera, 733, 734, 737, 739, 741
Agrotis segetum, 834
Air, 507
Air bag presses, 240
Air density, 366
Air-filled porosity (AFP), 456
Airflow, 225, 235, 360
Air flow rate, 357
Air freight, 427, 428, 507, 519, 520
Airfreighted, 507, 520
Airfreighting, 428
Air heaters, 344

Air humidity, 355, 356
Air pollution, 1047, 1060, 1075
Airports, 1009
Airport terminals, 768
Air quality, 765, 773, 774, 797, 1058–1061, 1075
Air specific heat, 366
Air temperature, 344, 346, 348, 353, 819, 822, 843
Air transport, 420
Air transportation, 426
Air velocity, 329, 357, 359, 360, 367
Ajowan (Ptychotis ajowan), 648
Alaska, 303
A. lawrencella, 436
Albania, 198
Albedo, 161, 162, 168, 185–187, 190
Albumin, 984
Alcea rosea, 409
Alcohol, 120, 215, 244, 247
Alcohol content, 210
Aldehyde, 120, 251
Alder, 716, 1186
Aldrovandi, 1270, 1271
Alexander the Great, 160, 649
Alexander von Humboldt, 441
Alfalfa, 217, 839, 1303
Algae, 514, 1033
Algeria, 166
Ali al-Masudi, 160
Alien, 829
Alien insect, 1175
Alien invasive species (AIS), 1172, 1174
Alien invertebrate, 1175
Alien pests, 1174
Alien species, 519, 837, 1175
Alien terrestrial invertebrates, 1175
A Life Cycle Assessment, 509
Alkaloids, 650, 656, 967, 981
All-America Selection (AAS), 415
Allelopathic compounds, 750
Allergenicity, 116
Alliaceae, 987
Alliums, 823, 976–978, 981, 987, 1234
Allocation of resources, 840
Allocation patterns, 838
Allometric relationships, 222
Allo-octoploid, 304
Allotment, 954, 962, 1001
Allotment gardening, 682, 1012, 1123
All Saints Day, 415
Allspice, 649
All's Well that Ends Well, 1247
Allyl (methyl-vinyl 2-propenyl), 977
Almassora, 166
Almeria, 367
Almeria-Spain, 346
Almond (Prunus amygdalus), 99, 101, 109, 110, 289, 605, 973, 1259, 1270
Almond (Prunus dulcis), 289
Alpha-carotene, 119
Alpha-linoleic acid, 966
Alphand, 679
Alpha-Tocopherol Beta-Carotene (ATBC), 980
Alpine flora, 804
Alpine tundra, 439
Alstroemeria pelegrina, 409
Alstroemerias, 409
Alternaria brassicae, 828
Alternaria brown spot, 166
Alternate bearing, 177
Alzheimer's disease, 267, 1054
Amazon, 270, 285
Amazon basin, 267, 270, 284, 438, 442
Amenity
  and environmental, 13
  grasses, 13
  grasslands, 13, 732, 734, 736, 752
  horticulture, 702, 788, 1119, 1120, 1122, 1123, 1235
  or ornamental horticulture, 1119
  plants, 446, 456
America, 124, 134, 135, 139, 264, 270, 282, 283, 412, 439, 648, 649, 661, 683, 821, 1173, 1176, 1206, 1219, 1258, 1289–1291, 1293, 1302, 1303
America irrigation technology, 1302
American Dietetic Association, 91
American Fruit Grower, 1239
American fruit industry, 1290
American Heart Association, 1050
American Horticultural Society, 1236
American Horticultural Therapy Association, 799
American mayapple (Podophyllum peltatum), 650
American Phytopathology Society, 1239
American Robin (Turdus migratorius), 1033
American Rose Society, 1236
American Social Science Association, 678
American Society for Horticultural Science (ASHS), 1077, 1239
American tropics, 720
Amerigo Vespucci, 1219

Amillaria spp., 231
Amino acids, 76, 210–213, 215, 249, 383, 627
Amino-N, 213
Amira, 308
Ammonia fertilizers, 1298
Ammonia (NH3), 457, 611, 614, 773, 1298, 1299
Ammonium (NH4+), 213, 215
Ammonium nitrogen, 1300
Ammonium sulphate, 1300
Amormorphallus paeoniifolius, 408
Amphibians, 835, 1026, 1027, 1031
A. muricata, 140
Amyloid β peptide, 988
Anacardiaceae, 147
Anacardium occidentale, 289
Anaerobic digester (AD), 476
Analgesics, 771, 981
Anchorage, 698
Andes, 439
Andrea del Verrocchio, 412
André Le Nôtre, 674, 1202, 1278
Androecium, 141
Anemone, 412
Angelonia angustifolia, 454
Angola, 438
Anicius Olybrius, 1218
Anigozanthos spp., 436
Animal Assisted Activity, 1015
Animal drugs, 867
Animal Feed, 715
Animal waste, 1145
Animal welfare, 1146
Anlage, 206
Anne de Bretagne, 1219
Annona, 140, 142, 143
Annonaceae, 139, 141
Annual bedding, 443
Annual crops, 613
Annual of Sicilian Agriculture, 1296
Annuals, 837
Annuals cornflower, 1032
Anoxia, 376
Antheraxanthin, 338
Anthesis, 98, 113
Anthochaera chrysoptera, 1032
Anthocyanidins, 119
Anthocyanin, 119, 165, 176, 212, 219, 236, 245, 250, 251, 307, 315, 351, 970
Anthony and Cleopatra, 1247
Anthracnose, 231, 845
Anthraquinones, 650
Anthropocene, 801
Anthropomorphic interpretations, 1027
Anthurium spp., 418, 453
Anthus pratensis, 1035
Antibacterial properties, 977
Antibiotics, 661
Anticancer, 650
Anticarcinogenic, 981, 988
Anti-establishment activism, 865
Antifungal properties, 978
Anti-inflammatory, 981
  responses, 982
Antimicrobial, 981
Antimutagenic, 981
Antiobesity, 981
Antioxidants, 76, 382, 383, 970, 978, 979, 981–983, 990, 1241
  activity, 843, 982
  capacity, 970, 980, 982, 983
  content, 846
  effect, 982
  enzyme, 985
Antioxidative capacity, 383
Antirrhinum majus, 408
Anti-transpirants, 186, 633
Antitumoral, 981
Anti-tumoral properties, 987
Antroposophic, 863
Antroposophism Movement, 862
Antroposopism, 862
Ants, 751, 1033
Ant species, 1175
Anxiety, 1013
Anxiety disorders, 1007
APETALA1 (AP1), 105
Aphid-borne viruses, 309
Aphids, 232, 306, 751, 832–834, 846, 847, 1300
Aphid-vectored viruses, 832
Apical dominances, 99, 102
Apical meristem, 102, 103
Apiculture, 721
Apis mellifera, 146, 180
Apocarotenoids, 980
Apomixis, 109, 113
Apoplast, 381
Apoplastic, 209
Apoptosis, 981
Apospory, 113
Apothecaries, 1235
Appalachian Mountains, 722
Appellation, 200

Apple, 1234, 1235
Apple (Malus domestica), 99, 100, 101, 109, 111, 112, 114, 118–120, 198, 235, 289, 344, 428, 474–476, 605, 613, 626, 822, 831, 844, 845, 850, 1183, 1233, 1257, 1260, 1264, 1271, 1273, 1283, 1286, 1288–1294
Apple-scab, 830
Application of organic manures and the application of manufactured fertilisers, 516
APPPC, 1181, 1184, 1185
Apprenticeship system, 1122
Appropriate level of risk (ALOP), 1173, 1181, 1184, 1188, 1191
Apricot, 1247
Apricot (Prunus armeniaca), 99, 100, 101, 109, 119, 821, 822, 1264, 1291, 1303
Aquaculture, 868
Aquaporin, 84
Aquatic ecosystems, 507
Aquatic life, 505
Aquifers, 805
Aquilegia, 455
Arabia, 266
Arabian Gulf States, 274
Arabic, 765
Arabica, 721
Arabica green bean, 1154
Arabidopsis, 84, 103, 106
Arab poetry, 1244
Arabs, 148, 160
Arachidna, 637
Aral sea, 606
Aranjuez, 674
Arborator, 673
Arboreta, 1001
Arboricoltura, 1295
Arboricoltura generale, 1296
Arboricultural, 708
Arboriculture, 13, 694, 702, 1119, 1285
Arbres fruitiers ou Pomonomie belge, 1288
Arcadia, 673, 674
Archaeology, 849
Architects, 783
Arctic, 804
Arctostaphylos uva-ursi, 447
Arduaine, 716
Areas of Outstanding Natural Beauty (AONBs), 795
A. reticulata, 140
Argentina, 139, 164–166, 170, 172, 199, 200, 317, 438, 697
Arginine, 209, 211, 215
A. rhodanthea, 436
Arid, 381
Arid climates, 778
Arid environments, 841
Arid regions, 341, 354, 440
Aristotle, 160, 1265, 1266, 1269
Arizona, 86
Armendariz and Morduch 2010, 1148
Armenia, 198
Armeria maritima, 455
Armillaria, 635
Armyworm, 751
Arnold Arboretum Horticulture Library, 1249
Aroids, 408
Aroma, 120, 212, 241, 242, 244, 248, 249, 385, 646, 650, 662, 663
Aroma-active compounds, 244
Aroma molecules, 211
Aroma note, 249
Aroma signal, 650
Aromatic, 843
Aromatic and medicinal, 13
Aromatic plants, 451, 457, 645, 646, 650, 662
Aromatic ring, 212
Aromatic wines, 240
Arrack, 722
Arrhenius, 819
Ars Poetica, 675
Art, 1197, 1205
Art deco, 764
Arteriosclerosis, 966
Arthropoda, 636, 637
Arthropod predators, 833
Arthur Tansley, 795
Artichokes, 78
Artificial environment, 1199
Artificial lighting, 336
Artificially raising CO_2 concentrations, 507
Artificial ripening, 12
Artificial substrates, 371
Artisan-farmer culture, 1256
Artistic, 1197
expression, 1197, 1205
Art Nouveau, 413
Art therapy, 954
Ascertain the impacts of a proposed process, 508
Ascomycetes, 1175
Ascorbic acid, 126, 190, 315, 335, 337, 353, 375, 845
Ash, 1260

Ash dieback, 1063
Ash trees, 1063
Asia, 76, 84, 99, 124, 128, 159, 160, 199, 268–270, 282, 284, 312, 314, 438–441, 453, 457, 466, 471, 648, 649, 662, 674, 703, 708, 721, 723, 733, 790, 826, 829, 1173, 1175, 1176, 1184, 1185, 1200
Asia and Pacific Plant Protection Commission (APPPC), 1184
Asia Minor, 97
Asian plum, 101
Asia Pacific region, 271, 286
Aspalathus callosa, 442
Asparagus, 78, 85, 606, 613, 981, 1264
Aspidistra, 767
A. squamosa, 140
Assam, 147
Assimilate supply, 340
Assyro-Babylon, 1301
Aster, 409, 766
Asteraceae, 436
Asthma, 775
Astilbe, 455
Astringency, 212, 239, 249, 250
Astringent, 250
Atemoya, 140
Aterra de Flamengo Park, 680
Atherosclerosis, 980, 984, 985
Atlantic poison oak (Toxicodendron pubescens), 722
Atria, 768, 777
Atriplex, 440
Attalea speciosa, 715
Attentional recovery, 1038
Attention deficit disorder (ADD), 1005, 1014
Attention restoration theory (ART), 1016, 1037
Attentiveness, 770
Aubert, C., 1149
Auction house, 421
Aurantioideae, 161
Australasia, 441, 733, 804, 806
Australia, 91, 98, 124, 136, 164, 170, 172, 199, 200, 219, 223, 229, 232–234, 238, 264, 266–268, 270–272, 281, 289–293, 304, 317, 318, 418, 421, 438–444, 446, 448, 450, 454, 456–459, 716, 719, 721, 733, 738, 744, 756, 795, 806, 831, 839, 862, 868, 1029–1034, 1040, 1118, 1119, 1121, 1125–1129, 1134, 1172, 1177, 1181, 1183, 1184, 1187, 1188, 1190, 1191, 1238, 1302, 1303
Australian Centre for International Agricultural Research, 1161
Australian Football League (AFL), 755
Australian horticultural education, 1129
Australian macadamia, 289
Australian Productivity Commission, 1128
Australian Quarantine Act (1908), 1178
Austria, 860, 1174
Autocatalysis, 117
Autogamy, 146
Automation, 1131
Autophosphorylation, 111
Auxin, 114, 115, 117, 181–183, 188, 191, 625–627
Auxin-ethylene, 117
Available Work Days (AWD), 831
Avicenna, 648
Avocado (Persea americana), 124, 125, 128, 143–147, 271, 272, 473, 480, 1271
Avocados, 1235
Awaji Island, 685
Awaji Yumebutai, 685
Axillary meristems, 105
Axonpus affinis, 734
Ayurveda, 648
Ayurvedic, 648
Ayurvedic medicine, 648
Azaleas (Rhododendron spp), 420, 621, 634
Azamboa, 160
Aztecs, 267, 284

B

Babaçu palm, 715
Baby carrots, 860
Babylon, 160, 678
Baby's breath, 409, 417
Bachelor's button, 409
Backcross introgression, 87
Bacteria, 831, 1030, 1175, 1179
  algae, 514
  plants, 514
Bacterial symbionts, 833
Bacterocera dorsalis, 845
Baghdad, 160
Bahamas, 168
Bahia, 285
Bahrain, 91
Bailey, L.H., 1288
Balconics, 456
Balcony plantings, 777
Balcony plants, 454
Balkan, 199

Balled and burlapped, 640, 641
Ball impact, 754
Ball-surface interaction, 753, 755
Baltic Sea, 675
Baltic States, 314
Bamboos, 764, 778, 1236
Banana, 120, 125, 127–131, 133, 134, 198, 264, 271, 274–280, 282, 285, 332, 345, 360, 369, 474, 610, 841, 844, 1181
  and plantains, 1234
  screenhouse, 345
Banana Cavendish, 473
Banana Musa AAA, 132
Banana Musa spp., 124
Bangladesh, 717, 721
Banksia, 418, 446, 1032
Banksia (Banksia hookeriana), 440
Banksia menziesii, 446
Banksia Production Manual, 458
Banksias (Banksia marginata), 409
Banksia spp., 442, 458
Banyan, 718
Baobab (Adansonia digitata), 715
Baptiste Van Helmont, J., 1301
Barbados, 168, 1032, 1033
Barberry, 1178
Barberry Berberis thunbergii, 457
Barcelona, 678
Bare-root, 640
Bare-rooted plants, 1177
Bargaining cooperatives, 1158
Bark, 378, 717, 722, 724, 726
Barley, 1206
Baroque, 411, 412
  painters, 1213
  park, 685
Bartholomew and Associates, 682
Bartlett, E., 1294
Bartolomeo Bimbi, 1214, 1282
Basf, 1298
Basil, 663
Basilio 2008, 1147
Basipetal gradient, 103
Basitonic, 102
Basket press, 240
Bassila, 722
Bastanbón, 160
Bat, 1031
Baths, 1028
Batonage, 241
Batrachochytrium dendrobatidis, 835
Bats, 724, 726, 1043
Battata Virginiana sive Virginianorum & Pappus, 1222
Batt, P.J., 1144, 1154, 1155, 1158, 1159
Baumea spp., 457
Bay, 663
B. cinerea, 226
Beans, 86, 1235
Bean sprouts, 1145
Bearded grapes, 1271
Beaujolais, 239
Beautiful garden art, 676
Becchetti, L., 1159
Bedding
  crops, 1226
  plant, 13, 408, 446, 452, 454, 456
Beech, 725, 840, 1185, 1295
Beech (Fagus sylvatica), 825
Beehives, 720
Beekeeping systems, 721
Bee-quarters, 1031
Bees, 109, 180, 751, 832, 1026, 1028, 1030, 1032, 1033
Beet, 77, 823
Beetle, 751, 1031
Beetroot, 1288
Beets, 85, 1300
Begonias, 443
Belgian Congo, 722
Belgium, 781, 1041, 1217, 1288
Belize, 272
Bell flower, 412
Bell pepper, 353, 846
  fruits, 356
Bemisia afer, 847
Bemisia tabaci, 847
Benchmarking, 701, 756
Beneficial insects, 751
Beneficial predators and parasitoids or important pollinators, 506
Ben Gairn, 316
Benign microbes, 823
Benign vertebrates, 835
Benin, 722
Benonite, 251
Bent neck, 356
Bentonite, 251
Benzene, 773
Benzene ring, 970
Benzene ring (C6H6), 250
Berberis, 627, 1031
Berberis thunbergia, 446
Berberis thunbergii, 457, 623

Berberis vulgaris, 1178
Berlin, 676
Berlin and Stuttgart Artificial Athlete, 753
Berlin Botanic Garden, 441
Bermuda grass, 748
Bernoulli equation, 359
Berries, 127, 201, 203–205, 207–210, 212, 218, 226, 229, 980, 986–988
Berries (avocado), 127
Berry, 133, 205, 207–214, 219, 227, 229, 245
  industry, 310
  maturity, 211
  metabolism, 211
  number, 205
  organisms, 243
  phenolics, 219
  ripening, 201, 210, 214
  set, 218
  size, 208, 209, 218, 219
  skin, 209
  softening, 208
  splitting, 227
  sugar, 227
  volume, 209, 210
  weight, 205
Berry polyphenols, 989
Best Bet Program, 448
Best practice, 512, 516
Beta-carotene, 843
Beuchelt, T.D., 1159
Beverage crops, 860
Beverages, 646, 722
Bhutan, 147
Bianchetti, 169
Biannual bearing, 104
Bible, 160
Bicarbonates, 746
Biddulph Grange, 672
Big Garden Bird Watch, 1036
Big step' innovation, 1134
Big vine, 223
Bioactive compounds, 967, 990
Bioactive ingredients, 659
Bioactive phytochemicals, 983, 986
Bioactives, 651
Bioactivity, 646, 980
Bioavailability, 967, 983, 984, 990
Biochar, 378
Biocontrol, 781
Bio control agents (BCAs), 850
Bioconversion, 242
Biodiesel, 269
Biodiversity, 5, 278, 605, 724, 732, 788–790, 792, 794–807, 809, 810, 860, 867, 1002, 1026, 1030, 1042, 1043, 1062, 1063, 1066, 1067, 1141, 1179, 1283
  conservation, 808
  loss, 803
Biodynamic
  agriculture, 862, 864
  farming, 863
Biodynamic Association, 864
Bio-Dynamic Farming and Gardening, 863
Bioenergy, 639
Bioflavonoids, 1241
Biofuel, 76, 266, 269
  production, 1147
Biofumigation, 233
Biogenic amines, 215
Biological, 517
  activity, 984
  control, 306, 750, 1175
  control agents, 518
  control, habitat manipulation, 517
  corridors, 724
  diversity, 788, 790, 809, 1172, 1266
  filtration, 513
  invasions, 1175
  nitrogen fixation, 516
  pest and disease control, 513
  processes, 514
  resources, 809
  rhythms, 818
  systems, 819
Biomass, 203, 233, 244, 245, 335, 338, 352, 360, 362, 363, 376, 377, 382, 514, 605, 607, 610, 774, 824, 836–838, 841
Biomes, 437, 459, 764, 788
Biophilia, 772, 954, 1036, 1037, 1042
Biophilia hypothesis, 1004
Bio-prospecting, 806, 809
Bioreactor, 651
Bioregulators, 109, 112
Biosecurity, 504, 1173, 1176, 1178–1180, 1182–1188, 1191
  capability, 1127
Biosurfactants, 513
Biotechnology, 10
Biotic stresses, 321
Birch, 716, 722, 726
  beer, 722
  sap, 722

Birds, 724, 726, 835, 839, 1027, 1028, 1041, 1043
  baths, 1031
  damage, 227
  feeders, 1033, 1036
  feeding, 1034
  food, 1042
  habitat, weed and insect control, 513
  life, 724
Birdsong, 1026
Birmingham Botanical Gardens, England, 1040
Biscogniauxia mediterranea, 825
Bitterness, 239, 249, 250
Biuret, 190
Blackberries, 310–313, 1290
Blackberry, 99, 310, 311, 313, 1267
Blackbirds, 1032, 1033
Black blight, 1300
Black-capped chickadees (Poecile atricapillus), 1033
Blackcurrant reversion virus (BRV), 316
Blackcurrants, 313–317
  clearwing, 316
  gall mite, 316
Blackman, F., 1301
Black raspberries, 307
Black rot, 831
Black scurf, 848
Black Sea, 199
Black Sigatoka, 279
Black spot, 231
Black stem rust, 1178
Blechnum gibbum, 764
Blissus leucopterus hirtus, 1029
Blister rust, 829
Blood oranges, 162, 165
Blood pressure, 770, 799
Blood Tree, (Harunga madagascariensis), 722
Blossom end rot (BER), 340, 352, 355, 356, 363, 364, 384, 843
  of tomatoes, 339
Blue baby syndrome, 505, 865
Blueberries, 310, 317–319, 321, 1234
Blueberry, 99, 320, 321, 973, 989, 1230
  anthocyanins, 989
  diet, 989
  extract, 987, 989
  juice, 989
  polyphenol, 989
  supplementation, 989
Blue Jay (Cyanocitta cristata), 1033
Blue water, 607, 608
B. napus, 828
Body mass index, 985
Bogota, 416
Bohemia, 1246
Bolivia, 87, 139, 289
Bologna University, 1271
Bolting, 823
Bonsai, 672, 1174, 1203
Bonseki, 1203
Borassus aethiopum, 722
Borate, 381
Borax, 750
Border, 446
Borneo, 286
Boron, 210, 634, 746
Bosch, C., 1298, 1299
Boston, 679
Boston Conference on Distribution, 1151
Botanical gardens, 13, 447, 799, 808, 1057
Botanic gardens, 13, 441, 447, 766, 790, 796, 803, 1040, 1075
Botanic Gardens Conservation International (BGCI), 796, 1133
Bot canker, 231
Both fresh and saltwater, 505
Botryosphaeria, 231
Botrytis, 227, 228
  bunch rot, 226
  cinerea, 226, 232
  grey mould, 355
Bottlebrush, 442, 1032
Bottle gourd, 1210, 1214
Bottling, 242
Bottom line benefit, 512
Boulevards, 678, 798
Bound sulfur dioxide, 251
Bouquet, 120
Bouquetier, 1274
Bower manuscript, 648
Bowling green, 736, 741
Box, 436
Box blight, 1186
Box (Eucalyptus angophoroides), 719
Boxwood (Buxus spp), 621
Boysen, 310
B. prionotes, 446
Brachyscome multifida, 448
Brain, 987
Brain function, 1005
Branding, 474
Branding factor, 685

Brasilia, Brazil, 683
Brassica, 611, 828, 831, 1213
Brassicaceae, 111, 828, 987
Brassicacea spp., 973
Brassicales, 134
Brassica napus, 828
Brassica rapa var. chinensis, 829
Brassicas, 77, 86, 605, 823, 828, 829, 836, 977, 982, 988, 1029, 1235
Brassica vegetables, 988
Brassinosteroids, 118, 208
Brazil, 8, 135, 136, 161, 163, 164, 268–272, 284, 285, 287, 289, 306, 414, 441, 675, 715, 717, 719, 781, 782, 795, 808, 822, 868
  fruit, 721
  nuts, 289
Bread
  fruit, 127
  fruit (Artocarpus altilis Fosb.), 126
Breast, 984, 987
Breast cancer, 987
Breast cancer cells, 987
Breeding, 430
Breeding performance, 835
Briggs, L.J., 1302, 1303
Britain, 828, 1178, 1179, 1183
British Association of Landscape Industries (BALI), 1133
British Colombia, 720, 722
British Trust for Ornithology (BTO), 1036
Brix, 175, 210
Broadacre City, 682
Broad beans, 1264
Broadleaved species, 778
Broccoli, 78
Broccoli florets, 976
Broken tulips, 412
Brokerage, 420
Brokers to grower, 420
Bromeliad, 271
Bronx-River-Parkway, 683
Bronze Age, 1206
Brookings, Oregon, 421
Bryant 1989, 1150
Bryant Park, 961
Bryophytes, 775
B. tabaci, 847
Buchloe dactyloides, 734, 737
Bud
  break, 98
  burst, 201
  differentiation, 104, 105
  fruitfulness, 225
  grafting, 626
Buddhist, 409
Budding, 625
Buddleia, 1031
Buddleia americana, 439
Buddleia davidii, 1031
Buddleia spp, 627
Building, 720
Building materials, 8
Bukina Faso, 721
Bulb crops, 1210
Bulbs, 1174
Bulgaria, 413
Bulk
  density, 698, 779
  produce, 468
Bullfinch, 1033
Bumblebee Conservation Trust, 1036
Bumble bees, 390
Bumblebees, 832
Bunch bunchstem necrosis (BSN), 215
Bunch end rot, 846
Bunch rot diseases, 232
Bundesgartenschau, 686
Bunjae, 672
Burbank, L., 1289, 1290
Burkina Faso, 840, 841
Burlap, 640
Burlapping, 640
Bürolandschaft, 767
Bush
  fruit, 832
  land, 802
  pickers, 442
Business
  as usual, 1140, 1151
  development, 1146
  development services, 1149
  environment, 1149
  models, 1161
  services, 1149
Butterflies, 832, 1026–1029, 1031, 1032, 1040–1042
Butterfly bush, 439, 1031
Butterfly Conservation Society, 1036
Butyrospermum paradoxum, 715
Butyrospermum parkii, 715
Buxus sp., 436
Buyer
  driven models, 1156, 1157
  power, 477

Byturus tomentosus, 309
Byzantine, 409, 411
Byzantine era, 1266
Byzantine period, 411
Byzantium, 411

C
C3 grasses, 733, 734, 737, 744, 747
C3 plants, 837, 838
C3 turfgrasses, 750
C4 grasses, 737, 744, 747
C4 plants, 837, 838
C4 turfgrasses, 733, 737, 744
C4 weeds, 838
Cabbage, 78, 85, 86, 353, 842, 847, 1264
Cabbage looper, 1029
Cabbage root fly (Delia radicum), 834
Cabernet Sauvignon, 212
Cacao, 1208
Cacatua galerita, 1029
Cacti, 439, 778, 841
Cactus pear, 841
Calçadão de Copacabana, 680
Calceolaria, 419
Calcium, 85, 109, 210, 211, 355, 384, 385, 634, 715, 967
Calcium carbide, 1298
Calcium carbonate, 190
Calcium chloride, 1300
Calcium cyanamide, 1298
Calcium deficiencies, 355
Calcium-dependant protein kinase, 84
Calcium nitrate, 185, 186, 190
Calcium oxalate, 355
Calcium sulphate, 385
California, 139, 164, 165, 167, 168, 170, 172, 289, 302–304, 312, 415, 417, 442, 444, 446, 456, 608, 621, 680, 681, 781, 821, 822
Californian almond, 291
Californian chaparral, 438
Californian extension system, 1128
California Spring Trials, 415
California USA, 418
California Visual Learning Test, 989
Callistemon, 1032
Callistemon spp., 442
Callose, 108
Calories, 967
Calvert Vaux, 679
Camellia, 409
Camellia sasanqua, 443
Camellia sinensis, 266, 289, 441
Cameroon, 284, 285, 715, 719, 1181
Campanian villas, 673
Campanula pyramidalis, 766
Campanulastrum americanum, 821
Canada, 91, 97, 136, 317, 378, 411, 421, 446, 723, 822, 828, 830, 868, 1029, 1030, 1032, 1033, 1120, 1173, 1238
Canadensis, 623
Canals, 714, 716
Canary Islands, 130, 136, 148
Canary Islands (Spain), 127
Canberra, Australia, 683
Cancer, 650, 966, 981, 982, 987, 1013, 1050, 1051
  incidence, 978
  prevention, 982
Cancer prevention, 982
Candida, 244
Candle, 722
Canellales, 143
Cangshan, 674
Canlas, D.B., 1147, 1148
Canna, 437
Canna (Canna x generalis), 437
C. annuum, 86, 87
C. annuum var. aviculare, 87
C. annuum x C. baccatum, 87
Canola, 233
Canopy, 203, 225, 226, 337, 358, 360, 366–368, 390
  architecture, 210, 225, 638
  development, 201
  health, 236
  management, 223, 225, 228, 235
  net radiation, 366
  resistance, 366
  structure, 330
  temperature, 221, 222, 236
  volumes, 236
Capacity building, 1118, 1124, 1157
Cape Biosphere Reserve, 442
Cape heaths, 442
Cape Hyacinths, 442
Cape Province, 438
Cape reed, 442
Capillary mats, 374
Capsaicinoids, 981
Capsaicins, 981
Capsicum/bell pepper aroma, 249
Capsicum peppers, 1216
Capsicum spp., 267, 649, 981

Capsid bugs, 306
Capsules (durian), 127
Carambola, 125, 128
Carassius auratus, 1030
Caravaggio, 1212
Caraway, 663
Carbohydrate, 76, 118, 147, 174, 177, 181–183, 203, 204, 206, 207, 213, 214, 225, 242, 244, 245, 247, 351, 362, 364, 388, 986
  partitioning, 182
  reserves, 741
Carbon (C), 633
  allocation, 382
  balance, 839
  efficient, 520
  emission, 520, 850
  foot prints, 426, 427, 509, 511, 611, 612
  sequestration, 603, 839, 1058
  sink, 513
  storage, 840
Carbon dioxide (CO2), 239, 246, 329, 360, 365, 427, 428, 451, 612, 613, 725, 790, 802, 803, 819, 820, 824–826, 836–838, 843, 846, 848, 850, 1299, 1301
  enrichment, 362
  fixation, 808
  foot prints, 428
Carbon dioxide equivalent (CO2e), 427, 507, 612
Carbon exporters, 203
Carbonic maceration, 238, 239
Carbon isotope discrimination (δ13C), 222
Carbon monoxide (CO), 451, 773
Carcinogenesis, 987
Carcinogens, 987
Cardamom, 267, 663
Cardinal, 1030
Cardinalis cardinalis, 1030
Cardiovascular
  diseases, 966, 978–980, 982, 984, 1007, 1013, 1050
  health, 984
  homeostasis, 985
  respiratory fitness, 1005
Carduelis carduelis, 1030
Carduelis tristis, 1030
Care Farming, 1015
Careless, 315
Cargill, 294
Caribbean, 264, 273, 274, 284, 438, 439, 721
Caribbean islands, 285
Carica, 137
Caricaceae, 134
Carica papaya, 137, 844
Carl Gottlieb Bethe, 676
Carl Linnaeus, 1172, 1233
Carl Per Thunberg, 441
Carmine production, 841
Carnations, 380, 412, 413, 419, 443, 453, 1246
Carnivore, 1029
Carob (Ceratonia siliqua), 648
Caro, N., 1298
Carotene, 335, 351, 980
Carotene (provitamin A), 126
Carotenoid pigments, 980
Carotenoids, 119, 175, 176, 182, 184, 335, 336, 351, 383, 979, 980, 982–984, 990
Carpodacus mexicanus, 1035
Carrizo citrange, 187
Carrots, 77, 85, 86, 361, 612, 823, 842, 982
Cartagena Protocol on Biosafety (CP), 1179
Carthage, 1264, 1265
Carya illinoinensis, 289
Casein, 250
Cashews, 289
Caspian, 199
Caspian seas, 100
Cassava, 1181, 1182
Cassia, 663
Castanea sp., 289
Castasterone, 208
Castilla elastica, 270
Castle of Mey, 716
Castle of Racconigi, 1258
Castor oil, 269
Casuarina spp., 723
Catalases, 353, 982
Catalpa, 639
Catalytic degradation, 477
Caterpillar, 1029
Catholic Relief Services (CRS), 1160
Cation exchange capacity (CEC), 191, 378, 628
Cato, 1259, 1262, 1264
Catonis, 1262
Cato the Censor, 1262
Cattle, 519, 716
Cattle fodder, 715
Cattleya, 418
Caucasus, 99, 100
Cauliflory, 128
Cauliflower, 78, 612, 842, 1213
C. aurantifolia, 162

C. aurantium, 162, 171
Causes of Plants, 1233
Cavendish, 128, 133, 134, 271, 276, 279, 844
Cavendish bananas, 133
C. chinense, 86
Ce-based resistance, 316
Cecidophyopsis ribis, 316
Cedar, 635, 1271
Cedar-apple rust, 635
Cedrus spp, 456
Celeriac, 842
Celery, 78, 384
Celery (Apium graveolens), 648
Cell signaling, 967
Cell-signaling action, 967
Cellulose, 126
Cell volume, 79
Cemeteries, 1001
Centaurea cyanus, 409, 1032
Center for Plant Conservation (CPC), 796
Center for Urban Horticulture, 1077
Center pivot, 81
Centipedes, 751
Centradenia, 448
Central Africa, 1182
Central African Republic, 719
Central America, 91, 135, 139, 143–146, 267, 269–271, 274, 284, 438, 719, 721, 795
Central England Temperature Archive, 820
Central Europe, 454
Centralised model, 1156
Centralised procurement, 1144
Central nervous system, 985, 986
Central Park, 679, 1122
Centre for redistribution, 1176
Centre of Phytosanitary Excellence (CoPE), 1182
Centrifugation, 240
Ceramic pots, 765
Cercis, 623, 627
Cercosporella rubi, 313
Cereals, 77, 973, 1144
Cerebral inflammation, 988
Certification, 1180
- process, 868
- programs, 424, 431
- schemes, 1188
- systems, 868

Certified EMS, 508
Certified pest-free, 1180
Cervix cancer cells, 987
Ceylon, 270
Chaenomeles japonica, 625
Chaffing, 737
Chalara fraxinea, 795
Chalcones, 119
C. halimii, 162
Chalk down-land, 1002
Chamber of Agriculture in Germany, 454
Chamelaucium, 444
Chamelaucium spp., 444, 446, 450
Chamelaucium uncinatum, 456
Chandigarh, Punjab, India, 683
Changi airport, 768
Changing climate, 310
Changing demographics, 18
Charaka Samhita, 648
Charcoal rot, 848
Chardonnay, 240, 246, 249
Charles Darwin, 441
Charles VIII, 1219
Checkland 1981, 1152
Chelsea Flower Show, 797
Chemical control, 781
Chemical growth inhibitors, 741
Chemical pesticides, 506
Chemical residues, 1145
Chemicals Regulations Directorate, 781
Chemie appliquée à l’agriculture, 1300
Cherimoya, 139–143, 146
Cherimoya (Annona cherimola Mill.), 128
Cherimoyas, 139
Cherries, 100, 101, 109, 120
Cherry, 100, 101, 413, 725, 839, 1234, 1264, 1269, 1286, 1289, 1291, 1292
Cherry (Prunus avium), 100
Cherry tomatoes, 353, 380, 382–384, 386
Chervil, 663
Chestnut, 99, 109, 110, 289, 1270, 1295
Chibanda, M., 1158
Chicago, 678
Chicory, 1235
Children’s health, 1050
Children’s Park, 679
Chile, 91, 139, 198–200, 271, 272, 303, 317, 318, 419, 438, 439, 830, 1035
Chilean Matorral, 438
Chile piquin, 87
Chili pepper, 77, 663
Chilled fruit, 276
Chilli, 267
Chilling
- hours, 821, 822
- injury, 353
- requirement, 98, 316
- units, 823, 845

Chilli pepper, 981
Chimeral, 151
Chimney flower, 766
China, 8, 76, 100, 101, 159, 161, 165, 166, 172, 199–201, 252, 264, 266, 268–270, 272, 289, 304, 306, 315, 317, 328, 409, 413, 414, 419, 436, 438, 439, 441, 613, 614, 648, 663, 672–675, 717, 721, 723, 765, 781, 782, 804, 828, 842–844, 863, 868, 1032, 1174, 1184, 1199, 1200, 1226, 1237, 1293, 1301
Chinese cabbage, 352, 829
Chinese medicine, 648
Chinese New Year, 414
Chinese Poetry, 1245
Chipmunks, 1029
Chiquita, 278, 294
Chives, 977
Chloride, 381
Chlorinated water, 779
Chlorine, 634, 746
Chlormequat, 453
Chlorophyl-a, 390
Chlorophyll, 119, 182, 184, 382
Chlorophyll a, 338
Chlorophyllase, 184
Chlorophyll b, 338
Chlorophyll content, 339
Chlorophyll degradation, 175
Chloroplast, 382
Chocolate, 281–284, 285, 287, 288
Chocolatl, 284
Choke throat, 844
Cholesterol-lowering, 966
Chongsheng, 674
Christian Cay Lorenz Hirschfeld, 679
Christmas, 414
Christmas Bells, 450
Christmas bush (Ceratopetalum gummiferum), 458
Christmas Day, 415
Christmas rose, 454
Christy, R., 1146
Chromista, 229
Chromoplasts, 119, 351
Chromosome doubling, 430
Chronic disease, 14, 966, 981, 982
Chronic inflammatory diseases, 966
Chrysanthemum, 349, 350, 408, 409, 413, 414, 419, 430, 443
Chrysanthemum (Chrysanthemum x grandiflorum), 408
Chrysanthemum City, 414
Chrysanthemum indicum, 380
Chrysanthemums, 350, 409
Chrysanthemum spp., 443, 1041, 1239
Chrysanthemum x grandiflorum, 409
Chuao, 285
Chylomicrons, 983
Chytridiomycosis, 835
Cicero, 1297
Cider, 1234
Cilantro, 663
Cinnamomum zeylanicum, 267
Cinnamon, 267
Citranges, 172
Citrate, 119, 211
Citreae, 161
Citric acid, 182
Citriculture, 190, 194
Citrinae, 161
Citrine, 429
Citron, 160, 169, 1273
Citrulline, 84
Citrullus, 84
Citrullus lanatus, 84
Citrumelo, 172, 173
Citrum piriforme, 1271
Citrus, 109, 111, 113, 119, 159–162, 166, 169–171, 173–184, 188–194, 212, 235, 271, 289, 625, 721, 766, 767, 841, 844, 1235, 1257, 1265, 1267, 1271, 1273, 1276, 1283, 1284
  aurantium, 160
  blight, 172, 173
  lemon, 110
  mealybug, 781
  medica, 160
  pollination, 180
Citrus fruit, 1284, 1295
Citrus limon, 160
Citrus macrophylla, 173
Citrus medica, 160
Citrus nematode, 233
Citrus sinensis, 160
Citrus spp, 289
Citrus Tristeza virus (CTV), 171
Citrus volkameriana, 172
City
  centres, 1009
  environment, 440, 803
  landscapes, 683, 783
  parks, 1002
  planning, 681

Civic decency, 679
Civic pride, 17
Civic spaces, 958
Civitas romanorum, 1262
C. jambhiri, 162, 171
Clarification, 240
Clarified juice, 240
Classrooms, 772, 783
Clay flower pot, 765
Clearwing, 316
Cleistothecia, 228
Clematis chiisanensis, 447
Clementine, 167, 168, 183, 194
Clementine mandarins, 166, 176, 180, 181, 184, 186, 188, 190, 192
Cleopatra mandarin, 171, 187
Climacteric, 117, 118, 477
Climacteric bananas, 476
Climate, 173
- change, 5, 18, 80, 233, 234, 320–322, 520, 694, 697, 708, 724, 801, 803–806, 808, 819, 821
- control, 328, 391, 392, 420
- mitigation, 513

Climate change, 820, 822–827, 829–833, 835–846, 848–850, 1047, 1049, 1058, 1059, 1061, 1063–1066, 1077, 1141, 1142, 1148, 1150, 1183
Climate strategy, 843
Climate systems, 1132
Climate warming, 831
Climatic
- change, 309, 804, 818, 824, 828, 830, 832, 833, 835–837, 842, 851
- disturbances, 1140
- niches, 832
- phases, 822
- stress, 835

Climbers, 764
Climbing plants, 436
Climograph, 346
C. limon, 162
Cling peaches, 1271
Clipping management, 740
Clone propagation, 103
Cloning, 1063
Clover mites, 751
Cloves, 267, 663
Clubroot, 828, 829
Clubroot disease, 828
Cluster capacity, 1160
Clustering Approach to Agro-enterprise Development, 1160
Cluster marketing, 1160, 1162
Cluster marketing groups, 1163
Cluster marketing groups (Cluster MGs), 1158
Cluster maturity, 1162
Clymenia, 161
C. maxima, 162
C. medica, 162
C. megalopetalum, 456
C. nobilis, 167
Coalbrookdale, 676
Coastal systems, 801
Coatings, 466
Coccinellids, 832, 833, 1031
Coccus hesperidium (soft brown scale insect), 781
Cochineal, 841
Cockatoos, 1029
Cocoa, 264, 266, 267, 270, 280–288, 860, 1159
Cocoa butter, 282, 283
Cocoa liquor, 283
Cocoa market, 287
Cocoa powder, 282, 283
Coconut (Cocos nucifera L.), 124
Coconut meat, 269
Coconut oil, 269
Coconut palm (Cocos nucifera), 722
Coconut palms, 719
Coconuts, 125, 127, 264, 267, 269–271, 844, 845
Cocos nucifera, 269
Codes of practice, 516
Codex Vindobonenis, 1218
Codiaeum variegatum, 776
Codron, J.-M., 1145
Coffea arabica, 266, 289
Coffee, 263, 264, 266, 289, 424, 426, 721, 844, 860, 973, 1154, 1159
Coffee rust, 795
Cognitive functioning, 1014
Coir, 378
Colaptes auratus, 1033
Cold
- acclimation, 820
- accumulation, 319
- chain handling, 479
- chilling, 821
- frames, 416
- pitting, 185
- room, 470
- shock, 353
- soak, 238
- storage, 428, 467, 469, 470

stress, 845
supply chain, 89
tolerance, 86, 87
Cold Tolerance, 86
Coleus x hybridus, 776
Colinearity, 304
Collaborative marketing, 1158, 1160
Collaborative marketing groups (CMGs), 1158
Collaborative marketing models, 1158
Collective marketing, 1160
Colletotrichum gleosporoides, 845
Colombia, 91, 139, 416, 419, 426, 439, 782
Colon, 987
Colon cancer, 987
Colorado, 426, 439
Colorado potato beetle (Leptinotarsa decemlineata), 1179
Colorectal cancer, 988
Color intensity, 350
Colour intensity, 183
Columba palumbus, 836, 1032
Columbia, 264, 269
Columbus, 268, 416, 649
Columbus, Christopher, 1219
Columella, 1257–1260, 1262, 1264–1267, 1270
Combretum glutinosum, 723
Combustion of natural gas, 507
Commercial industry, 850
Commission on Phytosanitary Measures (CPM), 1180
Commoditization, 474
Common Bushweed (Securinega virosa), 719
Common carp, 519
Common mandarin, 162, 166
Common oranges, 162
Common Plant Health Regime, 1183
Common reed (Phragmites australis), 457
Common scab, 848
Common sweet orange, 168
Commonwealth of Australia, 268
Commonwealth, Scientific and Industrial Research Organisation (CSIRO), 801
Communication, 1148
Communities, 788
Community, 17, 1055
cohesion, 5, 15, 801, 810
engagement, 1124
facilities, 962
forestry, 707
gardening, 1057
garden project, 1013
gardens, 686 , 790, 962, 1047, 1050, 1054, 1055, 1060, 1061, 1068, 1076, 1119, 1121
health, 1047, 1048
involvement, 705–707
spirit, 17
Community gardens, 1239
Com-munos, 957
Compaction, 504, 747
Companion crops, 233
Compatibility, 109
Competencies of extension, 1118
Complex spike, 132
Components of yield, 828
Compound fruits, 127
Compulsory competitive tendering, 1123
Computational Fluid Dynamics (CFD), 357
Computer-based monitoring, 371
Computerization, 420
Concentrate pesticide, 516
Concepcion, S.B., 1150
Condensation, 333, 334, 344, 355, 368
Condensation flux, 344
Condiments, 1226
Conductance, 344
Confucianism, 675
Congo, 270
Congo basin, 438
Coniferous trees, 438
Conifers, 408, 437, 456, 638, 1236
Conospermum spp., 440
Conservation, 722, 723, 727, 1034, 1035, 1042
Conservation agriculture, 6
Conservation dead wooding, 726
Conservation of wildlife, 1027
Conservation organisations, 1027, 1043
Conservation-oriented water, 723
Conservation or low soil tillage, 513
Conservation psychology, 1036
Conservatories, 416, 777
Constantinople, 411, 649, 1218
Consumer appeal, 621
Consumer demand, 1141, 1148
Consumer horticulture, 1077, 1118–1121, 1129, 1132
Consumers, 869, 870
Consumer trends, 88
Contact with nature, 953
Container-grown, 621, 640, 641, 849
Container-grown ornamentals, 628
Container-grown stock, 850
Containerised, 849
Containerized plants, 621, 631

Containerized transplants, 78
Container nurseries, 622, 629–632
Container production, 621, 622, 631
Container stock, 631
Contaminate surface water, 505
Contamination, 505
Contamination from nitrate and phosphate fertilisers occurs in aquatic ecosystems, 505
Contemporary gardens, 959
Continuous screw presses, 240
Contract farming, 1156, 1157, 1162, 1163
Controlled atmosphere, 477
Controlled atmosphere storage, 466, 467, 477
Controlled atmosphere systems, 477
Controlled environment, 416
Control measures, 517
Convenience attributes, 1145
Convenience factor, 479
Convenience food, 479
Convenience offerings, 477
Convention on Biological Diversity (CBD), 806, 1179
Cook and Chaddad 2004, 1158
Cool-Bot, 470
Cooling, 1059, 1060, 1067
Cool morning, 365
Cool season grasses, 734
Cool storage, 12, 477
Cool temperate forest, 437, 438
Cool temperate regions, 453
Cool transport, 469, 470
Cool transportation, 471
Co-operative Extension, 1124, 1125
Co-operative Extension Offices, 1125
Co-operative Extension System, 1121
Co-operative models, 1158, 1162
Co-operatives, 1149, 1158, 1159, 1162, 1163
Cootamundra wattle, 440
Copaifera demeusi, 722
Copper, 210, 211, 634
Copper hydroxide, 1300
Copper sulphate, 242, 1178, 1300
Coppicing, 639
Cordons, 203
Corn, 1299
Cornell University, 698
Cornflower, 409
Cornucopia, 410, 411
Cornus, 625, 640
Coronary heart diseases (CHD), 982, 984, 1013
Corylus sp, 289
Corymbia ficifolia, 437
Cosmetic act, 91
Cosmetic goods, 646
Cosmetics, 665, 715, 717
Costa Rica, 278, 285, 419, 439, 453, 781
Cost-benefit analysis (CBA), 1003
Cost benefit relationships, 848
Cost-efficient energy, 347
Cosystem, 518
Cote d'Ivoire, 283, 285
Cotoneaster, 636
Cotton, 264, 269, 270, 1159
Council Directive 29/2000/EC, 1183
Country in the city, 1122
Courgette, 1266
Courtyards, 456, 777
Cover cropping, 513
Cover crops, 233, 235, 513, 632, 1303
C. paradisi, 162, 172
Crabapples (Malus spp.), 635, 636
Cracking, 340, 350, 356, 364
Cradle to gate, 427
Craft, 720
Cranberry, 99, 611, 973
Cranfield University, England, 427
Creasing, 167, 186, 187, 190
Creative self-expression, 17
Creative therapies, 955
Creativity, 770
Crescent, 1205
C. reshni, 171
C. reticulata, 162
C. ribis, 316
Cricket, 755
Cricket grounds, 736
Cricket pitch, 755
Crime, 1047, 1054–1058
Crime and disorder, 17
Critical deficiency, 214
Croatia, 414
Cronartium ribicola, 315, 316, 829
Crop, 716, 719–721, 868
  duration, 846
  evaporation, 344
  fertility, 10
  history, 1222
  images, 1205
  nutrition, 10
  pathogens, 847
  photosynthesis, 337
  production, 10, 1147

protection, 517, 610, 723
protection, chemicals for, 322
protection, strategy for, 517
residues, 846
temperature, 344, 354
transpiration, 355, 367, 368
utilization, 475
yields, 1142, 1275
Crop coefficients, 81
Crop cover, 513
Cropland, 723
Crop management, 1233
Crop rotation, 806
Crop Water Stress Index, 221
Crossbreeding, 1289
Cross flow filtration, 241
Cross-pollinated, 179
Crown degradation, 840
Crown development, 701
Crown gall, 231, 636
Crown structure, 702
CRS-Philippines, 1160, 1161
Cruciferaceae, 233
Cruciferous, 78
Crucifers, 1235
Crusades, 765
Crushed grapes, 238
Crushing, 238
Cryphonectria parasitica, 825
Cryptochromes, 98
Cryptoxanthin, 119
Crystal Palace, 417
C. sinensis, 162, 172
C. solstitialis, 409
C. transvaalensis, 734
Cuba, 168
Cuckoo (Cuculus canorus), 818, 1035, 1040
Cucumber (Cucumis sativus), 77, 336–338, 340, 351, 353, 355, 361, 363, 376, 377, 384, 386, 388–390, 842, 846, 1208, 1212, 1214
Cucurbit, 77, 78, 86, 823, 1212, 1213
Cucurbitaceae, 387
Cucurbita pepo, 1214
Cucurbita pepo subsp. texana, 1219
Culinary herbs, 645
Cultivars of Scabious, 1032
Cultivation, 715, 721, 723
Cultural, 517
Cultural component, 1198
Cultural integration, 1012
Cultural management, 752
Cultural practices, 77, 517, 637
Culture of a nation, 17
Cumin, 663
C. unshiu, 162
Cupuacu (Theobroma grandiflorum), 717
Currant, 99, 313, 316, 626
Curtis Botanical Magazine, 1236
Custard apple, 140
Cut flowers, 329, 337, 408, 436, 446, 454, 456, 458, 766, 767, 771, 1129, 1174, 1176, 1226, 1235
Cutin, 970
Cutinases, 109
Cutworm, 751, 834
Cyanococcus, 317
Cyclamen, 335, 340, 419
Cyclamen persicum, 335
Cyclamen spp., 340
Cymose, 137
Cynodon dactylon, 733, 734, 737, 739
Cynodon hybrids, 737
Cyperaceae, 457
Cyperus spp., 457
Cyprinus carpio, 519
Cyrus the Great, 4
Cysteine, 249, 977
Cysteine sulphoxide, 977
Cysteine sulphoxide hydrolysis, 977
Cytokinesis, 112
Cytokinin 6-benzylaminopurine, 449
Cytokinins, 115, 206, 626, 627
Cytoplasm, 977
Czech Republic, 413, 842, 985

D

Dacelo novaeguineae, 1033
Dactylocterium australe, 734
Dados, 1210
Daffodil, 412
Dagger nematodes, 233
Dahlia (Dahlia pinnata), 408
Daily Contraction Amplitude (DCA), 222
Daily light integral (DLI), 335, 338, 348
Daktulosphaira vitifoliae, 1300
Dali, 674
Damping-off, 635
Damping-off pathogen, 635
Damson tree (Terminalia spp.), 648
Dante, 1268
Dark-adapted leaves, 389
Dark green vegetables, 988
Darwin, Charles, 114, 819, 1233, 1258, 1285, 1287, 1288, 1290, 1297

Dasineura tetensii, 316
Date palm (Phoenix dactylifera), 124, 127, 648, 1206
David Douglas, 441
David Livingstone, 441
David Ricardo, 794
Day and night temperatures, 349
Day-length, 98, 336, 449, 454
Day lily, 842
Day-neutral, 305, 449
Day-neutral flowering, 303
Deacetylase/carboxypeptidase, 84
Dead arm, 230
Dead wood, 724, 726, 727
De Candolle, 1288
De causis plantarum, 1233
Decentralisation, 1148, 1149
Deciduous, 98, 102
Deciduous fruit, 106
Deciduous senescence, 104
Deciduous species, 105
Deciduous trees, 98
Decurrent, 638
Deep flow technique, 378
Deep water culture, 378
Defective trees, 726
Defects, 476
Deficit Available Water (DAW), 217
Deficit irrigation, 80, 81, 84, 218, 219, 222, 375, 376
Definitions, 1119
Deforestation, 719
De gesloten kas, 347
Degradation, 719
Degradation of land and water quality, 507
Degree-day model, 834
Degreening, 189
Dehumidifying, 329
De la Quintinye, 1274, 1276, 1285
Della natura delle piante, 1268
Delphi, 1208
Demand for food, 1140
Demand-led extension, 1127
De Materia Medica, 1218, 1233
Dementia, 799, 988, 1005
Democratic empowerment, 687
Demonstration gardens, 1121
Dendometry, 222
Denmark, 313, 419, 860, 1014
Department of Agriculture Fisheries and Forestry (DAFF), 1188, 1190
Department of Agriculture (USDA), 868
Department of Environment, Food and Rural Affairs (DEFRA), 508, 509, 516
Depleted surface water and freshwater aquifers, 507
Depolymerization, 118
Depression, 6, 803, 1013, 1047, 1050, 1051
De re rustica, 1258, 1259, 1265, 1266
Der Wiener Dioskurides, 1218
De Sassure, N.T., 1299, 1301
De Serres, 1274–1276
Desert, 329, 437, 438, 723
Desert date palm (Balanites aegyptiaca), 721
Desertification, 723
Desiccation, 778
Design, 1123
Destructive Insects Act, 1179
Destructive Insects and Pests Act, 1178, 1179
Detergents, 514
Dethatching, 747
Developing world, 520
Development, 1148
Development of nuisance algae, 505
Dewberry, 311
Diabetes, 980, 981, 1123
Diacetyl, 246
Día de los Muertos, 414
Diagnostics, 1175, 1176, 1181, 1182, 1187, 1188
Di-ammonium phosphate (DAP), 215, 244
Dianthus caryophyllus, 380
Dianthus spp., 443
Diapause, 834, 848
Diaporthe perjuncta, 230
Diatomaceous earth, 241
Dichlorodiphenyltrichlorethane (DDT), 864, 1120
Dichogamous, 146
Die Mutations Theorie, 1287
Dies Rosationis, 411
Diet, 1064, 1066–1068, 1241
Dietary antioxidants, 982
Dietary fibre, 967
Dietary Guidelines for Americans (DGA), 91
Dietary health, 798
Dietary Supplement Health and Education Act, 661
Dietary supplements, 646
DIF-concept, 349
Diffenbachia, 782
Digestive illness, 1006
Digitara didactyla, 734, 739
Dihydroflavonols, 119

Dikili Tash, 198
Dill (Anethum graveolens), 662
Dilute pesticide waste, 516
Dimension, 1197
Dimorphism, 128
Dioecious, 136
Diplospory, 113
Directed attention fatigue, 1037
Direct marketing, 468
Direct seeding, 77
Discipline of Horticulture, 16
Discordant behaviour, 772
Disease, 1050, 1053, 1063, 1076, 1078
 control, 829
 emergence, 1174
 epidemics, 825, 1185, 1187
 incidence, 752
 management, 752
 management strategies, 847
 pandemic, 1177
 prediction, 826, 830
 resistance, 83
 risk, 827
 severity, 831, 836
Diseases of civilization, 966
Disocórides, 160
Disrupting bio diverse ecosystems, 505
Distraction therapy, 1012
Distribution, 466, 481
Distribution chain, 10, 421, 467, 1176
Distribution systems, 481
Diurnal cycling, 209
Diversification, 1119
Diversity, 442, 732, 802, 805
Diversity of produce, 1172
Diversity Review, 960
D-lactic acid, 247
DNA damages, 987, 988
Dodder (Cuscutum epythimum), 1271
Dole, 294
Dollar spot (Sclerotina homoeocarpa), 807
Dolomite, 744
Dolomitic limestone, 377
Domestic, 726
Domesticated species, 714
Domestic gardens, 1120
Domestic violence, 800
Dormancy, 98, 201, 205, 206, 444, 622, 623, 819–821, 834, 845, 849
Dormant period, 216
Double blossom rosette, 313
Doubled the supply of reactive nitrogen, 507
Double-layered female nodes, 132
Douglas fir, 1186
Doum palm (Hyphaene thebaica), 715
Dove, 1032, 1033
Downy mildew, 229, 1300
Dracaena draco, 768
Dracaena sanderiana, 455
Dragonflies, 1026, 1028
Dragon plant (Dracaena draco), 437
Drainage, 516
Drained Upper Limit (DUL), 217
Dream of the Red Chamber, 673
Drepanopeziza ribis, 316
Dried fruit, 1291
Drip, 80, 382
Drip irrigation, 6, 292, 374, 378, 380, 382, 846, 1303
Drip irrigation system, 382
Drip line irrigation, 514
Driscoll's Strawberry Associates, 308
Drooping she oak (Allocasuarina verticillata), 457
Drosophila suzukii, 306, 309
Drought, 77, 84, 85, 505, 519
Drought resistance, 850
Drought stress, 83, 355, 364, 387
Drought tolerance, 84, 85, 744
Drought tolerant, 84
Drought tolerant genotypes, 85
Drug, 91
Druid, 718
Druidism, 718
Drupes (mango), 127
Dryland salinity, 802
Dryocopus pileatus, 1033
Dual economies, 1153
Dualistic agrarian economies, 1157
Dualistic agribusiness systems, 1155
Dualistic chains, 1155
Dubai, 685
Du Breuil, 1285–1287, 1295
Duckweed, 408
Dulce et utile, 675, 676
Dumb cane (Dieffenbachia spp.), 458
Dune, 713, 723
Durian (Durio zibethinus), 125, 127, 128
Dutch, 427, 428
Dutch East and West India Companies, 416
Dutch elm disease (Ophiostoma ulmi), 831, 1063, 1185, 1186
Dutch-Flemish periods, 411, 412
Dwarf wheat, 7
Dye, 8, 722
Dyera costulata, 270

E

Early ripening, 318
Earthworm (Lumbricus terrestris), 1033
East Africa, 148, 719, 720, 722
East Asian brassicas, 86
Easter, 414, 415
Easter lily, 421
Eastern Asian, 199
Eastern Europe, 314, 315
Eastern teaberry, 454
Eastern U.S., 311
East India Company, 266
East Malling, 1287
E. balsamifera, 717
Ebb-flood benches, 374
Ebers papyrus, 648
E-books, 1249
E. coccinea, 442
Eco-friendly practices, 788
Ecological balance, 806
Ecological diversity, 788, 1146
Ecological factors, 1173
Ecological fitness, 821
Ecological footprint, 510
Ecological purity, 424
Ecological ranges, 829
Ecological systems, 867, 1132
Ecologists, 1134
Ecology, 702, 867, 870, 1030, 1034, 1035, 1123, 1285, 1299
ECOMAC II, 831
Economic, 1047–1049, 1058, 1064, 1066, 1070–1072, 1074, 1075
Economic benefits, 802, 808
Economic capital, 1150
Economic climate, 865
Economic development, 1143
Economic factors, 1173
Economic gain, 766
Economic growth, 76, 507, 1140, 1144
Economic impact, 1161
Economic migration, 505
Economic stability, 1047, 1049, 1058
Economic sustainability, 797
Economic value, 962
Economic viability, 1124
Economic yield, 846
Eco-regions, 788
Ecosystem resilience, 823
Ecosystems, 278, 439, 732, 788, 792, 801, 806, 809, 829, 860, 867, 1040, 1048, 1061, 1064, 1123, 1172, 1299
Eco-systems analysis, 2
Ecosystem services, 694, 704, 708, 792, 795, 797, 808, 1058, 1123
Eco-therapy, 1003, 1015
Eco-tourism, 8, 797, 807
Ecotypes, 734, 737
Ectotherms, 833, 835
Ecuador, 139, 280, 284, 285, 287, 413, 416, 419, 426, 439, 453, 722
Edaphoclimatic, 128, 129, 135, 140, 143, 145, 148, 173
Eddy covariance, 368, 369
Eddy covariance technique, 366
Eden Project, 763, 779, 797, 809, 960
Edwardian theme, 764
Effective alleviation, 1005
Effluents, 724
Eggplant (Solanum melongena), 78, 337, 350, 351, 352, 384, 390, 846, 1208, 1218, 1244
Egg white, 250
Egypt, 77, 80, 160, 198, 268, 270, 304, 409, 410, 413, 648, 672, 673, 715, 723, 827, 1199, 1206, 1208, 1210, 1235, 1237, 1301
Egyptian, 409, 411, 765
Eichhornia crassipes, 519
Eight Step Clustering Approach, 1160, 1162
E. lata, 231
Elective Affinities, 673
Electrical conductivity (EC), 362, 380, 382, 385
Electricity generation, 797
Electrochromatic glass, 777
Electrodermal activity, 770
Electrodialysis, 251
Electron delocalisation, 980
Electronic media, 1006
Electronic on-line delivery, 1249
Elementi di agricoltura, 1279
Elettaria cardamomum, 267
Elevated nitrogen deposits, 505
Ellagitannins, 309
Elm, 1063, 1260, 1270
Elm trees, 1063
El Niño, 6
El Pueblo de la Reyna de Los Angeles, 681
Elsinoe ampelina, 231
E. mammosa, 442
Embroidery, 1216
Embryogenetic, 113
Embryo rescue, 430, 444
Emerald ash borer (Agrilus planipennis), 1063
Emerald Necklace, 679

Emergent pests, 1176
Emerging pests, 1181
Emerging plant disease, 1185
Emerging risks, 1183
Emission rates, 838
Emperor Go-Mizunoo, 674
Emperor Ta-Yu, 160
Emperor Tiberius, 765
Emperor Xuan-zong, 1245
Employment, 505, 518
Empoasca fabae, 847
Emu grass (Podocarpus drouynianus), 458
Encarsia formosa, 781
Encrusting, 77
Endangered species, 796
Endangered Species Act, 806
Endeavour, 441
Endemic, 806
Endive, 1235
Endophytes, 230, 823
Endosperm, 112
Endothelial function, 984
Endothermic, 836
Endo-β-(1,4)-glucanases (EG), 118
Energy, 1059, 1060, 1071, 1076
  combustion, 507
Energy conservation and efficiency, 512
Energy consumption, 347, 348
Energy costs, 343, 512
Energy density, 967
Energy efficiency, 349, 357, 391, 512
Energy expenditure, 981, 986
Energy homeostasis, 986
Energy metabolism, 986
Energy prices, 1142, 1150
Energy production, 685
Energy reduction, 512
Energy refrigerated storage, 507
Energy requirements, 349
Energy saving, 361
Energy shortages, 5, 18
Energy transfer, 213
Energy transportation, 507
Energy use, 1059, 1060
Energy use efficiencies, 776
Engineering, 10, 1132
Engineers, 783
England, 340, 409, 413, 417, 675, 676, 679, 765, 808, 809, 830, 831, 1028, 1030, 1032–1035, 1040, 1121, 1199, 1200, 1245
England and Wales, 1127
Englischer Garten, 679
English, 411
English gardens, 1200
English Heritage, 1133
English ivy (Edera helix), 102
English landscape, 1200
English oak, 716
Enhanced memory, 988
Enrichment, 359
Enterocyte, 973, 984
Entomological Society of America, 1239
Entomology, 1181
Environment, 504, 506–509, 511, 512, 515–518, 520, 615, 1049, 1055, 1063, 1064, 1066, 1069, 1070, 1074, 1076, 1077
Environmental and ecological values, 14
Environmental attitudes, 1057, 1058
Environmental awareness, 17, 800
Environmental benefits, 808, 1172
Environmental burdens, 509, 510
Environmental care, 10
Environmental changes, 833, 1152
Environmental designs, 1055, 1057
Environmental destruction, 1047, 1048
Environmental footprint, 427, 429, 509, 510
Environmental health, 1002, 1047, 1049
Environmental horticulture, 13, 16, 849, 1071–1073, 1119
Environmental impact, 478, 505, 507, 508, 510, 512, 520, 604, 608, 609, 613, 1058, 1076
Environmental Impact Assessment (EIA), 508
Environmental inequality, 1052
Environmental justice, 1077
Environmental management
  tools and methodologies, 508
Environmental Management Systems (EMS), 508
Environmental movement, 506, 1120
Environmental protection, 511, 520, 715
Environmental restoration, 1048
Environmental sciences, 1128
Environmental services, 806
Environmental strategy, 843
Environmental stresses, 628, 750, 1052
Environmental sustainability, 870, 1124
Environmental threats, 504, 506, 507, 511, 516, 517
Environmental wellbeing, 962
Environment-friendly, 76, 1120
Environment impact assessment (EIA), 508
Environment surrounding extension, 1128
Enzyme, 970, 974, 976, 977, 986
Enzyme myrosinase, 975

Epazote, 663
Ephedra distachya, 409
Epicormic buds, 627
Epic Prospective Study, 987
Epicurean garden, 1261
Epicureanism., 1261
Epicuticular wax, 209
Epidemiology, 825, 830
Epinasty, 779
Epiphytes, 764
Epithelial proteins, 970
Epithiospecifier protein, 975
Era of Globalization, 1172
Erekh, 1206
Eremochloa ophiuroides, 734
Eremocitrus, 161
Eremophila (Eremophila glabra), 440
Erhai Lake, 674
Erica, 308
Erica caffra, 442
Ericaceae, 317, 634
Erinaceus europaeus, 1031, 1033
Erithacus rubecula, 1032
Ermenonville, 675
Ernest Wilson, 441
Erosion, 5, 504, 723, 810, 862
Erosion and depletion, 513
Erosion of resources, 819
Espalier, 638
Essen, 415
Essential minerals, 1241
Essential oil, 11, 645, 650, 717, 721
Establishment, 736, 737, 739
Esters, 120, 244
Esthetic, 1197–1199, 1201
Esthetic value, 1198
Estienne, 1274
Ethanol, 242, 244–247
Ethephon, 184, 453
Ethical employment, 520
Ethical Trade Initiative, 511
Ethiopia, 266, 431, 715, 718–720
Ethnoveterinary, 662
Ethyl acetate, 215, 244
Ethyl carbamate, 215
Ethylene, 114, 117, 118, 182, 184, 194, 208, 466, 477, 779
Ethylene binding inhibitor, 477
Ethylene climacteric, 117, 118
Ethylene control, 477
Ethylene in grape, 117
Ethylene synthesis, 384
Eucalyptus, 437, 438, 719–723, 1034
Eucalyptus macrocarpa, 440
Eucalyptus spp., 437, 723
Euclea racemosa, 442
Eucoreosma, 313
Eukaryotic cells, 245
Eumusa, 128
Euonymus, 621
Euonymus alatus, 627
Euphorbia pulcherrima, 349, 439, 447, 453
Euphorbia tirucalli, 717
EU Plant Health Directive, 1183
Eurasia, 439
Eurasian Blue Tit (Cyanistes caeruleus), 1033
Europe, 18, 78, 97, 198, 199, 201, 229, 232, 267–272, 274, 283, 284, 306, 308–310, 312–317, 334, 346, 348, 367, 378, 409, 411, 412, 415, 416, 421, 438–441, 446, 448, 451, 453–455, 457, 475, 606, 611, 614, 615, 648, 649, 662, 673, 674, 675, 678, 679, 696, 703, 707, 717, 721, 733, 737, 748, 765, 766, 804, 806, 826, 828, 829, 831, 834, 836, 839, 860, 862, 985, 1002, 1172–1176, 1178, 1179, 1182, 1183, 1185, 1186, 1218–1220, 1222, 1257, 1266, 1268, 1270, 1272, 1275, 1279, 1281, 1285, 1288–1293, 1295, 1296, 1298–1300, 1303
European and Mediterranean Plant Protection Organization (EPPO), 1182
European and Mediterranean Plant Protection Organization (EPPO) Standards, 1182
European apple canker, 830
European canker, 830
European Commission, 1183
European Directive on Traditional Herbal Medicinal Products, 651
European Environment Agency, 1003
European Food and Veterinary Organisation, 1183
European honeybee, 146
European pomology, 1291
European rabbit, 836
European Renaissance, 1206
European Spring Park Trials, 415
European Union (EU), 651, 865, 868, 1176, 1182–1184, 1186–1188, 1192, 1290
European Union (EU) Directives, 511
Europe hawthorn (Crataegus oxyacantha), 717
Eutrophication, 505, 507, 604, 610–612, 614, 615, 724, 1146
Eutrophication and acidification potential, 510

Eutypa die back, 231
Eutypa lata, 231
Eutypine, 231
Evaporation, 514
Evaporative cooling, 340, 343, 368, 696
Evaporative demand, 217, 219
Evapotransiration (ET), 81
Evapotranspiration, 81, 217, 222, 366, 367, 514, 607, 608, 746, 837
Evapotranspired, 371
Evelyn, 1257
Event management, 8
Events, 13
Evergreen Agriculture, 840
Evergreens, 620–622
Everlasting flowers, 436
Evolution and adaptation, 1288
Exacum, 419
Excessive nitrate, 505
Excurrent, 638
Exercise, 1049, 1052, 1053, 1069
Exhibitions, 797
Exine, 108
Exocarp, 133
Exogenous species, 1030
Exolite, 417
Exotics, 714
Exotic species, 519
Expansins (EXP), 118, 183
Export certification, 1184
Exposure to nature, 1003
Expression, 1197
Ex situ conservation, 796
Extension, 1118, 1120, 1124, 1125, 1132, 1134, 1148
Extension capability, 1127
Extension capacity, 1129, 1132, 1133
Extension delivery, 1118
Extension models, 1134
Extension practice, 1132
Extension practitioners, 1125
Extension programs, 1120
Extension provision, 1118
Extension publications, 1238
Extension reforms, 1127
Extension services, 1118, 1119, 1127, 1132, 1133, 1148
Extension skills, 1131, 1133
External landscaping, 13
Extinction, 832, 835
Extinction of species, 801
Extraction, 505, 518, 650
Extreme weather, 828, 832, 847, 850

F

F1 hybrids, 77, 87, 420
Fabaceae, 623
Fagus sylvatica, 840
Faidherbia albida, 720, 723, 840, 841
Fairtrade bananas, 278
Fairtrade Foundation, 294
Fairtrade (FT), 278, 424, 426, 474, 511, 520, 1145, 1158, 1159
Fairtrade International, 1159
Falkland Islands, 860
False Acacia, (Robinia pseudoacacia), 717
Family breakdown, 5
Fan and pad, 341–343, 633
Fan ventilation, 345
Farm business systems, 1133
Farm crops, 721
Farmer empowerment, 1127
Farmers, 716, 720, 721, 723, 1072
Farmers markets, 90, 1071, 1072
Farmer-to-Consumer Direct Marketing Act, 90
Farming, 716, 723, 1068
Farming models, 1157
Farming system, 862
Farm land, 713
Farmland, 714, 719, 720, 723, 724
Farms, 727, 1067, 1069, 1071
Far red light, 339
Fatigue, 770
Fatsia japonica, 768
Fatty acids, 76, 247, 966, 981
Faustino Malaguti, 1300
F. bucharica, 304
F. chiloensis, 303
Feather flowers, 442
Federal Food, 91
Federal garden show, 686
Federalist, 413
Feed, 715, 716
Feeders and bird baths, 1042
Feeding, 1028
Feeding activity, 832
Feeding of birds, 1032
Femminello, 169, 170
Fences, 717
Fencing, 720, 726
Fennel, 663
Fermentation, 211, 215, 236, 238–241, 244–250, 281, 282, 287
Fermentation kinetics, 215
Fermentation Management, 244
Fermenter type, 238
Ferme ornée, 675

Fernendo Po, 285
Ferns, 438, 764, 767, 775, 778
Fertigation, 193, 292, 377, 379, 386, 846, 1303
Fertigation management, 379
Fertigation systems, 377
Fertiliser burn, 744
Fertiliser distributor, 742
Fertiliser management, 450
Fertilization, 109, 112, 115, 117, 179
Fertilizer applications, 742
Fertilizer rate, 742
Fertilizers, 515, 633, 634, 797, 805, 864, 865, 867, 1299, 1300
Fertilizer tree, 840
Fertilizer use efficiency (FUE), 391
Festuca, 737
Festuca arundinacea, 734, 739
Festuca ovina, 734
Festuca rubra subsp. commutata, 739
Festuca rubra subsp. rubra, 739
Festuca rubra subsp.rubra, 734
Fibers, 966, 1198
Fibre, 8, 11, 126, 139, 266, 295
Fibreglass, 417
Ficus, 722, 767
Ficus benjamina, 768, 775
Ficus elastica, 270, 768
Field capacity, 1302
Field Capacity (FC), 217
Field-grown, 621
Field maple (Acer campestre), 1260
Field nurseries, 630–632
Field nursery crops, 629, 630
Field production, 622, 640
Field radiometers, 221
Fields and forests, 1007
Field to Fork, 511
Field vegetable, 823
Fig, 99, 1259
Fig (Ficus carica), 648
Figs, 198, 1264, 1283
Fiji, 269
Filmcoating, 77
Filtration, 241
Filtration system, 513
Financial services, 1148, 1149
Finch, 1035
Finland, 303, 337, 722
Fire blight, 636
Fireblight, 99
Fire blight (Erwinia amylovora), 99, 636, 831
Fireflies, 1028
Fire management programmes, 956
Firewood, 718
Firs, 439
Fir trees, 1295
Fishing, 519
Fitness, 835, 837
Fitzgerald River National and Park, 442
Five elements, 648
Flagship gardens, 1121
Flame weeding, 750
Flanders, 1217
Flat peaches, 100
Flavan-3-ols, 970, 983
Flavedo, 161, 162, 176, 184–187
Flavedo-albedo, 187
Flavone, 970
Flavonoid glycosides, 983
Flavonoids, 119, 250, 315, 650, 970, 984, 985, 988, 989
Flavonol methyl esters, 970
Flavonols, 315, 986
Flavor, 119, 476, 662, 663
Flavoring, 662
Flavour, 212
Fleas, 751
Fleece, 614, 615
Fleuroselect, 415
Flexible polyethylene, 417
Float hydroponics, 378
Floating reed beds, 457
Flood irrigation, 85, 194, 504
Floral arts, 1197, 1198, 1202, 1203, 1222
Floral design, 408, 1202
Floral differentiation, 151, 178
Floral displays, 1204
Floral identity, 179
Floral induction, 150
Floral inductive pathways, 106
Floral meristem, 179
Floral morphogenesis, 179
Floral walls, 436
Floriade World Horticulture Expo, 1042
Floricane, 307, 308, 313
Floricanes, 311, 312
Floricultural, 427
Floricultural crops, 420
Floriculture, 417, 421, 424, 428, 436, 450, 517, 843, 1231
Floriculture industry, 517
Florida, 87, 148, 163, 164, 166, 168, 172, 302, 304, 417, 721, 781
Florilegias, 1222
Florist retail, 421

Flotation, 240
Flower, 350
Flower abscission, 335
Flower certification, 424
Flower development, 106, 821
Flower differentiation, 105
Flower festivals, 954
Flower garden, 1274
Flower industry, 518
Flowering, 201, 820
Flowering gum, 437
Flowering hormone (florigen), 106
Flowering induction, 179
Flowering pathways, 105
Flowering perennials, 455
Flowering plants, 408, 776, 1274
Flowering shrubs, 1236
Flowering stages, 822
Flowering stimulus, 132
Flowering trees, 620, 621, 622
Flower initiation, 132, 133
Flower organ differentiation, 107
Flower organ formation, 106
Flower quality, 380
Flowers, 11, 368, 716, 717, 721, 850, 860, 1119, 1213, 1215, 1216, 1218, 1222, 1226, 1233, 1236, 1248, 1274, 1278, 1289, 1290
Flower shops, 408
Flower shows, 13, 1041
Flower structure, 106
Fluorescence, 364
Fodder, 716, 717
Fog cooling, 343
Fogging, 329, 341, 343, 357, 368
Fogging systems, 478
Fog system, 363
Foliage, 408, 436
Foliar pathogens, 828
Folic acid, 966
Folklore, 718
Fondateur du système physiocratique, 675
Food, 295, 408, 715, 719–722
Food additives, 646, 867
Food and Agricultural Organisation (FAO), 124, 198, 264, 1140, 1179
Food and Agriculture Organization of the United Nations (FAO), 1124
Food and Drug Administration (FDA), 90, 651
Food chain, 824, 1028
Food chains, 428, 506
Food demand, 1143
Food flavoring, 650
Food growing spaces, 962
Food industry, 511
Food insecurity, 1141, 1150
Food manufacturers, 1145
Food markets, 1140
Food miles, 504
Food plants, 1199
Food policy, 1150
Food preparation, 1145
Food prices, 1140, 1143, 1150
Food processing, 1144
Food producers, 16
Food-producing industry, 511
Food production, 18, 861, 1140–1142, 1145
Food production systems, 509
Food productivity, 1141
Food pyramid, 966
Food quality, 1145, 1152
Food retailer, 511
Food retailers, 1144
Food safety, 90, 91, 469, 470, 861, 1145, 1146
Food safety incidents, 1145
Food Safety Modernization Act (FSMA), 90
Food safety standards, 1145
Food security, 17, 18, 607, 801, 804, 825, 840, 859, 862, 1047, 1049, 1058, 1064, 1066, 1068–1070, 1140, 1150
Food service chains, 1143
Food service sector, 1144
Food storage, 1145
Food supplies, 16
Food supply, 724, 805, 836, 861
Food supply chain, 1143
Food system, 859, 866, 867, 870, 1151
Food web, 823, 824, 846
Football, 755
Football pitches, 748, 961
Footprint, 509
Footprint analysis, 511
Forced ventilation, 343, 357
Forest Development and Management, 808
Forest landscape, 803
Forest pathogen, 1185
Forest plantations, 13
Forest School concept, 1014
Forest tree, 839
Formaldehyde, 773
Formal garden, 1199
Formalism, 1199–1202
Formative pruning, 701
Former mining, 713
Fortification, 678
Fortunella, 161

Fox, 1033
Foxglove, 440
F. pratensis, 738
Fragaria, 304
Fragaria chiloensis, 101
Fragaria virginiana, 101
Fragaria x ananassa, 303, 842, 1280
Fragmentation, 802
Fragrance, 408
Fraise mowing, 748
France, 199–201, 223, 234, 239, 315, 413, 417, 674, 675, 678, 679, 724, 822, 825, 840, 842, 869, 973, 987, 1031, 1176, 1178, 1202, 1210, 1219, 1235, 1257, 1270, 1274, 1275, 1281, 1285, 1295, 1297
Francis Masson, 441
Franco Calabrese, 160
François Quesnay, 675
Frank, A., 1298
Frankincense (Boswellia sacra), 765
Frank Lloyd Wright, 682
Franz von Anhalt Dessau, 675
Fraxinus, 623, 697
Fraxinus excelsior, 623
Fraxinus spp., 1063
Freda Kahlo, 1215
Frederick Law Olmsted, 678, 679
Frederick the Great, 679
Free market, 1142
Free sulfur dioxide, 251
Freeze damage, 175
Freeze-sensitive, 170
French, 411
Frequency of application, 745
Frescoes, 1212
Fresh and processed food, 13
Fresh-cut, 480, 481
Fresh-cut processors, 479
Fresh-cut produce, 478
Fresh fruits, 275, 383, 384, 1159, 1291
Fresh market, 307, 311, 312
Fresh produce, 467, 470, 471, 473–475, 477, 609, 837, 860
Fresh produce supply chains, 477
Freshwater, 504, 505, 511, 518
Friction velocity, 360
Friedrich Bergius, 1298
Friedrich Ludwig von Sckell, 679
Frigoplants, 305
Fritz Haber, 1298
Frogmouth, 1033
Frogs, 1031
Frost damage, 170, 316
Frost-resistance, 124
Fructokinase, 352
Fructose, 119, 139, 207, 209, 210, 227, 242, 245, 351, 364, 722, 982
Fruit, 1234–1236, 1239, 1244
Fruit abscission, 188
Fruit acidity, 352, 844
Fruit and vegetables (FAV), 841, 844, 850, 860, 867, 869, 966, 1013, 1143, 1144, 1159, 1212, 1214, 1215, 1233, 1239, 1241, 1270, 1301
Fruit bud development, 98
Fruit colour, 183
Fruit colour-break, 183, 184, 188
Fruit colouring, 194
Fruit cracking, 356
Fruit crops, 274, 635, 823, 844
Fruit-derived antioxidants, 982
Fruit development, 112, 113, 115, 181
Fruit drop, 113, 177, 180–182
Fruit enlargement, 384
Fruit expansion, 363
Fruit fly, 845
Fruit garden, 1274
Fruit greenlife, 276
Fruit growing, 1289, 1297, 1302
Fruit growth, 182
Fruit industry, 1290
Fruiting cacti, 1290
Fruit irradiance, 335
Fruit landscape, 683
Fruitlet abscission, 174, 181
Fruitlet drop, 176, 180
Fruit load, 362, 390
Fruit malformation, 350
Fruit number, 182, 183, 376, 389
Fruit quality, 182, 188–191, 194, 236, 275, 356, 375, 383, 389, 390, 845
Fruit ripening, 105, 115, 118, 120, 182, 183, 383
Fruit rots, 230
Fruit russeting, 356
Fruits, 11, 12, 89–91, 266, 271, 329, 350–353, 355–357, 363, 364, 368, 375, 380, 383–385, 390, 466, 467, 470–475, 477–481, 606, 610, 715–717, 720, 721, 723, 726, 796, 843, 844, 850, 860, 1030, 1119, 1157, 1199, 1206, 1210, 1212, 1213, 1218, 1222, 1226, 1231, 1233, 1256–1261, 1264, 1265, 1267, 1269–1276, 1281–1287, 1289–1291, 1296

Fruit set, 179–181, 190, 363
Fruit setting, 390, 845
Fruit sink, 183
Fruit size, 182, 183, 189, 191, 194, 356, 384, 845, 846
Fruits of New York, 1234
Fruits of the State of New York, 1294
Fruit-tree, 727, 1279
Fruit tree culture, 1285
Fruit trees, 387, 727, 822, 1276, 1278, 1279, 1281, 1286, 1288, 1292, 1297, 1303
Fruit weight, 386
Fruit yield, 212, 380–382, 390
FT-certification, 1159, 1160
FT-cooperatives, 1159
FT-organic, 1163
FT-organic coffee, 1159
FT-organic markets, 1160
Fuchsia, 767
Fuel, 718–720
Fuelwood, 718, 719
Full-bodied wines, 240
Fumigants, 635
Fumigation, 629, 632
Functional attributes, 698
Functional foods, 91, 1066
Fungal pathogens, 206
Fungi, 726, 1175
Fungicides, 226, 228, 230, 231
Fungus, 720
Furniture, 720
Fusarium, 306
Fusarium circinatum, 1186
Future World Report, 801
Fuzian, 266
F. vesca, 304, 1280
F. virginiana, 303
F. virginiana glauca, 303
F. x ananassa, 303, 304
Fynbos, 438, 442

G

Galacturonic acid, 187
Galanthus, 766
Galanthus spp, 818
Gallo, 1272–1274
Galloylation, 983
Gametic sterility, 180
Gametophyte, 110
Gametophyte differentiation, 108
Gametophytic phase, 110
Garden advisory services, 1121
Garden architects, 680, 682
Garden centres, 420, 1035
Garden cities, 680
Garden design, 13, 1197, 1198
Gardeners, 1051, 1055, 1064, 1067, 1072
Gardener's Chronicle, 1239
Gardeners' settlements, 683
Garden exhibition, 685
Garden festivals, 797
Gardening, 408, 799, 1001, 1050–1055, 1057, 1067, 1069–1071, 1077, 1239
Garden of Alcinöus, 1242
Garden of Eden, 1244
Garden of Epicurus, 1261
Gardens, 685, 696, 703, 798, 1009, 1054, 1055, 1057, 1061–1063, 1069, 1071–1073, 1076, 1198–1202, 1241, 1274
Gardens by the Bay, 808
Gardens—formalism, 1199
Gardens of Babylon, 672
Garden tours, 8
Garlic, 842, 976, 978, 1266
Garonne river, 724
Gaseous deposits from the atmosphere, 516
Gaseous pollutants, 774
Gattchina, 676
Gaultheria procumbens, 454
Gay feather, 408
Gehlhar and Regmi 2005, 1144
Geitonogamy, 146
Gelatin, 240, 250
Gelling agent, 627
Gel mixtures, 77
Gene mapping, 318
Gene pools, 804
Generalists, 506
Generalists or specialists, 506
General Plan East, 683
General Zhang Qian, 199
Generic promotion, 1144
Genesis, 1265
Genetically modified organisms (GMOs), 1179
Genetic diversity, 788, 801, 804, 1062, 1063, 1222
Genetic engineering, 867
Genetic erosion, 804
Genetic fidelity, 622
Genetic heritage, 459
Genetic mapping, 430
Genetic modification, 10, 430, 477
Genetic pollution, 801, 804
Genetics, 83
Genetic variation, 430

Genista pilosa, 447
Genocide, 683
Genotype-environment interaction, 137
Genotypes, 387, 388, 391, 804
Genotypic diversity, 697
Georg Béla Pniower, 683
Georg Dionysus Ehret, 1215
Georg Eberhard Rumpf, 441
Georges-Eugène Haussmann, 678
Georgia, 165, 198
Georgia O'Keeffe, 1215
Georgics, 160, 1261, 1262
Geothermal energy generation, 519
Geothermal plant, 519
Geothermal power generation, 518
Geotropism, 1233
Geraldton wax, 444, 446, 449, 451, 458
Geraldton waxflower (Chamelaucium spp.), 440, 446, 456, 458
Geraniales, 161
Geraniineae, 161
Geranium, 381
Gerbera, 381, 443
German Horticultural Library, 1249
German Landscape Research, Development and Construction Society, 456
Germany, 199, 200, 318, 346, 347, 415, 417, 436, 447, 451, 453, 456, 672, 675, 676, 679, 683, 828, 862, 1145, 1287, 1303
Germination, 77, 109, 838
Gertrude Stein, 1248
Ghana, 284, 285, 287, 715
Gian Battista Ferrari, 1272
Gianpaolo Barbariol, 685
Giant Pineapple Plantation (GGP), 271
Giant reed (Arundo donax), 457
Giardini botanici (botanical gardens), 765
Giardini dei Semplici, 1275
Gibberellic acid (GA3), 77, 132, 151, 166, 168, 177, 191
Gibberellin biosynthesis, 177
Gibberellins, 115, 179, 180, 184, 206, 208
Gibberellin synthesis, 191
Gillyvors, 1246
Ginger, 267, 663
Ginkgo, 623, 1042
Giorgio Gallesio, 1258, 1282
Giovanna Garzoni, 1214
Giovanni Martini da Udina, 1212
Girdling, 181
Girdling roots, 1061
Girolamo Molon, 1258, 1289, 1290
Gladiolus, 380, 408, 410, 413, 418
Gladiolus x hybridus, 408, 418
Glass beads, 779
Glasshouse nursery, 778
Glasshouses, 330, 337, 346, 355, 367, 440, 453, 458, 614, 615, 766, 778, 843
Gleditsia, 623
Glen Ample, 308
Gliricidia sepium, 717
Global Climate Models (GCMs), 233
Global economic crisis, 869
GlobalGAP, 91, 511
Global Good Agricultural Practice, 91
Global health, 803
Global horticultural trade, 1178
Globalisation, 507, 837, 1174, 1183, 1191, 1291
Global market, 473, 1147
Global population, 76
Global radiation, 341, 367
Global retailers, 1143, 1145
Global trade, 459, 473, 1143
Global warming, 8, 210, 451, 478, 507, 509, 520, 732, 819, 821, 839, 842, 846, 1047, 1058, 1059, 1063, 1064, 1077, 1141, 1150
Global warming and climate change, 507, 519
Global warming potential, 507, 510
Glow-worms, 1028
Glucans, 227
Glucose, 119, 207, 209, 210, 227, 242, 245, 351, 364, 975, 986
Glucose homeostasis, 986
Glucosidase, 974
Glucosinolate, 973–976, 982, 984, 988
Glucosinolate biosynthesis, 974
Glucosinolate glucoraphanin, 976
Glucosinolates, 973–975, 977, 984
Glucotoxicity, 986
Glutamine, 211, 213
Glutathione, 249, 977, 984
Glutathione peroxidases, 982
Gluten meal, 750
Glycemia, 989
Glycerol, 227, 228, 243
Glycolytic pathway, 245
Glycoproteins, 984
Glycosides, 650, 981, 983
Glycosylated flavonoids, 970
Goats, 716, 726
Gobi, 438
Goddess Isis, 410

Gold Coast, 285
Golden Age, 416
Golden Carp, 1030
Goldfinch, 1030
Goldfish, 1030
Gold specks, 355
Golf courses, 446, 736, 789, 807
Golf greens, 736, 741, 753, 755
Golubka, 316
Good Agricultural Practice (GAP), 424, 511, 1124, 1146
Good Nutrient Management (GNM), 515, 516
Good practice, 516
Gooseberries, 99, 313–316, 626
Gossypium, 270
Gourd, 1219
Government service delivery, 1126
Grade standards, 468
Graft incompatibility, 625
Grafting, 78, 81, 103, 192, 235, 387, 625, 626
Grafting machines, 387
Grains, 1218
Granulation, 167
Granulocytes, 980
Grape berry, 207–209, 223, 227
Grape composition, 214, 215
Grapefruit (Citrus paradisii Macfad.), 124, 162, 166–169, 174–176, 185, 189, 190, 193, 844
Grape hyacinth, 409
Grape juice, 242
Grape production, 213
Grapes, 99, 110, 115, 119, 120, 198, 208, 648, 825, 844, 845, 860, 1210, 1233, 1234, 1243, 1264, 1269, 1283, 1292
Grapevine Leaf Roll Virus (GLRV), 232
Grapevine nutrition, 213
Grapevine reserves, 216
Grapevine root, 219
Grapevine rootzone, 218
Grapevine stomatal density, 203
Grapevine (Vitis spp.), 199, 201, 203–206, 212–216, 220, 221, 223, 228, 231–233, 1190
Grass cover, 756
Grasshoppers, 751
Grasslands, 5, 715, 801, 802, 838
Grass (Poaceae), 408, 439, 440, 732, 734, 736, 742
Grass roots activism, 685
Grass species, 849
Grassy stunt virus, 795
Great Britain, 270, 417, 732, 733, 737, 795, 864, 1035
Great Conservatory, 766
Great Giant Pineapple Plantation, 294
Great Recession, 1128
Great Sandy and Simpson, 439
Great Victoria, 438
Greco-Arab, 648
Greece, 101, 160, 198, 413, 1199, 1209, 1237, 1242
Greek, 160, 198, 409–411, 1257, 1262
Greek Revival, 413
Green Acres Program of New Jersey, 683
Green avenues, 798
Green bean, 375
Green bell pepper, 212
Green belt, 678, 679
Green bridges, 828, 836
Green-care farming, 1053
Green-care farms, 1048
Green cover, 446, 840
Green exercise, 798, 799, 1015
Greenfly, 1033
Greenhouse, 328, 330, 332–334, 336, 338, 340, 341, 343, 344, 347, 349, 351, 354–364, 367, 368, 375, 378, 381, 385, 386, 389–392, 453, 454, 458, 504
Greenhouse climate, 842
Greenhouse crops, 419
Greenhouse design, 391
Greenhouse gas, 6, 604, 606, 612
Greenhouse gas emissions, 516, 842, 1060, 1144
Greenhouse gases, 612, 613, 615, 725, 803, 819, 820
Green House Gases (GHG) emissions, 509
Green house gases (GHGs), 509, 510, 519
Greenhouse horticulture, 507
Greenhouse industry, 620
Greenhouse irrigation systems, 374
Greenhouse management, 381
Greenhouse microclimate, 329
Greenhouses, 328–332, 334, 335, 341, 343, 346–348, 351, 354, 357–361, 363, 365, 367, 368, 373, 375, 377, 387, 390–392, 416, 506, 507, 633, 641, 843, 1231, 1236, 1257, 1273
Greenhouse technology, 1131
Greenhouse tomato, 337
Greenhouse vegetable, 354
Green industry, 1198, 1199

Green infrastructure, 696, 708, 1001, 1006, 1119, 1122, 1123, 1129, 1131, 1132, 1134, 1135
Greening technologies, 1135
Green landscapes, 1003, 1011, 1017
Greenlife, 278
Green manure, 1298
Green networks, 1002
Green onions, 832
Green open space, 8, 13, 14, 17, 436, 789, 790, 797
Green revolution, 10, 18, 804, 864, 1141, 1142, 1150, 1151, 1299
Green roof gardens, 8
Green roofs, 456, 1002, 1050, 1059–1061, 1132, 1134
Green shoulder, 340
Green space, 18, 451, 678, 682, 685, 789–791, 798–800, 807, 960–962, 1001–1004, 1006–1011, 1014, 1017, 1123, 1133, 1237
Green space per inhabitant, 686
Green space theory, 678
Green square, 678
Greensward, 679
Green system, 678
Green Thumb, 686
Green tourism, 808
Green walls, 452, 1059, 1061, 1132, 1134
Green waste, 705
Green water, 607, 608
Gregor medel, 1233
Gregor Mendel, 1290
Grevillea, 1032
Grevilleas, 457
Grey mold (Botrytis cinerea), 781
Grey mould, 226, 227, 232
Grey water, 607
Greywater recycling, 513
Ground cover, 621, 1198
Ground cover plants, 764
Ground keepers, 836
Groundsel, 409
Growers, 16
Growers associations, 1158
Growing degree days, 99
Growing degree hours, 99
Growing media, 779
Growth analysis, 115
Growth inhibition, 365
Growth inhibitors, 365
Growth pattern, 733
Growth rates, 605, 638
Growth regulator, 624
Growth regulators, 191, 742
Growth retardants, 105
Growth stages, 825, 838
Guanabana, 140
Guanaja, 284
Guano, 1298
Guatemala, 143, 270, 272, 419, 439, 719
Guatemalan, 143, 145
Guava, 126, 127, 844
Guava (Psidium guajava L.), 124
Guiera senegalensis, 723
Guiseppi Arcimboldo, 1214
Gum, 722
Gum arabic (Senegalia senegal and Vachellia seyal), 648
Gustav Vorherr, 676
Gymnocladus, 623
Gymnorhina tibicen, 1033, 1034
Gymnosperms, 408
Gymnosporangium spp, 635
Gynodioecious, 137
Gynoecium, 141
Gyno-sterility, 110
Gypsophila paniculata, 409, 417
Gypsum, 744
Gypsum blocks, 219

H

Haber, 1298, 1299
Haber/Bosch process, 1299
Habitat deterioration, 610
Habitat loss, 802, 803
Habitat restoration, 796
Habitats, 1026, 1031, 1034
Habitat suitability, 1262
Habito and Briones (2005), 1147
Hamburg, 686
Hanahaku Japan Flora 2000, 685
Handling, 466, 467
Handling chains, 466
Hand pruning, 225
Hanging baskets, 436
Hanging Gardens of Babylon, 3, 1235, 1301
Hanseniapora, 244
Hanseniaspora uvarum, 244
Hans Sloane, 441
HAPIE Plants, 447
Haploidization, 430
Hardwood cuttings, 625

Hardy Amenity Plant Introduction and Evaluation scheme, 447
Hardy nursery plants, 378
Harrison Shull, G., 1290
Harvest index, 390
Harvesting, 640, 821
Harvesting the Sun, 7
Harvest quality, 353
Hatch Act, 1125
Hawaii, 135–137, 267, 290, 418, 803, 1030
Hawthorn (Crataegus spp), 635, 636, 721, 818
Hazard Analysis Critical Control Point (HACCP), 1145
Hazards, 1056, 1058, 1061, 1070
Hazel, 716
Hazelnut, 99, 109, 289, 973
H. brasiliensis, 270
Head cabbage, 1213
Healing, 646
Healing Gardens, 1015, 1017
Healing landscapes, 1052, 1053
Healing practices, 655
Health, 17, 646, 650, 654, 860
Health and safety, 1146
Health and well-being, 769, 771, 773, 794, 798, 953, 954, 1001–1004, 1017, 1238, 1285
Health attributes, 1145
Health awareness, 76
Health benefits, 784, 809, 1037, 1122
Health care, 14, 646, 655, 798, 1048, 1049, 1052–1054
Health of Communities, 1054, 1077
Health of the community, 1049, 1055
Health of the individual, 1047, 1049
Health policy, 1004
Health status, 658, 772, 1012
Healthy living, 966
Heart disease, 1050, 1053
Heart rate, 770, 799
Heat accumulation, 98
Heat delay, 352
Heat Extraction, 239
Heat flux, 368
Heather (Calluna), 634
Heaths (Erica spp), 634
Heating pipes, 344
Heat island, 796
Heat of fusion, 633
Heat-pulse, 366
Heat shock protein (HSP-70), 989
Heat stress, 234, 235, 352, 741
Heat summation, 175
Heat tolerance, 86–88
Heat treatment, 636
Heavy metals, 696
Hebe spp., 440
Hebrew bible, 1226, 1241, 1243
Hebrides, 441
Hedera, 1031
Hedera helix, 436
Hedgehogs, 1031, 1033
Hedge row, 446, 713, 717, 723, 724, 727
Hedges, 724, 1042
Hedrick, 1292–1294
Hedychium, 803
Helichrysum, 436, 440, 766
Helichrysum bracteatum, 440, 448
Helleborus niger, 454
Hemerocallus spp, 842
Hemiptera, 142
Hemp, 264
Hendrickson, A.H., 1303
Henry Doubleday Research Association (HDRA), 1033
Henry the Navigator, 649
Henry VIII, 1247
Henry Wickham, 270
Herbaceous, 1226
Herbaceous crops, 127
Herbaceous ornamentals, 352, 635
Herbaceous perennials, 1239
Herbal, 646, 1218, 1220, 1237
Herbal gardens, 646
Herball, 1220
Herbal medicines, 656
Herbals, 646, 1218, 1220, 1222, 1237
Herb gardens, 664, 800
Herbicide damage, 739
Herbicides, 864
Herbicide tolerance, 631
Herbivore, 1029
Herbivore repulsion, 967
Herbivory signal transduction, 974
Herbs, 662–664, 796, 1236, 1274
Herculaneum, 1212
Herkogamous, 141
Hermaphrodite, 109, 110, 133, 136–138, 141, 149, 150
Hermaphrodite flowers, 132
Hermaphroditic, 141, 149
Heterocyclic pyrane C ring, 970
Heterofermentative degradation, 247
Heterosis, 84, 420, 1290

Hevea, 270
Hevea brasiliensis, 289, 722
Hexose, 210, 383
Hibernation, 1031
Hibiscus, 436, 627
Hibiscus rosa-sinensis, 436
Hibiscus tiliaceus feed cattle, 717
Highbush, 317, 318
Highbush blueberry, 317, 318
Highbush breeders, 318
High density living, 436
Higher alcohols, 244
Higher Education (HE), 1128, 1129
Higher efficiency and lower costs, 513
Higher light intensity, 513
Higher temperatures, 175
High-Level Expert Forum on How to Feed the World to 2050, 1141
High quality, 641
High salinity, 77, 387
High value, 12
Himalaya, 147, 159, 439, 441, 450, 844
Himalaya blackberry, 311
Hinduism, 718
Hippeastrum, 455
Hippie revolution, 865
Hirundo rustica, 1040
Historia de plantes, 1233
Historic gardens, 849
Historic sites, 849
History of plants, 1233
H. macropylla, 448, 453
H. muercifolia, 448
Hobby gardeners, 850
Hockey, 755
Holland, 313, 409, 428, 429
Hollow fruits, 335
Hollyhock, 409
Holly (Ilex), 438, 440, 621
Holocene extinction, 801
Home gardeners, 1121
Homeotic genes, 106
Homer, 1233, 1241
Homogenetic sterility, 180
Homo sapiens, 788, 819, 1004
Honduras, 278, 284, 419, 781
Honey, 721
Honey bees, 146
Honeysuckle (Lonicera), 410, 439, 447, 765
Hong Kong, 808
Hop, 1274
Horace, 675
Horizontal screw presses, 240
Horse chestnut leaf miner (Cameraria ohridella), 1175
Horse racing, 753
Horse racing tracks, 736
Horse-surface, 753
Horsetails (Equisetum spp), 634
Horticultural activity, 518, 954
Horticultural and related education, 1128
Horticultural attractions, 1040, 1043
Horticultural biodiversity, 801
Horticultural businesses, 1071, 1072
Horticultural crop quality, 514
Horticultural education and training, 1128
Horticultural enterprises, 511
Horticultural events and festivals, 1040
Horticultural expertise, 1124
Horticultural extension, 1118
Horticultural farmers, 1143
Horticultural green revolution, 10
Horticultural industry, 504, 510, 511, 514, 517, 1064
Horticultural innovation, 1162, 1163
Horticulturalists, 783
Horticultural producers, 512
Horticultural production, 513, 520, 604, 794, 1148
Horticultural science, 809, 851, 1047–1049, 1052, 1058, 1059, 1061, 1062, 1064, 1066, 1070, 1071, 1076, 1077, 1131, 1226, 1227, 1230–1233, 1239
Horticultural shows, 1041
Horticultural statistics, 1127
Horticultural therapy, 799, 1015, 1053, 1077, 1238
Horticultural therapy trusts, 799
Horticultural trade, 519
Horticultural Trade Association (HTA), 1132
Horticultural value chain, 520
Horticulture, 504, 507–513, 516, 518, 520
Horticulture Australia Limited (HAL), 1128, 1130
Horticulture Collaborative Research Support Program (Horticulture CRSP), 471
Horticulture in Europe, 10
Horticulture practices, 800
Horticulture producers, 513
Horticulture's impact, 520
Horticulture therapy, 954, 959
Horticulture Week, 1133, 1239
Horti Lucullani, 673
HortTechnology, 1077

Hortus epicureo, 1261
Hortus pictus, 1271
Hospital gardens, 1012
Hospital patient recovery, 771
Hospitals, 770, 771, 783, 799
Hostas, 1236
Hot beds, 416
House, 720
House building, 720
Household, 719, 720, 722
Household gardens, 13
House plants, 764, 769, 1236
Hoverflies, 1032
Hover mower, 740
Howea forsteriana, 764, 768
H. quercifolia, 453
H. serrata, 448, 453
Huangdi Neijing, 648
Hugo de Vries, 1258, 1287, 1288
Human bonding, 957
Human carcinoma, 309
Human culture, 1198
Human diets, 321, 979
Human disease, 1050
Human health, 11, 14, 16, 76, 505, 516, 781, 798, 861, 863, 870, 1006, 1050, 1051, 1060, 1061, 1064, 1299
Human health and wellbeing, 801, 810
Human hygiene, 470
Human impact, 801
Human intervention, 956
Human life, 408
Human nutrition, 316
Human pathogens, 470
Human population, 802, 1047, 1049, 1064, 1066, 1076
Human productivity, 770, 772
Human resource capacity, 1132
Human resources, 1146
Human stress, 1050, 1051, 1076
Human survival, 13
Humidifying, 341
Humidity, 329, 344, 778
Humidity control, 478, 625
Humid tropics, 778
Hummingbird, 1032, 1041
Hungary, 413
Hunting, 1028
Hunting ground, 679
Hunt Morgan, T., 1288
Hunt of the Unicorns, 1217
Huxley Report, 795
Hyacinth, 410, 430
Hyakuda-en, 685
Hybridization, 804, 1290
Hybrid species, 1175
Hybrid vigour, 420, 1290
Hyde Park, London, 417, 679
Hydrangea, 439, 443, 446, 448, 453
Hydrangea macrophylla, 443
Hydrangea paniculata, 448, 453
Hydrangea spp., 439
Hydration, 77
Hydric stress, 319
Hydroculture, 779
Hydrogen, 634
Hydrogen sulfite (HSO3-), 251
Hydrogen sulphide, 238
Hydrolases, 109, 118
Hydrological amelioration, 797
Hydrolysis, 975, 976
Hydroponic culture, 378
Hydroponics, 8, 353, 379
Hydroponic systems, 378
Hydroxycinnamates, 208
Hydroxycinnamic acids, 986
Hygiene, 635, 636
Hymenoptera, 146
Hymenoscyphus pseudoalbidus, 1063
Hypanthium, 115
Hypertension, 1013
Hyphaene coraiacea, 722
Hypoclade, 204
Hypoxia, 375, 505

I

IAS, 1187
Ibn Al Awwam, 1257, 1266, 1267
Ibn Alò-Awwam, 1301
Ibn Butlan, 1218
Ibn Sara of Santarem, 1244
Ibn-Vahschiah, 1301
Ibn Wahsiya, 160
Iceberg lettuce, 832
Iceland, 441
Iconography, 1281, 1297
Ikebana, 409, 1203
Ikenobo, 1203
Ildefons Cerdà, 678
Ilex aquifolium, 440
Illinois General Assembly, 678
Imara, 308
Immune function, 979
Immune system, 979, 982

Impatiens, 443, 444, 448, 449, 809
Impatiens walleriana, 420
Import risk assessment (IRA), 1188
Inaequilateralis, 448
Inanna, 1206
Incident radiation (IR), 329
Inclusive design, 960
Income, 1051, 1052, 1055, 1067–1070, 1072, 1077
Incompatibility, 444, 1270
Incompatible pollen, 112
Incompatible signalling(s), 111
Increased algal production, 505
India, 8, 76, 135, 147, 148, 161, 264, 266–270, 272, 280, 306, 418, 436, 453, 469, 471, 648, 649, 716–721, 725, 795, 798, 827, 843–849, 863, 868, 869, 1199, 1206, 1210, 1237, 1301
Indian jujube (Ziziphus mauritiana), 721
Indian myna bird, 1030
Indian rubber, 270
Indian subcontinent, 438
India soil erosion, 723
Indigenous breeds, 804
Indigenous flowers, 1042
Indigenous species, 806
Indochina, 159, 438
Indo-Gangetic plains, 826, 846
Indole-3-butyric acid, 453
Indole acetic acid, 625, 626
Indole butyric acid, 625
Indonesia, 8, 159, 264, 266, 267, 269, 271, 284, 286, 287, 438, 439, 441, 717, 721, 802
Indoor plants, 443, 455
Indoor plants., 458
Indus, 3
Industrial, 722, 724
Industrial crops, 263, 266
Industrialization, 679
Industrialized, 727
Industrial oil, 269
Industrial Revolution, 795, 1171, 1174
Industrial sites, 713
Industry-driven extension, 1129
Industry-funded support, 1129
Industry levies, 1130
Inert gas, 242
Infection periods, 830
Infiltration rate, 746
Inflammatory biomarkers, 986
Inflammatory bowel disease, 980
Inflorescence, 203–207
Inflorescence axis, 107
Inflorescence emergence, 134
Inflorescence primordia, 205–207
Inflorescences, 207
Informal model, 1156
Infra-red radiation (IR), 333
Infra-red signalling, 449
Infrastructure, 782, 1147, 1148
Inga nobilis, 720
Inhibitor genes, 179
Injury potential, 753
Inner Niger Delta, 715
Innovation diffusion, 1118
Innovation intermediaries and brokers, 1133
Innovation intermediation/broking, 1134
Innovation systems, 1133
Innovation systems thinking, 1134
Inorganic fertiliser contamination, 513
Inorganic fertilisers, 505, 507, 513
Inorganic inputs, 507
Inorganic manufactured fertilisers, 516
iNOS activity, 985
Inputs, outputs and the potential environmental impacts, 509
Insecta, 636
Insect pests, 751
Insect-proof screens, 358
Insects, 724, 832, 839, 1026–1029, 1031–1033
Insect vectors, 848
Insolation, 176
Inspection, 1180
Inspect the environmental impacts, 508
Institute of Groundsmanship (IOG), 1133
Institute of Horticulture (IOH), 1133
Insulin, 986, 989
Insulin-Growth Factor-1, 989
Insulin resistance, 986
Integrated disease management, 827
Integrated management, 705, 706
Integrated Pest and Disease Management, 316
Integrated pest management (IPM), 6, 390, 516, 517, 752, 807, 850, 1131
Integrated plant management, 391
Intellectual property (IP), 421, 437, 458, 459
Intellectual property rights (IPR), 437, 459
Intelligent packaging, 481
Intensive crop production, 505
Intensive users of resources, 506
Intensive vegetable production, 77
Interactive landscapes, 1009
Inter-African Phytosanitary Council (IAPSC), 1181

Intercalary units, 151
Inter-Governmental Environment Summit, 795
Intergovernmental Panel on Climate Change (IPCC), 820
Interior design, 764
Interior environment, 775
Interior landscapes, 769, 779, 781, 783, 1002
Interior planted landscapes, 763
Interior plant hire, 1119
Interior plantings, 771
Interior plants, 769, 770, 774, 775, 778
Interior plantscapes, 783
Interlight, 338
Internal disorders, 846
Internal quality, 362
International Center for Tropical Agriculture (CIAT), 1160
International Convention, 1178
International Convention for the Protection of Plants, 1179
International Convention on Measures, 1180
Internationale Gartenschau, 686
International Federation of Agricultural Movements, 860
International Federation of Organic Agricultural Movements (IFOAM), 864, 867, 870
International Florist Organisation, 415
International Food Policy Research Institute (IFPRI), 1140
International Horticultural Congress, 10, 1077
International Hortifair, 415
International Institute of Agriculture (IIA), 1179
International Maize and Wheat Improvement Center, 864
International Organization for Standardization (ISO), 427
International Plant Protection Convention (IPPC), 1173, 1179–1181, 1184
International Society for Horticultural Science (ISHS), 7, 1077
International Society of Arboriculture, 701
International Standards for Phytosanitary Measures (ISPMs), 1180, 1181, 1184, 1191
International Standards Organisation, 509
International trade, 91, 507, 1173, 1180
International Union for the Conservation of Nature (IUCN), 1030, 1035
International Union for the Protection of New Varieties of Plants (UPOV), 437, 459
Internode appearance rate, 350
Internodes, 105
Interpersonal relationships, 17
Inter-specific hybridization, 318, 430
Inter-specific hybrids, 128, 456, 1286
Interstem, 625
Interstocks, 625
Intestine cancers, 987
Intine, 108
Introduced plants, 1172
Introduction of invasive species, 505
Introgression, 303
Invasion pathways, 1175
Invasive aliens, 846
Invasive alien species, 1179
Invasive pests, 1174, 1175, 1179
Invasive plants, 1183
Invasive species, 5, 801–803, 810, 837, 956, 1047, 1048, 1063, 1064, 1174, 1175
Inventive parks, 960
Inventory analysis, 509
Invernale, 169
Invertebrate pests, 832
Invertebrates, 724, 726
Inzenga, G., 1296
Ion exchange, 251
Ion sensors, 380
Ion translocation, 354
Ipomea batatas, 1222
IQF markets, 308
Iran, 160, 198, 268, 413, 438, 721, 723, 1205
Iraq, 160, 198, 268, 409, 413, 723, 1205, 1206
Iraq intercrop, 723
Ireland, 826, 1173, 1178
Iris, 409, 412, 414
Irish potato blight, 795
Iron, 210, 211, 634, 715, 967
Irradiance, 775–778, 782
Irradiation, 349, 350
Irrigated, 366, 723
Irrigated grapevines, 222
Irrigation, 80, 83, 100, 216, 217, 219, 220, 234, 292, 362–366, 368, 369, 371, 373, 376, 381, 382, 385, 386, 392, 450, 456, 458, 504, 514, 605, 607, 611, 629, 632, 727, 736, 739, 744–746, 777, 779, 782, 805, 810, 825, 831, 849, 864, 1121, 1132, 1199, 1272, 1297, 1301, 1303
Irrigation canals, 723
Irrigation design, 746, 1131
Irrigation efficiency, 513, 745
Irrigation frequency, 746

Irrigation loss, 367
Irrigation management, 77, 81, 219, 220, 371, 374, 375
Irrigation methods, 513
Irrigation protocols, 321
Irrigation scheduling, 220, 222, 223, 374, 607
Irrigation strategies, 218
Irrigation systems, 513
Ischemic stroke, 1013
Isinglass, 250
ISO 14001, 508
Isobutylidenediurea, 742
Isobutyl methoxypyrazine, 212
Isopentenil diphosphate (IPP), 119
Isoprenoid polymers, 979
Isoprenoids, 119
Isothiocyanates, 973, 974, 982, 984, 988
Israel, 77, 80, 91, 130, 136, 148, 150, 170, 272, 344, 345, 369, 371, 418, 444, 1205, 1243, 1303
Israelsen, O.W., 1302
Istanbul, 411
Istar, 1206
Isthmus of Suez, 648
Italian prune, 109
Italian Renaissance, 1212, 1235
Italy, 160, 165, 198–200, 436, 673, 675, 850, 985, 987, 1176, 1208, 1209, 1212, 1235, 1257, 1258, 1260, 1265, 1270, 1271, 1273, 1274, 1283, 1290, 1291, 1296, 1300
i-Tree, 704
Ivy, 410, 436, 1035
Ivy (Hedera helix), 457
Ivy tree (Schefflera octophylla), 721

J
Jaboticaba (Mirciaria cauliflora Berg), 128
Jacques Le Moyne de Morgues, 1215
Jaggery, 722
Jamaica, 167, 168, 285
James Cook, 441
James Cunningham, 441
James Hobrecht, 678
James Joyce, 1226, 1241, 1248
Jane Austen, 1247
Jane Austin, 1241
Japan, 78, 84, 97, 136, 165, 166, 172, 198, 274, 306, 317, 328, 409, 413, 414, 417, 421, 436, 438, 439, 441, 446, 453, 457, 672, 674, 675, 722, 804, 863, 868, 1041, 1184, 1199, 1200, 1203, 1226, 1241, 1293
Japanbreite, 675
Japanese barberry, 446
Japanese beetle, 636
Japanese germplasm, 86
Jardin à la Française, 674
Jardin fruitier, 1274
Jardin potager, 1274
Jarrah (Eucalyptus marginata), 719
Jasminum, 767
Jasmonate, 974
Java, 286
J. De La Quintinye, 1279
Jean Baptiste de La Quintinie, 1257, 1276
Jean Bourdichon, 1219
Jean-Charles Alphand, 678
Jean de la Quintinye, 1274
Jean Jacques Rousseau, 675
Jedermann Selbstversorger, 682
Jelutung, 270
Jewish Hanukkah, 415
Job dissatisfaction, 773
John Bennet Lawes, 1300
John Evelyn, 1279
John Gerard, 1220
John H. Rauch, 678
John Tradescants the Younger, 1172
Jordan, 91, 1205
Joseph Banks, 441, 1172
Joseph Berkeley, 441
Joseph Dalton Hooker, 1172
Joseph Henry Gilbert, 1300
Joseph Hooker, 441
J.S.Fry & Sons, 284
Judas Tree, (Cercis siliquastrum), 717
Jugland regia, 289
Juice clarification, 240
Jujube, 99
Juliana Anicia Codex, 1237
Juliana Anicia Codex (JAC), 310, 1218
Jumping Bean Tree (Spirostachys africana), 717
Juncaceae, 457
Juncus spp., 457
Juneberry, 99
Juniper, 717
Junipero Serra, 681
Juniperus, 625
Juniperus procera, 450
Just in time logistics, 471, 477
Jute, 264
Juvenile, 102, 103, 113
Juvenile phase, 102, 136, 622
Juvenility, 102, 103

K

Kaferstein 2003, 1145
Kahili ginger, H. gardnerianum, 803
Kaine and Cowan 2011, 1152
Kalahari, 438
Kalanchoe spp., 337
Kale, 86
Kalmia, 634
Kangaroo paws, 436
Kangaroo paws (Anigozanthos spp.), 440, 444, 458
Kansai airport, 685
Karnak, 160
Kearney, T.H., 1302
Kelly 2003, 1147
Kenya, 266, 419, 427–429, 431, 517–520, 611, 615, 715, 720, 1182
Kenyan flower industry, 518
Kenyan roses, 429
Ketelaar 2007, 1145
Ketone, 251
Kew, 799, 808
Kew Garden, 1279
Kew Magazine, 1236
Key and Runsten 1999, 1157
Keystone species, 829
King Charles I, 679
King protea, 440
King Solomon, 409
Kirstenbosch Botanic Garden, 808
Kitab al-felahah, 1266
Kitchen gardens, 1199
Kitul palm (Caryota urens), 722
Kiwi, 120
Kiwi fruit, 99, 109, 113, 480
Kloeckera apiculata, 244
Knossos, 765
Knowledge generation, 1134
Knowledge/technology transfer, 1118
Koala fern (Caustis blakei), 458
Koishikawa Korakuen, 676
Kookaburra, 1033
Korakuen, 676
Korea, 78, 91, 165, 328, 409, 672, 722, 863, 1184, 1216
Kumquat, 161
Kuopio Ischemic Heart Disease Risk Factor Study (KIHD), 984
Kweli, 308
Kwongan, 438
Kyoto, 674

L

Label recommendations, 1145
Labour, 505
Labour costs, 426
Labrador, 441
Laccase, 227
Lacewings, 751
Lachenalia, 440
Lachenaliai bulbifera, 442
Lachenalia spp., 440
Lack of soil fertility, 504
La classe des propriétaires, 675
Lactic acid, 241, 246
Lactic acid bacteria, 241, 246
Lactobacillus, 246
Lactones, 120, 242
Lactuca sativa, 380
Lactuca sativa var. capitata, 338
Lady beetles, 751
Ladybirds, 1028, 1031
Ladybugs, 1031
Lafinita, 272
Lajos Mitterpacher, 1279
Lake Naivasha, 428, 517–519
Lake Naivasha Water Resource Users' Association (LNWRUA), 519
Lakes, 714
Lake sedimentation, 519
Lamm, F.R., 1303
Lampung, 271
Land, 504, 513
  and soil degradation, 504
  consolidation, 727
  degradation, 802, 839, 1141
  embellishment, 675
  grabbing, 685
  improvement, 840
  management systems, 793
  markets, 1146
  resources, 18
  tenure, 1146, 1147, 1157
  use, 5, 615, 794, 1147, 1201
Landesverschönerung, 676
Land-grant institutions, 1125
Land grant system, 1125
Land grant universities, 1121, 1238
Land-management systems, 792
Landolphia kirkii, 270
Landraces, 805
Landscape, 13, 614, 615, 622, 628, 630, 633, 635, 638, 641, 672, 673, 675, 782, 788, 792, 795, 798, 808, 839, 840, 867, 1200, 1216, 1222, 1239

architecture, 678, 681, 702, 707, 1123, 1197, 1201, 1202, 1237
contractors, 420
designs, 1048, 1058, 1059, 1071, 1076
horticulture, 1001, 1119
industry, 622
of exemplification, 683
plants, 769, 1226
restoration, 13
technicians, 782
trees, 698
Landscape designs, 1120, 1232
and construction, 1119
Landscape gardens, 1235
Landscape horticulture, 1235
Landscaping, 408, 1226, 1235
Language of flowers, 412, 413
La Niña, 6
Lannea stuhlmanni, 720
Lapeirousia silenoides, 450
Lapis specularis, 416, 765
La Quintinye, 1278
Larch, 716
Large white butterfly, 1029
Larkspur, 410
Laryngitis, 775
Larynx, 987
Late blight, 826, 827, 846
potatoes, 635, 848
Latent heat, 340
Latham, 309
Lath houses, 641
Latin America, 17, 127, 266, 269, 273, 274, 276–278, 439, 723, 869
Lauraceae, 143
Laurales, 143
Laurel (Cordia alliodora), 719
Lavandula, 1031
Lawn
mower, 736
Lawn bowls, 755
Lawns, 732–734, 736, 737, 740, 749, 751, 752, 849, 1029, 1236
bowls, 755
Laws of Hammurabi, 1241, 1243
Layerage, 626
Layering, 626
Leach into the ground and runoff into water courses, 505
Leaf, 717, 724
abscission, 203
analysis, 191, 807
area, 222
blight, 848
conductance, 385
irradiance, 335
litter, 1031
photosynthesis, 376
photosynthetic, 339
primordia, 105, 206
senescence, 201
temperature, 221, 222, 344, 346
water potential, 219, 220, 221
Leaf area index (LAI), 337, 843, 846
Leaf-curling midge, 316
Leafhopper, 751, 847
Leafy greens, 467
LEAFY (LFY), 105
Leafy vegetables, 77, 982
Lean manufacturing, 471
Leaves, 715–717, 719, 1031
Lebanon, 1205
Leberecht Migge, 682
Lechenaultia biloba, 440
Le Corbusier, 683
Leek, 78
Leeward ventilation, 360
Legionella pneumophilia (Legionnaires' disease), 778
Legislation, 511, 516
Legislative requirement, 508
Legumes, 837
Leisure, 715
and recreation, 1119
provision, 958
Le Language des Fleurs, 412
Lele 1981, 1158
Lemon, 160, 162, 168–170, 173, 175, 189, 190, 193, 844, 1273, 1276
Lenné, 676, 679
Le Nôtre, 1279
Lent, 415
Leonardo da Vinci, 412
Leonhard Fuchs, 1220
Lepidoptera species, 845
Leptomastix dactylopii wasp, 781
Leptospermum scoparium, 440
Leptospermum spp., 444
Leptosphaeria maculans, 825
Leschenaultia, 440
Les Grandes Heures d'Anne de Bretagne, 1219
Lesion nematodes, 233
Les vignobles et les arbres à fruit à cidre, 1285
Lettuce, 85, 335, 337, 338, 352, 353, 355, 361–363, 365, 380, 384, 608, 805, 1235

Leuconostoc, 246
Leucospermum conocarpodendron, 442
Leucospermums, 442
Leukemia cells, 987
Levels, 505
Levy-funded research, 1128
Lewis, 1153
Lewisia cotyledon, 455
Liatris spicata, 408
Liberalising international and national regulatory framework, 508
Liber de agri cultura, 1262
Liberia, 270
Liberty Hyde Bailey, 1233
Libya, 723, 1209
Liébault, 1274
Liebig's Law of the Minimum, 391
Liechtenstein, 860
Life cycle, 513, 517
Life Cycle Assessment (LCA), 509, 511
Life expectancy, 14, 17
Life-span, 506, 724, 726
Lifestyle, 18, 807
  horticulture, 1119, 1131
Light absorbance, 337
Light brown apple moth (LBAM), 227, 232
Light compensation point, 339, 362
Light duration, 334
Light-emitting diode (LED), 338
Light extinction coefficient, 387
Light intensity, 334–337, 339, 341, 348, 350, 391, 777, 845
Light interception, 389
Light quality, 339
Light saturation point, 339
Light transmission, 333, 334
Light use efficiency, 361, 387
Lignin, 970
Ligno tubers, 443
Lilac, 416
Lilac (Hardenbergia comptoniana), 457
Lilium, 766
Lilium candidum, 412
Lilium spp., 443
Lily, 409, 410, 412–414, 430, 443
Lily of the valley, 412
Lime, 162, 170, 171, 175, 190, 744, 844, 1273, 1276, 1300
Lime (Citrus aurantifolia (L) Swingle), 124
Lime (fruit), 721
Limiting factor, 391, 1147, 1301
Limoni, 169, 170
Limonin, 164, 167
Limonium sinuatum, 409, 417
Lindley Library, 1249
Liners, 630
Linking the Environment and Farming (LEAF), 511
Linneo, 1271, 1284
Lipid, 650, 986
Lipophilic, 970, 980
Lipoproteins, 983, 985
Litchi, 128, 271
Litchi (Litchi chinensis Sonn.), 124
Liver, 984
Liver cancer cells, 987
Livestock, 715, 716
Livestock and arable agricultural land, 509
Livestock feed, 715, 716
Living walls, 457
Livio, 1264
Lizard, 1031
Llimes, 175
Lobelia erinus, 448
Local ecosystems, 505
Local food, 869
Locally grown, 90
Local markets, 310
Local sourcing, 870
Logan, 310
Logistics, 469, 481
Lolium, 737
Lolium multiflorum, 739
Lolium perenne, 733, 734, 737, 739
Lolium perenne/Festuca pratensis, 738
Lolium perenne (perennial ryegrass), 738
Lombardy-Venetia, 1257
London, 678
Longan (Dimocarpous longan Lourr.), 127
Long-canes, 308
Long-day annual, 449
Long days, 305
Longevity, 17
Long-term prevention of pests, 517
Lonicera, 766
Lonicera periclymenum, 1031
Lonicera spp., 439, 447
Look to the Land, 863
Lord & Burnham Co, 417
Lorikeets, 1032
Los Angeles, 681
Loss of biodiversity, 6, 506
Lost Gardens of Heligan, 1040
Lost nutrients and organic matter, 504

Lotus, 409
Louis XII, 1219
Louis XIV, 1202, 1274, 1276, 1279, 1285
Lowbush, 317
Lowbush blueberries, 317
Low-density lipoprotein (LDL), 985
Low-energy greenhouse, 347
Low-energy precision application, 81
Lower Limit (LL), 217
Low fertility, 77
Low/high tunnels, 416
Low irradiance, 362
Low-pressure irrigation, 374
Low temperatures, 151, 175
Loxigilla barbadensis, 1033
L. perenne, 738
L. subsp. melo, 1212
Lucerne, 217, 839
Lucius Junius Moderatus Columella, 1258
Lucretius, 1257, 1261
Lucullus, 673, 1264
Ludwig Mitterpacher, 1257
Luigi Savastano, 1295
Luigi Vilmorin, 1288
Lung cancer cell, 987
Lupin, 412, 839
Lutein, 338
Luther Burbank, 1289
Lychee, 1245
Lycopene, 119, 335, 336, 339, 351–353, 375, 383, 844, 984
Lycopersicon esculentum, 1030
Lysimeters, 81, 366, 367, 368

M

Macadamia, 125, 272, 289–294
Macadamia integrifolia, 290
Macadamia (Macadamia spp.), 124
Macadamias, 289
Macadamia tetraphylla, 290
Machine
  harvest, 225
  harvested, 311, 315
  harvesters, 321
  harvesting, 77, 308
  transplanting, 78
Macro and micro-nutrients, 513
Macroeconomic and political stability, 1146
Macro-environment, 1150
Macronutrients, 627, 634
Macrophomina phaseolina, 848
Macrosporogenesis, 108
Macular degeneration, 979, 982
M. acuminata, 128
Madagascar, 135, 148, 717
Madder, 1274
Madeira, 139
Madonna lily, 412
MADS domain, 107
M.A. Du Breuil, 1285
Magdeburg, Germany, 683
Magnesium, 210, 634, 967
Magnolia, 439
Magnolia grandifolia, 437
Magnoliales, 139, 141, 143
Magnoliid, 143
Magpie, 1032–1035
Mains water supply, 513
Maintenance of mobility, 1005
Maiolini, 169
Maize, 274, 720, 826, 841, 1212, 1290, 1299
Major retailers, 475
Making Connections survey, 960
Malacca tree (Emblica officinalis), 648
Malate, 119, 211
Malawi, 840
Malaya, 266
Malay Archipelago, 147, 170, 441
Malaysia, 136, 159, 264, 266, 267, 269, 270, 284, 286, 436, 802
Malaysian Cocoa Board, 287
Male gametogenesis, 108
Mali, 715, 720
Malic, 119, 207
Malic acid, 209–211, 241, 246
Malolactic fermentation, 246
Malolactic fermentation (MLF), 241, 246
M. Alphonse Du Breuil, 1278, 1285
Mal secco disease, 170
Malus, 109, 623
Malus angulosa, 1271
Malus domestica, 289
Malus medica, 160
Malus spp., 766
Mammals, 1027, 1033
Mammary carcinoma, 987
Managed turf, 752
Management, 725, 726
Management of fruit, 1258
Management practices, 636, 806
Management skills, 1125
Management system, 725
Manchuria, 98, 101
Mandarin, 162, 166–168, 175, 184–186, 190, 193, 844

Mandarin oranges, 844
Mandarins, 162, 165, 167, 168, 175, 180, 181, 183, 184, 188, 189
Mandatory inspections, 1190
Mandatory testing, 1190
Manganese, 210, 211, 634
Mangifera, 147
Mangifera indica, 717
Mango, 124, 126–128, 147–152, 271, 473, 845, 1235
Mango in India, 127
Mango (Mangifera indica L.), 124
Mango (polyaxials), 128
Mangosteen, 127
Mangosteen (Garcinia mangostana), 125
Mangrove, 722
Manila Galleon, 124
Manilkara achras, 719
Manorina melanocephala, 1032, 1034
Mansfield Park, 1247
Manuscripts, 1218
Maple, 437, 722
Maples (Acer spp.), 625
Maravilla, 308
Marc, 240
Marcii Porcii Catonis, 1262
Marco Polo, 649
Marginal water, 81
Marienstein Farmers' Conference, 863
Marigold, 348
Marigolds, 414, 443
Markelova and Mwangi (2010), 1162
Marker-assisted, 86
Marker Assisted Selection (MAS), 87, 103
Marketability, 380, 444
Marketable yield, 384
Market access, 1131
Market chain, 664, 1160
Marketing, 77, 88, 420, 479, 481, 641, 850
Marketing capacity, 1161
Marketing chains, 466, 467, 850
Marketing clusters, 1160
Marketing contracts, 1156
Marketing cooperative, 1159
Marketing decisions, 426
Marketing/distribution/consumer, 466
Marketing groups, 1146
Marketing plan, 630, 1160
Marketing strategy, 235
Marketing structures, 1158
Market intelligence, 475, 477
Market niches, 734
Market price, 468
Market share, 472, 860
Mark Twain, 1248
Martech 2005, 1145
Martin Wagner, 682
Marula, (Sclerocarya birrea), 717
Massachusetts, 679
Mass market, 477
Mass selection, 1288
Master Gardener programs, 1121
Master Gardeners, 1121
Mating success, 772
Matthiola incana, 417
Maturity, 102, 103, 104
Maturity index, 182
Mauria spp., 722
Mauritania, 723
Mauritius, 1030
Maximum Daily Shrinkage (MDS), 222
Mayas, 284
Maze test, 989
Measure or predict its impact, 508
Mechanical damage, 206
Mechanical harvesting, 308, 376
Mechanical thinning, 114
Mechanisation, 235, 1131
Mechanised hedge pruning, 225
Media, 160
Media and fertilisers, 1131
Medical apple, 160
Medical care costs, 17
Medicinal, 646, 722, 843, 1274
Medicinal and aromatic plants, 12, 646, 648, 650, 651, 653, 654, 658, 664, 665, 667, 668, 846, 847, 860
Medicinal and aromatic products, 653
Medicinal and pharmaceutical products, 8
Medicinal herb garden, 1274
Medicinal herbs, 645, 648, 1218, 1272
Medicinal plants, 648, 844, 1226, 1274
Medicines, 408, 646, 715, 717, 718
Mediterranean, 101, 130, 160, 162, 165, 169, 173, 180, 183, 192, 199, 268, 302, 304, 331, 332, 340, 346, 362, 367, 375, 376, 418, 438, 446, 456, 457, 663, 733, 831, 1182, 1214, 1244, 1291
Mediterranean Basin, 78, 966
Mediterranean Biomes, 763
Mediterranean diet, 966
Mediterranean lemon, 169
Mediterranean oak decline, 831
Mediterranean region, 456

Mediterranean Sea, 77, 648
Mega-cities, 804
Megaspore mother cell, 108
Mega-sporogenesis, 111
Meiosis, 108
Meiotic diplospory, 113
Melaleucas, 457
Melampsora spp., 825
Melbourne's, 719
Meloidogyne, 233
Melons, 78, 84, 85, 87, 351, 353, 390, 480, 1208, 1214, 1230, 1266
Membership-based organisations, 1149
Memory and learning, 988
Memory retention, 17
Mendel, 83, 1297
Mendoza and Vick 2010, 1149
Meng haoran, 1245
Mental ability, 773
Mental fatigue, 799, 1008, 1051
Mental health, 1009, 1016
Mental health disorders, 1007
Mental illness, 810, 1123
Mental skills, 770
Mercapturic acid pathway, 984
Meristem, 98, 99, 102–106, 627, 636
Meristem culture, 636
Meristem identity genes, 107
Meristems, 105
Mesnager, 1274
Mesoamerica, 139
Mesocarp, 113
Mesopotamia, 160, 672, 1199, 1205, 1237, 1243, 1301
Messe Essen, 415
Messmate Stringybark (Eucalyptus obliqua), 719
Metabolic syndrome, 986, 988
Metabolic waste, 650
Metabolism dysfunction, 986
Metals, 724
Metamorphoses, 1212
Metaphycus alberti wasps, 781
Metarhizium anisopliae, 824
Methane, 365
Methemoglobinemia, 505
Methoxypyrazines, 249
Methyl bromide, 1176
Methylene urea, 742
Metrics, 1227
Mevalonic acid, 652
Mexican, 143, 145
Mexican Day of the Dead, 414
Mexican oregano, 663
Mexico, 87, 91, 124, 136, 139, 143, 152, 267, 272, 284–286, 312, 313, 419, 438–440, 663, 1030, 1173, 1178, 1219
México, 161
Mexico, Brosimum alicastrum, 720
Mica, 416
Mice, 1035
Michelangelo Merisi, 1212
Microbial activity, 513, 631, 742
Microbial communities, 610, 829
Microbial phytotoxins, 750
Microbial safety, 470
Microbiological contamination, 1145
Microcitrus, 161
Microclimate, 328, 329, 342
Microclimatic, 358
Microfinance, 468, 1148, 1149
Microflora, 513
Micro-irrigation, 1303
Microlaenia stipoides, 734
Micro-mapping, 292
Micromorphometry, 222
Micronutrients, 627, 634
Micro propagation, 450, 622, 626, 627, 639, 1176, 1177
Microsporangia, 108
Microsporangium, 108
Microsporogenesis, 107, 108
Middle Ages, 411, 412
Middle East, 271
Middle Eastern, 77
Migration, 832, 835, 1027, 1029, 1034
Migration patterns, 803
Migrations, 834
Mildew, 355
Milieu Programma Sierteelt (MPS), 424
Milk, 250
Millet, 841
Millipedes, 751
Mindanao, 1152, 1161
Miner, 1032, 1034
Mineralization, 824
Minerals, 76, 139, 966, 967
Mineral theory, 1300
Minimalist' style, 764
Minimizes risks to human health, beneficial and non-target organisms, 517
Minimum air temperatures, 349
Minimum pruning, 225
Minimum tillage, 839

Ministry for Primary Industries, 1191
Minneapolis, MN, USA, 419
Minoan, 765
Miombo (Brachystegia), 721
Missouri Botanical Garden, 796, 1249
Mist irrigation, 633
Mites, 206, 306, 1030
Mitigation, 850
Mitigation practices, 850
Mitochondria, 986
Mitotic diplospory, 113
Mitotic divisions, 108
Mitsukuni, 676
Mitterpacher, 1280, 1281
M. javanica, 233
Model-based irrigation, 373
Model of Flower Development, 107
Modified, 1026
Modified atmosphere packaging, 466
Mole crickets, 751
Molecular assisted breeding, 430
Molecular biology, 11, 391
Molecular markers, 87, 309, 318
Mollusca, 637
Molon, 1289–1291
Molon, G., 1289
Moluccas, 441
Molybdenum, 634
Molybdenum (Mo), 187
Monatsblatt, 676
Mongolia, 98, 438
Monoaxial, 127, 128
Monocarpic, 104
Monocultures, 278
Monopodial, 103
Monopodial branching, 102
Monopsony buyer, 1157
Monosporascus root rot, 84
Monoterpenes, 120
Monsoon periods, 847
Monsoon system, 847
Monstera deliciosa, 768
Monstrifica barbis insignita, 1271
Monterey Cypress (Cupressus macrocarpa), 716
Monterey Pine (Pinus radiata), 716
Mood scores, 771
Morbiana, 1203
Morocco, 91, 163, 164, 166, 168, 419
Morphogenesis, 334
Morphogenetic, 98
Morphogenetic phase, 178
Morrill Act, 1125
Mosaic art, 1209
Mosaics, 1209, 1222
Mosquitoes, 751
Mosses, 439, 849
Mother Earth, 1261
Mothering Sunday, 414, 415
Mother's day, 415
Mottlecah, 440
Mountain ash, 818
Mountain ash (Sorbus spp), 636
Mountain scree, 955
Moustier (2012), 1155, 1159, 1161
Mouth cancers, 987
Mowing, 750
Mowing frequency, 741
Mowing heights, 739, 741
Mowing regimes, 740
Mozambique, 717, 721, 722
MS medium, 627
Mt Fuji, 672
Mud, 1031
Mulberry, 109, 716, 1274
Mulberry trees, 723
Mulching, 846, 1031
Mulla mulla, 454
Multinational retailers, 477
Multiple retailers, 469, 474
Multispectral imagery, 236
Multi-storey plantings, 777
Munich, Bavaria, 679
Municipal parks and gardens, 1122
Muriate of potash, 744
Murray Prior et al. 2006, 1155
Murray-Prior, R., 1152–1155, 1158–1163
Musa, 128, 129
Musa acuminata, 128
Musa balbisiana, 128
Musaceae, 128
Muscari racemosum, 409
Muscle tension, 799, 1005
Mushrooms, 843, 1119, 1226, 1235
Music therapy, 954
Muskau, 681
Muskmelon, 721
Muskmelons (Cucumis melo), 380, 1212
Must, 238, 242
Mustard, 233
Mutation, 1288
Mutation induction, 430
Mutualism, 823
Myanmar, 147, 266, 439

Mycoherbicides, 750
Mycorrhizae, 823
Mycorrhizal colonisation, 824
Mycorrhizal fungi, 824
Myristica fragrans, 267
Myrosinase, 973, 974, 976
Myrosin cells, 974
Myrtle, 410, 1267
Myrtle (Myrtus communis), 457
Myzus persicae, 834, 847

N
Nabataeans, 1266, 1301
Nageire, 1203
Naktuinbouw, 1188
Naphthalene acetic acid, 625
NAPPO, 1181
NAPPRA, 1188
Narcissus, 409, 414, 430, 766
National Agricultural Library of the United States, 1249
National Audubon Society, 1036
National Botanic Gardens of Wales, 763
National Cherry Blossom Festival, 414
National Flower, 413
National Forum on Biological Diversity, 788
National Library in Beltsville, 1249
National Marine Fisheries Service, 806
National Organic Program (NOP), 868
National parks, 956
National Parks, 795
National Plant Protection Organisation (NPPO), 1180
National Socialist Generalplan Ost, 683
National strength, 685
National Trust for Scotland, 1133
National Trust UK, 1133
Native, 1029, 1030
Native and non-native species, 1035
Native bush, 5
Native plants, 1063
Native species, 1062
Natural antioxidant defenses, 988
Natural cycle, 1043
Natural England, 795, 805, 1003
Natural Environmental Research Council (NERC), 732
Natural Environment and Rural Communities Act, 795
Natural environments, 799, 1002, 1003, 1007, 1049, 1073, 1078, 1140, 1238
Natural grasslands, 732
Natural green spaces, 799–801, 803, 809
Natural habitats, 1030
Natural History Museum Library, 1249
Naturalism, 1199–1201, 1203
Natural landscape, 1009, 1012, 1200, 1261
Natural law, 675
Natural products, 659, 661
Natural resources, 16, 18, 505, 604, 801, 805, 1150
Natural selection, 1034
Natural spaces, 962
Natural systems, 801, 809
Natural ventilation, 357, 359
Natura morta, 1212
Nature-based tourism, 807
Nature Conservancy, 795
Nature conservation, 788
Nature et Progrès, 864
Nature reserves, 1002
Naturescaping, 1041
Nature's Choice, 511
Navel oranges, 162, 164, 165, 172, 180, 187, 188, 192
Navel rind stain, 187
Navel sweet orange, 164
Nazca period, 1216
Ncukana, L., 1152
N. de Bonnefons, 1279
Neanderthals, 409
Near East, 723
Nebuchadnezzar II, 678
Nectar, 1030, 1032
Nectarine, 99
Neem Tree, 717
Negative environmental effect, 508
Neglected and underutilised species (NUS), 125
Nematicides, 233
Nematodes, 171–173, 232, 233, 306, 636, 751, 824, 832, 1175
Nematus ribesii, 316
Nemesia, 419
Neolithic, 198
Neolithic people, 3
Neolithic Revolution, 1206
Neonectria galligena, 830
Neotropics, 139, 140
Neoxanthin, 338
Neoxantin, 119
Nepal, 147
Nephrolepis exaltata, 764
Nerium oleander, 364

Nero, 416
Nest boxes, 1028, 1036, 1042
Nesting boxes, 1031
Nests, 1031
Netherlands, 413, 765, 1042
Net-houses, 328, 331
Net photosynthesis, 391
Net radiation, 346
Net radiometers, 332
Network theory, 1176
Neurodegenerative diseases, 966, 988
Neurogenesis, 989
Neuro-hormonal imbalanc, 1007
Neuronal cell apoptosis, 988
Neuronal death, 988
New English Weekly, 863
Newfoundland, 441
New Guinea, 438
New Guinea hybrids, 443, 444, 448
New Mexico, 86, 438
New Ornamental Plants, 447
New South Wales, 290, 291
New testament, 1226, 1241, 1243
New World, 199, 200, 238
New York Botanical Garden, 808
New York City, 679, 1122
New York ICE Futures market, 1154
New York State Agricultural Experiment Station, 1291
New York State experiment station, 1234
New Zealand, 98, 145, 199, 314, 316–318, 440, 441, 716, 756, 806, 1119, 1127, 1177, 1181, 1183–1187, 1190, 1191, 1238
NGIA, 1130, 1131
Niacin, 139
Nicaragua, 272, 1159
Nickel, 634
Niger, 723, 840, 841
Nigeria, 8, 264, 269, 270, 284, 285, 662, 715
Nikolay Vavilov, 6
Nile, 3
Nile River, 715
Nilsson et al. 2012, 1158
Nitidulid, 142
Nitrate, 335, 611, 865, 1298, 1300
Nitrate (NO3-), 213
Nitric oxide, 984
Nitric oxide synthase, 985
Nitrogen, 79, 211, 238, 240, 365, 377, 457, 611, 612, 614, 631, 634, 723, 724, 742, 777, 824, 834, 974, 1298–1300
Nitrogen-based fertilizers, 1299
Nitrogen cycle, 507
Nitrogen fertilisers, 505
Nitrogen fertilizers, 744
Nitrogen fixation, 805, 840
Nitrogen-fixing microbes, 837
Nitrogen-fixing rhizobia, 824
Nitrogen fraction, 211
Nitrogen (N), 213, 742, 743
Nitrogenous compounds, 215
Nitrous oxide, 451, 612, 613
Nitrous oxides (NOx), 773, 837
Noble rot, 228
Noble symmetry, 1201
Noise, 769
Noise barriers, 457
Noise pollution, 803
Non-climacteric, 117, 184
Non Climacteric, 118
Non-climacteric fruits, 119, 208
Non-climacteric ripening, 182
Non-digestible fibers, 967
Non-native species, 1029
Non-Saccharomyces yeasts, 244
Nonselective herbicides, 632
Non-tillage method, 194
Norisoprenoids, 212
Norman Borlaugh, 7
Nortes, 152
North Africa, 101, 303, 314, 438
North America, 101, 199, 229, 232, 271, 272, 274, 284, 310–315, 317, 426, 439, 441, 637, 678, 679, 697, 703, 707, 722, 725, 804, 806, 827, 829, 984, 1121, 1175, 1185, 1290, 1300
North American, 317, 426
North American Free Trade Agreement, 91
Northern Africa, 97
Northern China, 98
Northern European, 426
Northern Hemisphere, 97
Northern highbush, 319
Northern Ireland, 830
Northern papaya, 99
North Pakistan, 721
Norths, 152
Northwest Flower and Garden Show, 416
Northwest U.S., 312
Norway, 303, 309
Novel packaging, 477
NPPO, 1180, 1181, 1184, 1187, 1188
Nucleic acids, 211

Nucleus-estate model, 1156
Nurseries, 420, 1035, 1129, 1275, 1297
Nurseries and garden centres, 1040
Nursery and Garden Industry Australia (NGIA), 1130
Nursery catalogues, 1239
Nursery crops, 620, 621
Nursery industry, 1063, 1120
Nursery plants, 436
Nursery production, 629, 1119
Nursery stock, 1188, 1190
Nursery trade, 1177
Nut, 112, 715
Nut crops, 272, 289
Nuthatch, 1033
Nutlets (litchi or longan), 127
Nutmeg, 267
Nutraceutic, 383
Nutraceutical, 309, 352, 375
Nutrient cycling, 840
Nutrient degradation, 839
Nutrient depletion and erosion, 513
Nutrient film, 361
Nutrient film technique (NFT), 378
Nutrient leaching, 807
Nutrient management, 247
Nutrient mobility, 213
Nutrient reserves, 214
Nutrients, 513, 516
Nutrient status, 215, 216
Nutrient supply, 216, 777
Nutrient–use efficiency, 838
Nutritional status, 191
Nutritional value, 480, 1144
Nutrition retention, 480
Nutritious fruits, 481
Nuts, 127, 266, 843, 860, 966, 973, 1119, 1226, 1264, 1290
Nuts crops, 289

O

Oak-derived, 248
Oak (Quercus sp.), 1186
Oaks, 437, 438, 715, 718, 724, 726, 1185
Oases, 723
Oats, 839, 981
Obesity, 14, 798, 985, 986, 1006, 1050, 1123
Obesity crisis, 479
Obesity related disease, 1006
Occupational therapy, 954
Oceania, 124, 269
Octoploid, 303, 304
Octoploid strawberry, 304
Odyssey, 1242, 1256
Oenococcus oeni, 246
Office environments, 767, 771, 772, 774, 778
Office-landscape, 767
Offices, 769, 771, 772, 778
Office work, 767
Ohio, 416
Ohio Florists Association, 416
Oidium, 228
Oidium magifera, 845
Oil, 266, 717
Oil crops, 295
Oil palm, 264, 269–271, 286, 287, 802, 845
Oil palm (Elaeis guineensis), 269, 289
Oil seed rape, 828, 836
Oilseed rape (Brassica napus), 825
Oktoberfest, 676
Olchondra 2010, 1148
Old World, 199, 200
Oleander, 364
Olea oleracea, 110
Oleria, 440
Olfactory nerve, 650
Olitor, 673
Olive (Olea europea), 101
Olives, 99, 101, 109, 410, 438, 605, 860, 1234, 1241, 1243, 1257–1259, 1262, 1264, 1265, 1270, 1291, 1302
Olivier de Serres, 1257, 1270, 1274, 1275
Olmsted Brothers, 682
Oman, 91
Omnivore, 1029
O. Montalbani, 1271
Oncidium, 418
On Farming, 1262
Onions, 77, 81, 84–86, 475, 612, 842, 846, 977, 978, 982, 986–988, 1234, 1266
On-line gardening information, 1132
On Medical Matters, 1218
Ontogenetic stages, 102
O. oeni, 247
Oomycetes, 1175
Open field, 839
Open field cultivation, 78
Open-field production, 77
Open-field vegetable, 86, 92
Open green spaces, 790
Open plan offices, 767
Open-pollinated, 87, 318, 420
Open pollination, 77
Open space planning, 686
Open space programme, 683

Open spaces, 679, 681, 683, 685–687, 790, 798, 1119
Open space structure, 682
Open systems, 1152
Open trade policies, 1146
Ophiostoma novo-ulmi, 831
Ophiostoma ulmi, 1185
Optimal air temperatures, 349
Optimum air temperatures, 348
Optimum day temperatures, 348
Optimum water content, 365
Opuntia, 841
Opuntia ficus indica, 841
Opuntia spp., 841
Opuntie (prickly pears), 1289
Opus ruralium commendorum, 1268
Orangeries, 416
Oranges, 160–162, 175, 176, 184, 187, 189, 190, 193, 1273
Orchard, 613, 614, 713, 1274, 1275, 1297, 1303
Orchard crops, 605, 612
Orchard management, 1298
Orchards, 721, 790, 1276
Orchard topography, 292
Orchid, 409, 413, 414
Orchids, 443, 453, 768
Oregano, 663
Oregon, 310–312
Organic, 426, 1033, 1120
Organic acids, 118, 119, 208, 209, 211, 243, 247
Organic agriculture, 859–861, 865, 867–869
Organic certification, 861, 867–870, 1159
Organic composting, 1239
Organic farmers, 1072
Organic farming, 863, 864, 866, 867, 869, 1235
Organic Farming and Gardening, 864
Organic fertilizers, 1298
Organic food movement, 865, 867
Organic foods, 12, 90, 860, 865, 869
Organic gardeners, 1029
Organic gardening, 1076
Organic gases, 773
Organic horticultural products, 870
Organic horticulture, 806, 860, 869
Organic ingredients, 869
Organic markets, 860, 1158
Organic material, 216
Organic matter, 365, 513
Organic matter decline, 504
Organic movement, 861–864
Organic principles, 861, 865
Organic production, 391, 867–869
Organic products, 90, 861, 866, 868, 869, 1145
Organic rice, 1149
Organic standards, 511, 868
Organic systems, 863
Organisation for Economic Co-operation and Development (OECD), 1140
Organoleptic, 166, 351, 383
Organoleptic properties, 966
Oriental vegetables, 1241
Origin, 90
Orinoco Valleys, 270
Orius, 142
Ornamental, 337, 362, 365, 436, 627, 1177, 1226
Ornamental crops, 1226
Ornamental gardens, 962, 1235
Ornamental horticulture, 781, 1002
Ornamental plants, 349, 374, 408, 439, 440, 444, 446, 456, 642, 860, 1191, 1236, 1239
Ornamental plant trade, 1176
Ornamental pot plants, 378
Ornamental production, 631
Ornamentals, 11, 12, 361, 365, 381, 439, 440, 620, 621, 796, 850, 1191, 1199, 1226, 1235
Ornamental shrubs, 1274
Ornamental species, 1176
Ornamental trade, 1174, 1176
Ornamental trees, 1267
Ornamented farm, 675
Orthophosphates, 457
Oryctolagus cuniculus, 836
Osaka Bay, Japan, 685
Osmanthus fragans flowers, 717
Osmolytes, 382
Osmotic adaptation, 364
Osmotic potentials, 386
Osteoporosis, 1006, 1013
Othello, 1247
Otiorhynchus sulcatus, 834
Ottoman victory, 649
Outcrossing, 319
Outdoor concerts, 797
Outdoor nurseries, 777
Outdoor Recreation Resources Review Commission (ORRRC), 683
Outreach, 1118

Outreach projects, 1121
Ovary, 987
Ovary abscission, 179
Over-exploitation, 505
Overhead irrigation, 374
Overhead nozzles, 374
Overhead surface, 374
Overhead surface irrigation, 374
Overpopulation, 801
Oviposition, 834
Ovule abortion, 110
Ovule differentiation, 107
Ovule formation, 108
OXFAM, 1140
Oxidation, 240, 242, 247
Oxidative damage, 1005
Oxidative enzyme, 227
Oxidative stress, 84, 984
Oxides of nitrogen, 614
Oxygen, 630, 634, 1299, 1301
Oxygen deficiency, 365, 376
Oxygen enrichment, 376
Oxygen permeation rates, 252
Ozone, 825, 838
Ozone (O_3), 837, 838
Ozothamnus diosmifolius, 440

P
Pacific, 124, 268, 282, 284, 1176, 1184
Pacific Coast, 310
Pacific Islands, 267, 269
Packaging, 467, 469, 471, 479, 481
Packhouses, 514
Packhouse water, 515
Packing, 466, 473
Paclobutrazol, 177, 449, 453
Pad and fan, 368
Padova, Italy, 685
Pain relief, 1005
Paintings, 1210, 1212, 1222
Pain tolerance, 770
Palace of Versailles, 766
Palatability, 116
Palladio, 160, 1265
Palladium, 1270
Palladius, 1257
Palm oil, 264
Palms, 410, 621, 723, 764, 766, 767, 1236
Palm wine, 722
Palm wine-making, 722
Paludina vivipara, 1033
P. americana var. americana, 143
P. americana var. drymifolia, 143
P. americana var. guatemalensis, 143
Pampas, 439
Panama, 91, 439
Panama disease, 285
PanAmerican Seed Company, 420
Pancreatic islets, 980
Pan evaporation, 218, 219
Pan Evaporation, 217
Panicle, 150, 151
Pansies, 443, 454, 455
Pansies (Viola tricolor), 459
Pansy, 412, 419
Papas, 1222
Papaveraceae, 112
Papaya, 127, 134–139, 148, 271, 272, 844
Papaya (Carica papaya Linn), 124
Papaya Ring Spot (PRV), 137
Papaya Ringspot Virus (PRV), 136
Papayas (monoaxial), 128
Paper Birch (Betula papyrifera), 720
Paper daisy, 440
Papua New Guinea, 267, 284, 286, 1154
Papyrus, 409, 410
Paradise apple, 1286
Parasite, 1035
Parasitic wasps, 751
Parasitoids, 832, 833
Paris, 678
Park des Buttes Chaumont, 679
Park design, 960
Parkia biglobosa, 717
Parkland, 13, 713, 715
Parkland management, 830
Park landscapes, 1001
Parks, 8, 436, 685, 696, 789, 798, 958, 962, 1009, 1047, 1050, 1054, 1059, 1061, 1062, 1069, 1073, 1075, 1076, 1201
Parks and gardens, 790, 850, 1119, 1120, 1132
Parks and gardens maintenance, 1122
Parks and landscapes, 1129
Parks and open spaces, 1007
Parks and recreation department, 686
Parks management, 1232
Park trees, 702
Parkways, 683, 1201
Parsley, 361, 663
Partenocarpic, 117
Partenocarpy, 109
Parthenocarpic, 115, 133, 137, 166, 181, 350, 390
Parthenocarpic mandarin, 181

Parthenocarpy, 111, 115, 180
Partial contracts, 1156
Partial roof shading, 342
Partial root zone drying, 375, 607
Partial Rootzone Drying (PRD), 218, 219, 386
Participatory Guarantee Systems (PGS), 869
Particulate matter, 779, 1061
Partnership, 705, 706
Parus caeruleus, 1035
Parus major, 1033
Pasadena, California, 414
Paspalum notatum, 734
Paspalum vaginatum, 734
Passer domesticus, 1032
Passion fruit, 128, 841
Passionfruit aroma, 249
Passion fruit (Passiflora edulis), 127
Pasture, 714
Patagonia, 438, 839
Pathogen growth, 825
Pathogen patterns, 831
Pathogens, 825, 847
Pathogen taxa, 1175
Patrick Dougherty, 1202
Paul Cezanne, 1215
Pavlovsk, 676
Payback period, 508
P. betulaefolia (rootstock), 1293
P. brassicae, 828
P. communis, 1293
P. cynaroides, 442
PDCI, 1150
P. domestica, 101
Pea, 83, 842, 1233
Pea aphid, 833
Pea aphid (Acyrthosiphon pisum), 833
Pea aphids, 833
Peach, 99, 100, 101, 106, 110, 115, 117, 119, 1212, 1234, 1235, 1264, 1270, 1281, 1286, 1291, 1299, 1303
Peaches, 109, 118–120, 822, 1233, 1234, 1271, 1283, 1292
Peach fruit development, 115
Peach (Prunus persica), 100
Peanut, 1208
Pear (Pyrus caucasica), 99
Pear (Pyrus communis), 99
Pears, 99, 101, 109, 111, 115, 118, 120, 480, 605, 625, 626, 822, 831, 1233, 1234, 1257, 1260, 1264, 1267, 1269, 1271, 1273, 1274, 1283, 1286, 1288, 1290–1294
Peas, 1235
Peat, 634
Pecan, 99, 289, 973
Pectate lyases (PL), 118
Pectin, 126
Pectinases, 109
Pectinmethylesterase, 187
Pectin methyl-esterases (PME), 118
Pectins, 240
Pectolytic enzymes, 240
Pecuaria Development Cooperative Incorporated (PDCI), 1149
Pedanios Dioscorides, 1218, 1233, 1237
Pediococcus, 246
Peel oil, 189, 190
Peel pitting, 167, 185, 186
Peel senescence, 188
Peepal (Ficus religiosa), 718
Pehibaye palm (Bactris Gasipaes), 720
Pelargonium x hortorum, 382
Pelleting, 77
P'en ching, 672
Pencil cedar, 450
Penman-Monteith equation, 367, 838
Penman-Monteith model, 369
Penman-Monteith (PM), 366
Pennisetum clandestinum, 734, 737
Pennsylvania, 416
Penstemon fruticosus, 447
Peony, 409
People management, 773
Peppers, 77–79, 85–88, 267, 335, 337, 340, 343, 350, 352, 360, 361, 363, 367, 369, 376, 388–390, 649, 721, 1234, 1235
Perceived risk, 1187
Perchlorates, 190
Perennial crops, 613
Perennial fruit, 821
Perennial grass, 81
Perennial plants, 837
Perennial ryegrass, 748
Perennials, 849
Perennial vegetables, 127
Perennial weed, 750
Performance, 714
Performance assessment, 752
Performance Assessment of Sports Surfaces (PASS), 756
Performance indicators, 472, 756
Perfumes, 8, 1273, 1290
Pericarp, 133

Pericarp cracking, 350
Periodic drought, 84
Periparus ater, 1033
Perlite, 378, 380, 779
Permafrost, 439
Permanent settlements, 3
Permanent wilting, 1303
Permanent Wilting Point (PWP), 217, 633
Perrault, Charles, 1278
Persea americana, 143
Persia, 160, 268, 1199, 1217, 1235
Persian walnut, 721
Persica alba, 1271
Persica lutea, 1271
Persimmon, 99
Persistence, 734
Personal control, 1006
Personal hygiene, 1145
Personal Protective Equipment, 516
Perspicua gemma, 416
Peru, 87, 91, 139, 166, 199, 289, 439, 444, 606, 608, 1216
Pest and disease management, 458
Pest control, 1042
Pest damage, 845, 848
Pest eradication, 1184, 1185
Pest free, 1184
Pest-free areas, 1184
Pest-free status, 1180
Pesticide, 516, 517, 609, 610, 797, 805, 861, 864
Pesticide contamination, 516
Pesticide drift, 806
Pesticide pollution, 518
Pesticide residues, 10, 610, 615
Pesticides, 510, 516, 517, 609, 610, 734, 807, 810, 850, 864, 867
Pesticide use, 610
Pest impact, 1179
Pest introductions, 1191
Pest invertebrates, 833
Pest management, 725
Pest population, 848
Pest prevalence, 1184
Pest risk analysis, 1188
Pest Risk Assessment, 1187
Pests, 825, 847, 848
Pests and diseases, 517
Pests and pathogens, 513
Pest species, 834
Pest threats, 1191
Pest vertebrates, 836
Petén, 719
Peterhof, 676
Peter Joseph Lenné, 676
Peter Martyr D'Angheria, 1219
Petrik method, 807
Pets, 1027
Petunia, 335, 348, 350, 421, 449
Petunias, 443, 454
Petunia x hybrida, 337, 348, 362, 420
P. ferruginea, 449
Pharaoh Thutmose III, 409
Pharmaceutical compounds, 798
Pharmaceutical drugs, 659
Pharmaceutical industry, 1298
Pharmaceutical interest., 1238
Pharmaceutical Products, 717
Pharmaceuticals, 11, 646, 648, 650, 653, 655, 658, 659, 717, 797
Pharmacognosy, 653
Pharmacological benefits, 654
Pharmacology, 1274
Phase I enzymes, 984
Phase II enzymes, 976
Phases of intensive growth, 98
Phase transition, 103
Phellodendron amurense, 803
Phenol, 383
Phenol biosynthesis, 970
Phenolic, 119, 210, 236, 239, 245, 249–251
Phenolic compounds, 843
Phenolics, 211, 212, 240
Phenological, 449
Phenological change, 819
Phenological indicators, 234
Phenological stages, 222
Phenological triggers, 835
Phenology, 98, 104, 803, 821, 822, 832
Phenols, 383, 970
Phenophases, 820, 822, 823
Phenylalanine, 212, 970
Phenylalanine ammonium lyase (PAL), 119
Phenylpropanes, 656
Phenylpropanoids, 650
Philadelphia Flower Show, 416, 1041
Philadelphus, 766
Philippines, 124, 135, 198, 264, 269, 276, 286, 453, 716, 720, 1146–1150, 1152, 1160, 1161
Phloem mobility, 213
Phoenicians, 148
Phoenix reclinata, 722
Phoma terrestris, 85

Phoma tracheiphila, 170
Phomopsis, 230, 316
Phomopsis cane and leaf blight, 230
Phomopsis viticola, 230
Phosphate, 611
Phosphorous, 79, 210, 351, 457, 611, 634, 715, 824, 967, 1300
Phosphorus (P), 743
Photo-assimiliates, 388, 390
Photographs, 1222
Photomorphogenesis, 336, 775
Photoperiod, 98, 132, 178, 349, 449, 782, 819
Photoperiodic, 336
Photorespiration, 362
Photosynthate, 174, 183
Photosynthesis, 339, 364, 385, 632, 775, 823, 837, 850, 1300, 1301
Photosynthesis rate, 391
Photosynthesis reduction., 360
Photosynthetic acclimation, 362, 363
Photosynthetic Active Radiation, 331
Photosynthetic activity, 337
Photosynthetically active region, 775
Photosynthetic capacity, 337, 362, 390
Photosynthetic disorder, 182
Photosynthetic gain, 362
Photosynthetic photon flux (PPF), 337
Photosynthetic potential, 741
Photosynthetic rate, 387, 388
Photosynthetic rates, 361
Photosystem II, 364, 376
Photosystem II (PSII), 389
Phototropism, 1233
Phtophthora spp., 171
Phyllosphere, 774
Phylloxera (Daktulosphaira vitifolii), 232, 1300
Phylloxera free status, 1178
Phylloxera (Phylloxera vastatrix), 1178
Phylloxera-resistant rootstocks, 229
Phylloxera vastatrix, 1180
Physical, 517
Physical activity, 1008, 1013
Physical and mental health, 13
Physical health, 14
Physiocracy, 675
Physiological benefits, 1005
Physiological disorder, 185–187, 215, 355, 476
Physiological disorders, 188, 355, 845
Physiological drop, 113
Physiological fruit disorders, 184, 194
Physiological health, 1004
Physiological motor performance, 1005
Physiological races, 827
Physiological stresses, 696, 823, 837
Physiological well-being, 799
Phythophthora, 781
Phythophthora cinnamomi, 144
Phythopthora root rot, 145
Phytochemical-rich diet, 966
Phytochemicals, 76, 382, 966, 967, 981–983, 987, 990
Phytochromes, 98, 339
Phytoene (C40), 119
Phytomer, 105
Phytomers, 102–104
Phytomonitoring, 222
Phytonutrients, 381, 967
Phytopathology, 1181
Phytophthora, 306, 513, 635, 1186, 1188
Phytophthora cinnamomi, 231, 825, 831
Phytophthora citricola, 825
Phytophthora infestans, 826, 848, 1178
Phytophthora ramorum, 1186
Phytophthora root rot, 171, 172
Phytophthora rubi, 309
Phytophthora spp., 169
Phytoplasma, 637
Phytoplasmas, 1175
Phytoremediation or bioremediation, 514
Phytosanitary, 1173, 1177–1179, 1181–1184, 1187, 1188, 1190, 1191
Phytosanitary agreements, 1178, 1181
Phytosanitary certification, 1176, 1180, 1181, 1184, 1188
Phytosanitary certification (ISPM No. 12), 1184
Phytosanitary inspections, 1187, 1190, 1191
Phytosanitary legislation, 1178
Phytosanitary measures, 1181
Phytosanitary risks, 1176
Phytoseiulus persimilis, 781
Phytosterols, 966
Phytotherapy, 654
Phytotoxic effects., 381
Pica melanoleuca, 1035
Pica pica, 1032, 1035
Picea spp., 439
Pichia, 244
Picoides pubescens, 1033
Pierce's disease, 232
Pierre Charles L'Enfant, 678
Pierre Joseph Redouté, 1215

Pierris brassicae, 1029
Pierris rapae, 1029
Pieter Aertsen, 1213
Pietra dura, 1210
Pietro de Crescenzi, 1257, 1268–1270, 1302
Pigmentation, 350
Pigmented oranges, 162
Pimelea physodes, 440, 442, 449
Pimenta dioica, 649, 719
Pineapple, 127, 128, 264, 271, 844
Pineapple (Ananas comosus Merr.), 124
Pine bark, 380
Pine nuts, 289
Pine pitch canker (Fusarium circinatum), 1189
Pine pitch canker (PPC) (Fusarium circinatum), 1188
Pines, 439, 723, 1188
P. infestans, 826, 827, 1173
Pinolene, 186
Pinot noir, 227
Pinstrup-Andersen and Watson, 1150
Pinus patula, 450
Pinus radiata, 716
Pinus spp., 439
Piperales, 143
Piper nigrum, 267, 649
Pipits, 1035
Pirotte et al. 2006, 1159
Pistachio, 99, 109, 147, 973
Pistachios, 289
Pistacia vera, 289
Pistacia vera L., 147
Pitch canker-, 1186
Plagiotropic, 102
Planned management, 702
Planning and policy, 805
Planococcus citri, 781
Planorbis corneus, 1033
Plantains, 125–129, 274
Plantains (Musa), 126
Plantains (Musa AAB), 125
Plant architecture, 334
Plantation-based, 1154
Plantation crops, 266, 294, 613, 844, 1147
Plantations, 12, 612, 614, 790, 843, 1267
Plant Available Water (PAW), 217
Plant-based industries, 1172
Plant Breeder Rights (PBR), 421, 459
Plant breeding, 10, 1235
Plant Breeding Rights, 437
Plant compactness, 79
Plant defense, 967
Plant development, 967
Planted balconies, 454
Plant growth, 1226, 1233, 1245, 1257, 1302
Plant growth regulators (PGRs), 191, 194
Plant health, 1131, 1180, 1183, 1188
Plant Health Directive, 1183
Plant Health Directive (77/93/EEC), 1183
Plant Health (Great Britain) order 1993, 1183
Plant health legislation, 1179, 1186
Plant hunting, 441
Plant husbandry, 1132
Plant iconography, 1205, 1220, 1222
Plant images, 1205
Planting density, 629
Planting schemes, 1042
Plant/insects-pathogens interactions, 973
Plant introduction schemes, 446
Plant landscaping, 773
Plant-microorganism interaction, 967
Plant nursery, 1275
Plant nutrition, 1235
Plant Passport, 1183
Plant Patent Act 1930, 1290
Plant Patents, 421
Plant pathogens, 825, 826, 830
Plant pests, 1174, 1187
Plant propagation, 1232
Plant protection, 1173, 1179, 1184
Plant quality, 361, 459
Plant Quarantine Act, 1178
Plantscapes, 763, 765, 767, 768, 773, 775, 777–779, 781–784
Plant sciences, 1230
Plant sculpture, 1204
Plant selection process, 1122
Plant stress, 364, 843
Plant trade, 1173, 1174, 1185, 1191
Plant Variety Protection, 421
Plant water status, 220
Plasma antioxidant capacity, 982
Plasmodiophora brassicae, 828, 829
Plasmopara viticola, 229, 825, 1300
Plastic cover, 330
Plastic-covered houses with simple or no heating systems and a low level of technical complexity, 506
Plastic covers, 333
Plastic films, 328, 329, 331, 339
Plastic greenhouses, 331, 350
Plastic pots, 380
Plastic sheets, 846
Plastic tunnels, 357, 392

Plasticulture, 80, 81
Platanus, 639
Plate glass, 417
Plato, 1266
Player fatigue, 755
Player opinion, 755
Player-surface, 753
Player-surface interaction, 753
Playgrounds, 958, 1201
Playing quality, 752, 753
Playing surface, 747, 755
Playing surface performance, 757
Play spaces, 962
Pleasure gardens, 1199
Plectranthus, 448
Pliny, 1257
Pliny the Elder, 160, 1265
Pliny the Younger, 416
Plugging, 737
Plug growers, 419
Plug Revolution, 420
Plumcots, 1290
Plums, 99, 101, 109, 110, 1233, 1234, 1260, 1284, 1286, 1289, 1290, 1292
Plum trees, 1303
Plywood, 273
P. magnifica, 409
P. mahaleb, 1286
Poa, 737
Poa annua, 748
Poaceae, 732
Poa pratensis, 734, 737, 739
Poa trivialis, 739
Podargus strigoides, 1033
Pod borer, 286
Pod rot and trunk canker, 282
Poinsettia, 349, 439, 443, 446, 447, 453, 455
Poinsettias, 350, 453
Polack 2012, 1147
Poland, 307, 315, 782, 985
Poliaxial, 128
Policy, 1147
Policy and targets, 508
Political environment, 1152
Polka, 308
Pollarding, 639
Pollen, 1030, 1032
Pollen differentiation, 108
Pollen germination, 109
Pollen mother cell, 105
Pollen stability, 87
Pollen transfer, 109
Pollen viability, 109
Pollination, 109, 110, 115, 179, 181, 390, 820, 832, 1033
Pollination activity, 795
Pollination period (EPP), 110
Pollinator attraction, 967
Pollinators, 109, 408, 832
Pollinizers, 109, 110
Pollution, 504, 505, 515, 604, 792, 1141
Pollution risk, 807
Polyaxial, 128
Polycarbonate, 417
Polycarpic, 104
Polyethylene, 329, 332, 334, 380, 629
Polyethylene screen, 345
Polygalacturonases (PG), 118
Polynesia, 1030
Polynesian Tapa cloth, 722
Polyphenol, 973, 982
Polyphenol glycoside, 983
Polyphenolic, 250
Polyphenols, 315, 967, 970, 973, 982–990
Polyphenols transporters, 983
Polyphenol supplementation, 989
Polyploidization, 430
Polysaccharides, 227
Polythene, 614, 615
Polytunnels, 614, 615
Polyvinylpolypyrrolidone (PVPP), 250
Pome fruit, 99, 1293
Pome-fruits, 112, 113, 115, 116, 120, 822
Pomegranates, 99, 721, 1259, 1270
Pomelo, 167, 168
Pomology, 1256–1258, 1270, 1273, 1274, 1282, 1283, 1285, 1289, 1291, 1292, 1298
Pomona Italiana, 1258
Pomonas, 1234, 1257, 1261
Pompeii, 1212, 1265
Poncirus, 161
Poncirus trifoliata, 172, 1276
Ponds, 1029–1031, 1033, 1036
Poor nutrition, 5
Popillia japonica, 636
Poplars, 716, 723, 724
Poplars (Populus spp), 639
Poppy, 412
Population, 5, 16, 76, 513, 518, 685, 1150
Population build-up of resistant target pests, 506
Population growth, 481, 727, 1140, 1144
Population size, 836

Populus, 625
Porous nets, 339
Porous pipes, 374
Porous screen, 328, 331
Porous screens, 329, 359
Portfolio diet, 966
Portion-sized packages, 481
Portugal, 139, 199, 317, 451, 649, 1244
Posidonia, 380
Posidonia oceanica, 380
Positive mood state, 1005
Post-dormancy chilling, 822
Post-emergence, 750
Post-emergence herbicides, 632
Post-emergent herbicides, 631
Postharvest, 458, 466, 468, 1131
Postharvest biology, 476, 1231
Postharvest care, 466, 467, 481
Post-harvest damage, 825
Postharvest deterioration, 353
Postharvest disease, 466
Postharvest Education Foundation, 471
Postharvest handling, 12, 468–471
Postharvest life, 473, 476, 477
Postharvest losses, 466, 467, 470, 478
Postharvest practices, 468, 471
Postharvest quality, 17, 477
Post harvest storage, 477, 1144
Postharvest technologies, 467, 468, 471–473, 476
Postharvest Training and Services Center, 471
Post-transplanting stress, 79
Post-veraison, 226, 228, 230
Potamogetonaceae, 380
Potassium, 79, 85, 210, 211, 364, 377, 634, 967, 974
Potassium bitartrate, 242
Potassium hydrogen tartrate, 251
Potassium (K), 743
Potassium permanganate, 477
Potato, 473, 612, 826, 827, 846, 1178, 1208, 1220, 1222
Potato apical leaf curl, 847
Potato blight, 826, 1173, 1178
Potato blight (Phytophthora infestans), 1173
Potatoes, 475, 477, 605, 836, 846
Potatoes of Virginia, 1222
Potato late blight, 827
Potato murrain, 826
Potato (Solanum), 1173
Potato (Solanum tuberosum), 1173, 1220
Potherbs, 645
Pot ornamentals, 374
Pot plants, 374, 408, 443, 446, 447, 452, 454, 456, 766, 774
Potsdam, 676
Potted ornamentals, 13, 329
Potted plant production, 419
Potted plants, 459
Poverty, 1066, 1068–1070
Powdery mildew, 228, 229, 845
Powdery mildew management., 229
PPC, 1188
P. polyandra, 161
PPPO, 1181
Prairie grasslands, 1002
P. ramorum, 1186
Prantilla 2011, 1148
Prebiotic benefits, 973
Predators, 832, 1027
Pre-emergence, 750
Pre-emergent herbicide, 631
Preferred suppliers, 1156
Pregnancy, 1047, 1052
Premium tier, 474, 475
Premium Waxflower, 458
P. repens, 442
President's Council on Environmental Quality, 788
Press, 238, 239
Pressing, 239
Preventing Chronic Diseases A Vital Investment', 981
Pre-veraison, 209, 229
Prey densities, 833
Price volatility, 1141
Priestley, J., 1301
Primarily poplars, 723
Priming, 77
Primocane, 307, 308, 312, 313
Primocanes, 311, 312, 313
Primofiore, 169
Primofire, 170
Primordia initiation, 225
Primrose, 412, 455
Primula vulgaris, 455
Prince Hermann von Pückler-Muskau, 681
Private garden, 679
Private gardens, 954
Private-good, 1127
Private space, 962
PRIVATE TYPE=PICT;ALT=We Do Our Part, 802
Private Voluntary Standards (PVS), 511

Privatisation and commercialisation, 1127
Proanthocyanidins, 986
Proanthocyanidins (PAC), 973
Process attributes, 1145
Processing, 313, 466
Processing companies, 1159
Processing industry, 314
Processing market, 311, 312
Processors, 315, 869, 1156
Procurement, 475
Produce handlers, 479
Produce retailers, 478
Producer organisations, 1158
Producers, 869
Product handling, 468
Production, 420
Production and productivity, 1142
Production costs, 392
Production horticulture, 1119, 1120, 1129, 1131, 1135
Production industries, 849
Production nursery, 781
Production schedule, 453
Production scheduling, 1144
Production system, 603
Productivity, 772, 773, 1071, 1075
Productivity Commission, 1128
Productivity per unit area, 17, 307
Product label, 868
Products quality, 77, 329, 330, 334, 350, 352, 355, 365, 381, 386, 387, 389, 391, 475, 1157
Product system throughout its life cycle, 509
Pro-environmental attitudes, 1039
Professional capacity, 1129
Professional development, 1133
Professional Gardeners' Guild, 1133
Profitable, 513
Proflora, 416
Program-team model, 1134
Pro-inflammatory cytokines, 980, 986
Proleptic, 102
Proline, 88, 209, 211, 215, 970
Pro-lycopene, 119
Promenade, 678
Promoter genes, 179
Promotion, 641
Promotional campaigns, 475
Promotion fatigue, 473
Pro-oxidant activity, 980
Pro-oxidants, 980
Propagation, 622, 1174
Property crime, 1057
Property rights, 1146
Property values, 1047, 1071, 1075
Prose and poetry, 1226
Prostate, 984
Prostate cancer, 987, 988
Prostate cancer cells, 987
Proteacea, 440
Proteaceae, 418, 443
Protea cynaroides, 409, 440
Proteas, 409, 418, 440, 442
Protected areas, 13
Protected cropping, 843, 1131
Protection, 724
Protein, 242, 250, 251, 715
Protein fining, 242, 250
Protein quality, 846
Proteins, 211, 213, 247, 250, 251, 362, 970, 983, 984
Protein synthesis, 191
Protozoa, 1030
Provenance, 622
Pro-vitamin A, 979
Prowse 2012, 1156, 1159
P. rubi, 309
Prune, 1061
Prunes, 1269, 1290, 1291
Pruning, 105, 225, 388, 389, 638, 639, 701, 725, 1061
Prunus, 110, 1289, 1290
Prunus africanum, 717
Prunus ceradosa, 719
Prunus domestica, 101
Prunus salicina, 101
P. serotina, 1293
Pseudochromosomes, 304
Pseudomonas, 775
Pseudomonas putida, 774
Pseudotsuga menziesii, 1186
Psidium guajava, 844
P. sinensis, 1293
Psychological Benefits, 1008, 1026
Psychological function, 1017
Psychological health, 770, 783, 1003, 1005
Psychophysiological stress, 772
Psychophysiological stress recovery theory, 1003
Psychosomatic illness, 1005
Psychrometric constant, 366
Ptilotus, 454
Ptilotus exaltatus, 454
P. trifoliata, 172

P. trifoliate, 161
Public extension systems, 1127
Public gardens, 960, 1199
Public-good goals, 1127
Public greenspace, 962
Public health, 966, 1047–1050, 1052, 1054, 1058, 1066, 1070, 1078
Public horticulture, 962, 1119, 1122, 1123, 1132, 1135
Public landscapes, 959, 1119, 1122
Public open space, 1123, 1132
Public parks, 679, 954
Public-Private Partnerships (PPP), 1148, 1150
Public recreation, 436
Public safety, 1047, 1054, 1055
Public sector extension, 1118, 1126
Public sector services, 1127
Public space, 960, 962
Public squares, 678
Puccinia graminis, 1178
Puddle, 1031
Puffiness, 187
Puffing, 163, 165, 170, 186, 190
Pulmonaria sp., 455
Pulmonary health, 966
Pummelo, 160, 175, 180
Pummelo (Citrus grandis (L.) Osbeck), 124
Pummelos, 175
Pumpkin, 1214
Purchasing decisions, 476
Putting greens, 736
PVS, 511
Pycnidia, 230
Pyra moscatella augustan, 1271
Pyranometers, 332
Pyra viridia, 1271
Pyra zucchella, 1271
Pyrenopeziza brassicae, 828
Pyrgeometers, 332
Pyrolysis, 378
Pyrus, 1292
Pyrus pyrifoglia, 99
Pyrus pyrifolia, 99
Pyrus ussuriensis, 99
Pyruvate, 245
Pythium, 306, 635, 781
Pythium ssp., 354
Pythium ultimum, 365

Q

Qi, 648
Quality, 83, 420, 481
Quality assurance, 10, 1188, 1226
Quality assurance procedures, 1156
Quality assurance programs, 1145
Quality attribute, 474
Quality defects, 475
Quality fruit, 844
Quality grades, 823
Quality index, 382
Quality management, 471
Quality of life, 982, 1048, 1058, 1071, 1123, 1262
Quality position, 474
Quality products, 330, 1152
Quality retention, 466
Quality tiers, 475, 476
Qualup bell, 440, 442, 449
Quantify and compare a product's impact with another, 509
Quantify and control environmental impact, 508
Quantitative polygenic trait, 98
Quantitative trait loci, 86
Quarantine, 452, 459, 1178–1184, 1187, 1190, 1191
Quarantine assessment, 1191
Quarantine legislation, 1178
Quarantine pests, 1176, 1184
Queen Elisabeth I, 679
Queensland, 267, 268, 290–292, 442, 457
Quercetin, 986
Quercus, 627
Quercus spp, 437, 715, 724
Quesnay, 675
Quince rootstocks, 626
Quinces, 99, 118, 626, 1264, 1270, 1286
Quinone reductase, 987
Quinones and Seibel 2000, 1149
Quiscalus lugubris, 1033
Quito, 416
Qu'ran, 1226, 1241, 1243

R

Rabbit, 836
Rabbiteye blueberries, 317, 319
Rachel Carson, 864, 1120
Radiant energy, 348
Radiant heating, 451
Radiation, 328–336, 338–340, 382
Radiation intensity, 339, 392
Radiation models, 81
Radiative cooling, 344
Radiative heating, 344

Radioactive contaminants, 724
Radionuclide-contaminated solutions, 724
Radish, 233
Rain erosion, 841
Rainfall patterns, 844
Rainforest Alliance, 278, 511
Rainforests, 269, 290
Rain gardens, 1059, 1132
Rainshelters, 329
Rain tree (Samanea saman (Jacq.)), 716
Raisins, 1243, 1264
R. allegheniensis, 311
Ramayana epic, 673
Raphael Sanzio, 1212
Rapid establishment, 734
Rare species, 809
R. argutus, 311
R. armeniacus, 311
Raspberries, 99, 307–310, 1234, 1290
Raspberry beetle, 309
Raspberry bushy dwarf virus (RBDV), 309
Ratoon, 133
Rats, 1035
Ravana's vimana, 673
Ravaz, 1297
RDC, 1128
Reaction rate, 348
Reactive oxygen species, 982
Readily Available Water (RAW), 217
Real estate, 1075
Reardon and Huang (2008), 1158
Recherches chimiques sur la vegetation, 1299
Reconstituted panels, 273
Recreation, 798, 800
Recreational activities, 1015
Recreational and leisure, 13, 16
Recreational benefit, 800
Recreational drugs, 650
Recreational facilities, 961
Recreational gardening, 1012
Recreational spaces, 962
Recreation and leisure, 672
Rectifying Health by Six Causes, 1218
Red apples (Malus rubra), 1271
Red barberry, 457
Redcurrants, 313, 314
Red/far-red ratio, 339
Red fox, 1028
Redgauntlet x Hapil, 304
Red grapes, 244
Red List, 1035
Red pepper, 649
Red raspberry, 307, 310, 311, 834
Red Sea, 648
Red spider mite, 781
Reduced biodiversity, 505
Reduced fertility, 504
Reduce diabetes, 979
Reduced natural fertility, 513
Reduce the use of water and fertilisers, 513
Reducing biodiversity, 507
Red wine fermentation, 238
Red wines, 219, 227, 238, 239, 250, 251
Reeds, 440
Reel mowers, 740
Reference Evapotransiptation, 217
Reflectance, 330
Reflected energy, 341
Reflecting, 1043
Reflection, 1038
Refrigeration, 18, 466, 478
Regeneration cycles, 840
Regional Regoverning Markets Programme Communities (RECs), 1181
Regional green space, 683
Regional plant protection organisations (RPPO's), 1181
Regoverning Markets Programme, 1150
Regreening, 174, 189
Regulated Deficit Irrigation (RDI), 218, 514
Regulated pests, 1183, 1184, 1187
Rehabilitation, 798
Rehder, 1293
Rehder, Alfred, 1293
Reichenbach in Pomerania, 676
Relative Humidity (rH), 343, 354, 367, 778, 843, 1075
Religion, 718
Renaissance, 411, 412, 1201, 1217, 1220, 1235, 1237, 1256, 1257, 1270, 1274
Renaissance art, 1209
René-Louis Girardin, 675
Renewable energy, 392
Renovation, 747, 748, 750
Reproductive ability, 848
Reproductive growth, 838
Reproductive phase, 105
Reproductive sinks, 823
Reptiles, 835
Republic of Korea, 304
Research, 1148
Research and development (R&D), 1128, 11150
Reservation, 683

Reserve forests, 798
Residential gardening, 1121
Residential landscapes, 1121
Residue, 513
Resin, 722
Resins, 650, 722
Resistance genes, 826
Resistant varieties, 517
Resource-efficient, 76
Resource-intensive crop, 842
Resource management, 508, 702
Resource poor, 1153, 1154
Resource rich, 1153, 1154
Resources management, 1135
Resource-use efficiency, 838, 842
Responsible, 426
Responsive Element-Binding Protein, 989
Restaurants, 89
Restionaceae, 440, 442
Restios, 440
Restoration, 1043
Restoration potential, 1009
Restorative process, 1038
Restrictive water management, 365
Resveratrol, 986
Retail and shopping areas, 768
Retail consumers, 10, 18
Retail customers, 849
Retail environment, 783
Retailers, 420, 473–475, 480, 869, 1156
Retailing, 1119
Retail markets, 474, 622, 1155
Retail outlets, 1144
Retail produce, 478
Retail quality, 479
Retail stores, 18, 89
Retinol Efficiency Trial, 980
Reverse osmosis, 251
Reyes 2002, 1147
R. flagellaris, 311
R. grossularia, 314
Rhamnus frangula, 1031
Rhinitis, 775
R. hirtellum, 314
Rhizobacteria, 77
Rhizoctonia, 306, 635, 781
Rhizoctonia solani, 354, 848
Rhizomatous grasses, 748
Rhizome, 131
Rhizosphere, 774, 824, 829
Rhizosphere bacteria, 824
Rhodendron, 625
Rhododendron, 439, 443, 627, 634, 819, 1185, 1186, 1239
Rhododendron indica, 774
Rhododendron maddeni, 450
Rhododendron spp., 439, 443
RHS, 1121
Ribbed apples, 1271
Ribes, 314–316, 829
Ribes dikuscha, 316
Ribes nigrum, 313
Ribes spp., 626
Riccardi, G.L., 1280
Rice, 274, 716, 720, 721, 826, 1159
Rice blast (Magnaporthe oryzae), 825
Rice flower, 440
Richard II, 1245, 1246
Ricinus communis, 269
Riesling, 228, 236
Rikka, 1203
Rind breakdown, 187
Rind colour-break, 183
Rind disorders, 188
Rio de Janeiro Botanic Garden, 808
Rio de Janeiro, Brazil, 680
Riparian buffers, 714, 724, 727
Ripeness, 479
Ripening, 115, 116
Ripening phase, 376
Ripening syndrome, 117, 118
Risk analysis, 1180
Risk of land and water pollution, 505
River lime (Nyssa ogeche), 721
River Red Gum, 719
River Red Gum (Eucalyptus camaldulensis), 716
Rivers, 721, 724
Riverside, 680
Riyadh, 764
R. multiflorum, 313
R. nigrum var. sibiricum, 313, 315, 316
Roads, 714, 722
Roadsides, 713, 716
Roadside verges, 1002
Robert Fortune, 266
Robert Marsham, 818
Roberto Burle Marx, 680
Robert Schmidt, 683
Robin, 1033
Robinia, 721
Robinia pseudoacacia flowers, 721
Robins, 1032
Robotics, 235

Rockwool, 361, 378, 380, 779
Roda, 1258
Rodale Press, 864
Rodents, 1043
Rola-Rubzen et al. 2012, 1161
Roman, 198, 409, 411, 675
Roman Empire, 160, 411, 1209, 1265, 1266
Romans, 198, 416, 765
Rome, 413, 673
Romeo and Juliet, 1247
Roof coverings, 715
Roof gardens, 436, 456, 790
Roof greening, 452
Roofing, 720
Roof top gardens, 451, 686, 1237
Rooftops, 1239
Root and bulbs, 77
Root asphyxiation, 376
Root crops, 860
Root cuttings, 625
Root Environment, 779
Root growth, 201
Rooting environment, 696, 698, 699
Rootknot nematodes, 233
Root pathogens, 781
Root pruning, 640
Root rot, 169, 309
Root rot diseases, 171
Roots, 717, 724, 726, 1144
Roots and bulbs, 77
Root-stock, 621
Rootstocks, 103, 105, 141, 144, 145, 161, 171, 172, 187, 207, 385, 387, 622, 625, 626, 1233, 1286
Root system, 219
Root system management, 725
Root temperature, 353, 363
Root vegetables, 842
Root vigor, 84
Root zone, 173, 362, 365, 699, 745, 746, 748
Rootzone water, 220
Rosa, 766
Rosaceae, 99, 102, 106, 112, 626, 636, 1294
Rose Bowl parade, 414
Rosemary, 663
Rose (Rosa x hybrida), 408
Roses, 337, 339, 353, 356, 361, 364, 368, 371, 379, 389, 409–413, 419, 428, 430, 431, 436, 443, 444, 449
Roses (Rosa), 765
Rossellinia necatrix, 144
Rotary drum vacuum (RDV), 241
Rotary mower, 740
Rotary sprinklers, 744
Rotation cropping, 513
Rothamsted, 1300
Rothamsted station, 1300
Rottger and Da Silva (2007, 1146
Rough lemon, 171
R. oxyacanthoides, 314
Royal Botanic Garden, Edinburgh, 447, 1249
Royal Botanic Garden Kew, 441, 1249
Royal Botanic Gardens, 799, 808
Royal Botanic Gardens at Kew, 804
Royal Botanic Gardens, Edinburgh, 766
Royal Botanic Gardens, Kew, 764, 766
Royal Botanic Gardens, Sydney, 1040
Royal gardener, 679
Royal Horticultural Society, 1040, 1236
Royal Horticultural Society in London, 1249
Royal Horticultural Society (RHS), 1121
Royal Prussian Garden Administration, 679
Royal Society, 1279
Royal Society for the Protection of Birds (RSPB), 1003, 1035, 1036
R. petraeum, 313
RPPO, 1181
R. rubrum, 313
R. sativum, 313
RSPB, 1036
R. spicatum, 313
Rubber, 263, 264, 270, 271, 289, 844
RuBisCo, 362
Rubus, 307, 310, 625
Rubus idaeus, 307, 311, 834
Rubus spp., 307
Rubus subg. Rubus, 310
Run-off, 516
Rural Advisory Services (RAS), 1118, 1125, 1128, 1133
Rural architecture, 676
Rural development, 1135
Rural entity, 16
Rural environment, 13
Rural Industries Research and Development Corporation (RIRDC), 448
Rural landscapes, 802, 810
Rural RDC, 1128
Rural retreat, 679
Rural trees, 713
R. ursinus, 311
Ruscus, 346
Ruscus hypophyllum, 346
Rush, 457

Rus in urbe, 1122
Russeting, 356, 363
Russia, 266, 306, 307, 315, 439, 674, 1141
Russian Federation, 8, 97
R. ussuriense, 317
Rusts, 635
Rusty apples (Malus ferrugineum), 1271
Rutaceae, 161
Rutgers Cooperative Extension, 1124
Rutgers New Jersey Agricultural Experiment Station (NJAES), 1124
Rwanda, 470

S
Sabah, 286
Saccharomycecs cerevisiae, 244
Saccharomyces, 244
Saccharomyces cerevisiae, 244
Sacred oak, 718
Safe food supply, 90
Safe play areas, 17
Safety, 517, 1055, 1057, 1061, 1070
Saffron, 663
Sahara, 438, 717
Sahel, 715
Salad leaves, 605
Salads, 475, 476, 663, 860
Salicylate, 974
Salinity tolerance, 84, 386
Salinization, 382, 504
Salinization of land and water, 507
Salix, 625, 640
S-Alk(en)yl-cysteine suphoxides, 976
S-alk(en)yl L-cysteine sulphoxides, 976
Salmonella, 832
Salmonella typhimurium, 832
Salt, 696
Saltbush, 440
Salt index, 744
Salt sensitivity, 85
Salt stress, 85, 385, 387
Salt tolerance, 83, 85, 86, 144
Salt tolerant, 86
Salvia, 348, 449
Salvia splendens, 348
Sandersonia aurantiaca, 450
Sand pears, 1293
San Francisco, 680
Sanitation, 5, 752
Sansevieria trifasciata, 764, 768
São Paulo, 685
Sao Tome, 285
Sap flow, 366
Sap flow gauges, 366
Sap flow rate, 366
Sapindales, 147
Sapodilla (Manilkara zapota), 128
Saponins, 650, 981
Saps, 717, 722
Satellite imagery, 703
Sat-nav Britain, 2
Satsuma, 165–167, 190
Satsuma mandarins, 162, 165, 175, 176, 180, 183, 186, 189, 194
Sauvignon Blanc, 228, 236, 248, 249
Savanna, 439
Savannah habitats, 798, 1004
Savastano, 1296, 1297
Savoy cabbage, 842
Sawfly, 316
Sawn wood, 273
Saxifraga x arendsii, 455
Scabiosa, 1032
Scaevola aemula, 440
Scaevolas, 440
Scaevola saligna, 448
Scaevola (Scaevola coriacea), 459
Scale insects, 751
Scandinavia, 313, 315, 603
Scarab, 751
Scarification, 623
S. cerevisiae, 244, 247
S. cheesmanii, 85
Schefflera sctinophylla, 768
Schima wallichii, 719
School and hospital gardens, 954
School and youth gardening education, 1057
School gardening, 1012
Schrekenberg and Mitchell, 1148
Science and technology, 1225
Science-based advice, 1132
Science-driven extension, 1135
Sciences and technologies, 1230
Scientific Revolution, 860
Scilly Isles, 716
Scions, 387, 625, 626, 1186, 1210, 1233, 1267
Sckell, 679
Sclerocarya birrea, 720
Sclerophyll forests, 438
Sclerotinia sclerotiorum, 828
Sclerotinia wilt, 848
Sclerotium rolfsii, 848
Scotland, 309, 716, 828
Scottish Biodiversity Strategy, 788

Scott Report, 795
Scouring, 181
Screen, 345, 360
Screen constructions, 328
Screen covers, 359
Screenhouses, 328, 331, 332, 344–346, 359, 360, 366, 368, 369, 392
Screens, 331, 332, 344, 359, 360
Screen transmittance, 332
Sculptures, 1205, 1222, 1279
Sea Guarrie, 442
Sea of Galilee, 369
Seasonal changes, 1043
Seasonal cycles, 823
Seasonal demand., 473
Seasonality, 90
Seasonings, 645, 646
Seattle, 416
Secondary metabolites, 650, 967
Security, 513
Sedentary lifestyle, 1006, 1007
Sedge, 408, 457
Sedges, 439
Sedges (Cyperus spp.), 631
Seed, 714, 724, 1174
Seed-borne pathogen, 831
Seed coating, 77
Seed development, 112
Seed dispersion, 967, 1029
Seed encapsulation, 450
Seed enhancement, 77
Seed heads, 1031
Seeding rates, 738
Seeding technologies, 77
Seedless cultivars, 180
Seedlessness, 350
Seedling density, 840
Seedling emergence, 77
Seedling establishment, 840
Seedling performance, 78
Seedlings, 721
Seed maturation, 105
Seed mixtures, 737
Seed physiology, 623
Seeds, 720, 724, 726
Seiko Goto, 676
Selective herbicides, 836
Self-awareness, 1005
Self-concept, 1005
Self-esteem, 1005
Self-fertility, 319
Self–fertilization, 110
Self-harm, 772
Self-identity, 1039, 1043
Self-incompatibility (SI), 110, 111
Self-incompatible, 112, 167
Self pollination, 109
Self-sufficiency garden, 682
Semi-arid, 329, 381, 385, 606
Semi-hardwood cuttings, 624, 625
Semillon, 228
Semi-natural amenity grasslands, 732
Semi-natural grasslands, 732
Semi-natural landscapes, 1006
Seneca, 416
Senecio vulgaris, 409
Senegal, 715
Senescence, 102, 104, 105, 188
Senna (Cassia angustifolia), 648
Sense of place, 768
Sensor-based control, 373
Sensory attributes, 1145
Sensory perceptions, 1007
Sensory Trust, 960
Septoria leafspot, 316
Septoria ribis, 316
Sequestered carbon, 613
Sequestration of carbon, 797
Serbia, 307, 312
Serengeti, 724
Serrurias, 442
Serrurias trilopha, 442
Service wood, 720
Sesame, 1206
S. esculentum, 87
Sesquiterpene lactones, 384
Sesquiterpenes, 120
Setting objectives, 508
Settling, 240
Seven Greens, 808
Seychelles, 809
Seychelles (Impatiens gordonii), 809
Shade, 715, 716, 719, 723, 724, 1198, 1199
Shade houses, 416
Shade nets, 339
Shade perennials, 1236
Shade screens, 340
Shade tolerant, 775
Shade trees, 621, 622
Shading, 340, 359
Shakespeare, 1226, 1241, 1245
Shakkei, 674
Shanidar IV, 409
Shea butter, 269

Shea-nut, 715
Sheath blight, 825
Sheet glass, 416, 417
Shelf life, 336, 385, 421, 476, 480
Shelter, 715, 716, 723, 724, 1198
Shennong, 648
Shepherd 2005, 1144, 1146
Shepherd and Galvez 2007, 1145
Shepherd and Tam 2008, 1145
Shikimic acid, 653
Shipping, 466
Shiraz, 209, 212, 227
S. hirsutum, 86
Shoot apical meristem, 102
Shoot density, 741
Shoot extension, 849
Shoot growth, 741
Shoot inclination, 105
Shoot primordia, 205
Shopping centers, 1201
Shopping malls, 768, 1009
Short-day, 305
Short day onions, 86
Short-day plants, 449
Shoulder check, 363
Shrubs, 621, 622, 638, 694, 764, 849, 1198, 1226, 1236
Shugakuin Rikyu, 674
Siberia, 101
Sicily, 1266
Sick building syndrome, 773
Siedlungsverband Ruhrkohlenbezirk, 683
Signaling cascades, 990
Signalling pathways, 98, 824
Silent Spring, 506, 864, 1120
Silicon, 634
Silk Road, 648
Silleptic, 102
Simcha Blan, 1303
Singapore, 91, 270, 436, 781
Singapore Botanic Garden, 808
Singer, 1153
Singh 2007, 1155, 1156
Sir Joseph Banks, 441
Sisal, 264
Site Preparation, 629, 736
Sitta canadensis, 1033
Skin conductance, 1005
Sky gardens, 783
Slash and burn, 606
Slatyer, R.O., 1302
S-locus, 111
Slovakia, 413
Slug, 834
Small American bell-flower, 821
Small and medium sized enterprises (SMEs), 508
Small fruits, 1290, 1292, 1293
Smallholder agriculture, 1142, 1146
Smallholder chains, 1162, 1163
Smallholder coffee cherry, 1154
Smallholder farmers, 1141, 1146, 1150, 1152, 1155–1160, 1162, 1163
Smallholder farming, 1149, 1151, 1157
Smallholder horticultural farmers, 1160, 1162, 1163
Smallholder horticultural producers, 1160
Smallholder producers, 1147–1150, 1160
Smallholder resources, 1163
Smallholder vegetable farmers, 1152
Small-scale agriculture, 519
Small step' or incremental improvements, 1134
Small white butterfly, 1029
SMART targets, 703
Smit and Nasr 1992, 1147
Smith-Lever Act, 1125
Smith River, California, 421
Smoke bush, 440
Snail, 1031
Snake melons, 1210, 1212
Snakes, 1036
Snap bean, 77
Snapdragon, 408, 449
Sneezewood, (Ptaeroxylon obliquum), 717
Snout apples, 1271
Snowdrop, 818
Soak hoses, 744
Soccer pitches, 755
Social acceptance, 1124
Social and health, 14
Social and Therapeutic Horticulture, 1015
Social capital, 1159
Social class, 679
Social communication, 1006
Social construction, 1028
Social Functioning, 1054, 1055
Social Horticulture, 14, 17, 849, 1053
Social impact, 1172
Social infrastructure, 505
Social interactions, 770, 962, 1047, 1049
Social networking, 958
Social responsibility, 424
Social services, 17

Social spaces, 958
Social standing, 17
Social Sustainability Toolkit, 960
Social & Therapeutic Horticulture, 1015
Social welfare, 18
Socio-economic composition, 18
Socio-economic performance, 17
Socio-economic status, 16
Sod, 737, 738, 739
Sod farm, 737
Sodic clay soils, 1120
Sodium, 381, 382
Sodium arsenite, 230
Sodium chloride, 382, 383, 386, 744
Sodium-dependent glucose transporter (SGLT1), 983
Sodium sulfate, 1300
Sods, 736
Softening rate, 118
Soft fruits, 614
Softwood cuttings, 624, 625
Soil, 513, 723
Soil acidification, 839
Soil aeration, 698
Soil aggregation, 513
Soil amendments, 476, 629, 632, 744
Soil and water protection, 723
Soil Association, 864, 868
Soil Association of South Africa, 864
Soil-borne diseases, 387, 752
Soilborne diseases, 829
Soil borne microbes, 850
Soil borne pathogens, 828, 848
Soilborne plant pathogens, 829
Soil-borne viruses, 832
Soil bulk density, 696
Soil characteristics, 698, 699
Soil compaction, 698, 699, 752
Soil contamination, 802
Soil degradation, 513, 723
Soil drainage, 628
Soil erosion, 723, 792, 796, 802, 839
Soil fertility, 723, 752, 823, 849, 863
Soil fungi, 751
Soil health, 6
Soil heat-flux density, 366
Soilless culture, 365, 377, 378, 380, 386
Soilless culture sweet pepper, 353
Soilless culture systems (SCSs), 363
Soilless media, 361
Soil-less production, 1177
Soilless systems, 380, 382
Soil management, 1299
Soil mechanics, 736
Soil microbes, 807, 823
Soil microbiota, 696
Soil moisture, 365, 377, 828
Soil moisture content, 753
Soil moisture deficits, 365
Soil moisture sensors, 456
Soil nutrients, 513
Soil organic matter, 513, 516
Soil-plant-atmosphere continuum, 364
Soil profile, 736, 747
Soil quality, 605, 862
Soils, 1132
Soil salinity, 77, 839
Soil sampling, 216, 745
Soil solarisation, 750
Soil solarization, 632
Soil stability, 605
Soil stabilization, 723
Soil sterilants, 629
Soil structure, 840
Soil structure, texture and fertility, 513
Soil temperatures, 819, 828
Soil testing service, 634
Soil texture, 218, 746
Soil water balance, 235
Soil water fraction, 220
Solanaceae, 111, 387, 1222
Solanaceous, 77, 78
Solanum melongena, 351
Solanum pennellii, 85
Solanum pseudocapsicum, 769
Solanum tuberosum, 1220
Solar energy, 347, 361
Solarization, 629
Solar radiation, 334, 340, 341, 343, 349, 352, 359, 361, 367, 392, 733, 744
Solar radiation (GSR), 332
Soluble solid, 210
Soluble sugar, 177
Solute accumulation, 371
Song of Songs, 1244
Song thrush, 818
Sorbus intermedia, 818
Sorghum, 841
Source-sink, 203
Source-sink balance, 363
Sour cherries, 101
Sour cherry (Prunus cerasus), 101

Sour orange, 160, 171, 187
Soursop, 140
South Africa, 98, 137, 150, 163–166, 170, 199, 200, 272, 289, 317, 418, 438–442, 448, 456, 719, 808, 862, 1153
South Africa's fynbos, 442
South America, 98, 189, 268, 271, 284, 285, 304, 314, 419, 426, 438–442, 456, 662, 708, 721, 733, 822, 1173, 1176, 1298
South American poinsettias, 440
South Amcrican rainforest, 439
Southeast Asia, 270
South East Asia, 17, 438, 719
South-eastern Asia, 125
South-Eastern Europe, 99, 101
South-East Europe, 99
South Ecuador, 720
Southern China, 100
Southern Hemisphere, 98, 839
Southern highbush, 317, 319
Southern Iran, 719
Southern Mexico, 284
South Korea, 868
South-Western Siberia, 99
Sowing, 738
Space utilization, 468
Spain, 139–143, 145, 161–166, 168, 170, 172, 187, 199, 200, 268, 272, 304, 317, 318, 328, 341, 367, 451, 605, 606, 608, 614, 649, 674, 676, 696, 719, 721, 985, 1208, 1219, 1235
Spanish, 160, 191
Spanish dehesa agroforestry, 715
Sparkling wine, 236
Sparrows, 1032, 1033
Spathiphyllum wallisii, 764, 775
Spathodea campanulata, 803
Spatial development model, 685
Spear and bud rot, 846
Species composition, 714
Species survival, 801
Specimen plant collections, 849
Spectral composition, 338
Spectral quality, 334
Specularia, 416
Specularium, 765
Sphaeropsis sapinea, 825
Sphaerotheca mors-uvae, 315, 316
Sphagnum bogs, 634
Spices, 263, 266, 267, 426, 645, 648, 649, 662–664, 843, 844, 1199
Spice trade, 267
Spider mites, 637
Spider plant (Chlorophytum comosum), 763
Spiders, 1036
S. pimpinellifolium, 85, 86
Spinach, 81, 86, 335, 981, 1145
Spiritual health, 17
Spiritual needs, 17
Spiritweed, 720
Spittlebugs, 751
Split-root, 385
Split root fertigation (SRF), 386
Splitting, 184, 185, 190
Sporobolus virginicus, 734
Sporophytic tissue, 108
Sport and amenity grasslands, 757
Sport and play, 961
Sports facilities, 1002
Sports fields, 736, 789
Sports turf, 13, 732, 734, 736, 737, 740, 741, 744, 749, 751, 752, 756, 807, 849
Spotted wing fruit fly, 306
Sprigging, 737
Sprinkler, 745
Sprinkler irrigation, 633
Sprinklers, 744, 1303
Sprinkler systems, 746
Spruce (Picea abies), 825
Spruces, 439
SPS, 1179
Squash, 1214
Sri Lanka, 269, 270, 722, 781, 795
SSR markers, 304
Stack effect, 357
Stadtpark, 679
Stand Establishment, 77
Staple food, 976
Star anise, 663
Starch, 120, 147, 203
Starch metabolism, 201
Starch synthase, 120
Starfruit (Averrhoa carambola L.), 125
State Extension Leaders' Network, 1125
Statice, 409, 417
St Augustine grass, 1029
St. Barnaby's thistle, 409
Stem dieback, 316
Stem water potential, 220
Stenotaphrum secundatum, 733, 734, 1029
Stenotaphrum species, 733
Steppes, 439

Sterols, 229
St George's Chapel Windsor Archives, 1249
Still life, 1212
Still life painting, 1197
St. Louis, Missouri, 680
Stock, 412, 417
Stock plan, 623
Stock plant, 636
Stolonising, 738
Stolonizing, 737, 738
Stomach cancers, 987
Stomata, 85
Stomatal closure, 352, 364
Stomatal conductance, 210, 221, 222, 357, 361, 364, 367, 837
Stomatal function, 356
Stomatal morphology, 356
Stomatal resistance, 355, 368
Stomata regulations, 364
Stone fruit, 99, 112, 113, 115, 116, 845, 1281, 1283, 1293
Storage, 466, 467, 473
Storm protection, 723
Storm water, 1058, 1061, 1076
St. Petersburg, Russia, 676
Stratification, 623
Straw bale cultural, 361
Strawberries (Fragaria vesca), 99, 101, 109, 118, 302, 303–306, 335, 390, 475, 476, 613, 839, 842, 973, 1280, 1293
Streams, 724, 1031
Street fittings, 685
Street scapes, 456, 457
Street trees, 696, 697, 702, 1002, 1007, 1050, 1054, 1063, 1073
Streptomyces scabies, 848
Stress Available Water (SAW), 217
Stress Deficit Irrigation (SDI), 218
Stresses, 17, 844
Stress hormone, 1005
Stress management, 211
Stress reactions, 1007
Stress recovery, 770
Stress-related depression, 17
Stress tolerance, 85, 823
Stress tolerant plants, 457
Strobilurins, 229
Strobus, 829
Strong Republic Nautical Highway (SRNH), 1147
Structural decline, 839
Structural soils, 698, 699
St. Valentine's Day, 413, 415
Stygmasterol, 109
Stylbenes, 986
Suberin, 970
Sub-irrigated systems, 381
Sub-irrigation, 373, 382
Sub-irrigation systems, 373, 374, 382
Sub Rosa, 411
Sub-Saharan Africa, 715, 717
Subsistence farming, 869
Substrate cultures, 353, 365
Substrates, 374, 630
Subsurface, 374
Sub-surface drip, 744
Subsurface Drip Irrigation (SDI), 1303
Subsymbolic, 1009
Subtropical, 329
Sub-tropical forest, 437
Subtropical moist deciduous forest, 438
Sub-Tropics, 98
Suburban landscapes, 1119
Sucrose, 119, 120, 210, 351, 364, 627, 722
Sudan, 715
Sudano-Sahelian region, 715, 723
Sudden oak death (Phytophthora ramorum), 1185
Sugana, 308
Sugar, 120, 189, 210–212, 228, 263, 264, 266, 335, 337, 353, 356, 375, 845, 970, 976, 986
Sugar apple, 140
Sugarbeet, 981
Sugarcane, 264, 268
Sugar Gum (Eucalyptus cladocalyx), 719
Sugar loading, 210
Sugars, 119, 120, 209, 211, 212, 223
Sugar transport, 211
Sulawesi, 282, 284, 286
Sulfenic acids, 977
Sulfite (SO32-), 251
Sulforaphane, 984
Sulfotransferases, 984
Sulfur, 634
Sulfur dioxide, 247, 251, 252
Sulfur dioxide (SO_2), 238, 251
Sulphated ketoxime, 973
Sulphate equivalents, 614
Sulphoraphane, 976
Sulphur, 210, 229, 614, 742, 967, 973–977, 982, 990
Sulphur compounds, 977
Sulphur dioxide, 240, 451, 614

Sulphur dioxide (SO_2), 837
Sulphur oxides (SOx), 773
Sumac (Rhus spp.), 625
Sumatra, 271, 286, 721
Sumer, 198
Sumeria, 648
Summer bedding, 444
Sunburn, 169, 176, 352, 823
Sunflower, 449, 1032
Sunflower seeds, 1029, 1032
Sun scald, 339, 352
Supermarket chains, 1176
Supermarkets, 18, 89, 468, 472, 866, 869, 1152, 1156, 1273
Superoxide dismutases, 353, 982
Superphosphate, 744
Supplemental assimilation lighting (SAL), 336
Supplementary lighting, 454
Supplier codes, 1157
Suppliers, 473, 476
Supply and demand, 866
Supply chain, 10, 88, 89, 276, 277, 307, 420, 427, 431, 468, 471, 477, 608, 861, 869, 1134, 1146, 1152, 1154, 1176
Supply chain. Amenity horticulture, 797
Supply Chain Management, 469, 477
Suppressants, 631
Suppression, 631
Suppressive soils, 829
Surface sealing, 696
Surveillance, 1183–1185, 1187
Sushruta Samhita, 648
Sustainability, 278, 424, 790
Sustainability of agricultural and food systems, 507
Sustainable, 278, 426
Sustainable Agribusiness Transformation, 1151
Sustainable agriculture, 863, 866
Sustainable city, 17
Sustainable design, 960
Sustainable development, 13, 810
Sustainable Development, 806
Sustainable environment, 18, 808
Sustainable farming, 1146
Sustainable greenhouse production, 391
Sustainable horticulture, 806
Sustainable husbandry, 841
Sustainable landscapes, 1123
Sustainable management, 702, 790
Sustainable production, 1141
Sustainable solutions, 18
Sustainable space, 800
Sustainable turf management, 807
Sustainable urban drainage, 694
SUVIMAX cohort study, 966
Swallows, 1040
Swede, 828
Sweden, 453, 674, 869
Sweet cherry, 112
Sweet Chestnut, 721
Sweet corn, 77
Sweet orange, 160, 162–165, 167, 171, 172, 174, 179–181, 188, 190, 844
Sweet oranges, 162, 168, 180, 189, 190
Sweet pepper, 335, 337, 355, 384
Sweet potato, 1222
Sweet potato (Ipomoea batatas), 1220
Sweetsop, 140
Switzerland, 309, 831, 860, 987, 1174, 1183
Sycamore, 716
Sylva\
a discourse of forest-trees, 1279
Symbiosis, 1042
Symbolic imagery, 1009
Symmetry, 1203
Symplastic, 209
Symplastic network, 98
Synanthedon tipuliformis, 316
Syncarp, 141
Syncarpium, 127
Synchitrium endobioticum, 848
Synthetic chemicals, 864, 865
Synthetic fertilizers, 1298, 1299
Synthetic pesticides, 1300
Syria, 77, 101, 723, 1199, 1205, 1209
Syringa (lilac), 625
Syringa vulgaris, 416
Systematic Management, 704
Systematic pomology, 1294
Systemic competiveness, 1157
Systems theory, 1151
Systems thinking, 1124
Syzygium aromaticum, 267
Szechuan pepper, 663

T

Table grapes, 198, 219
Tables of Health, 1218
Table-top growing, 306
Table-top structures, 305
Table wines, 248
Tactile quality, 480
Tacuinum Sanitatis, 1218

Tadao Ando, 685
Tagetes erecta, 414
Tagetes patula, 348
Tahiti, 441
Taiwan, 87, 88, 438, 439
Tajikistan, 199
Taj Mahal, 1210
Takeout food, 479
Tamarind, 721, 723
Tamarindus indica, 715
Tamarix, 719
Tamius striatus, 1029
Tang dynasty, 1245
Tangelos, 167, 168, 175, 180
Tangerine, 167
Tangors, 175, 180, 190
Tannins, 208, 209, 212, 250, 722
Tanoak, 1185
Tanoak (Notholithocarpus densiflorus), 1186
Tanzania, 266, 419, 471, 715, 720, 724, 1159
Tapestries, 1216, 1222
Tapestry, 1216
Tapetum, 108
Taqwim al-Sihha bi al-Ashab al-Sitta, 1218
Targetes sp., 451
Tarragon, 663
Tartaric, 209
Tartaric acid, 207
Tartrate, 211
Tasmania, 98
Taste, 476
Taxus, 625, 640
Tea, 263, 264, 266, 289, 426, 438, 441, 841, 844, 860, 973, 981
Teatree, 440, 444
Technical competency, 1157
Technical expertise, 1129
Technology and management strategies, 520
Technology-based advice, 1134
Technology transfer, 1124
Telomere length, 1005
Telopea speciosissima, 440
Tem blight, 848
Temperate fruits, 97–99, 117, 821
Temperate Houses, 766
Temperate regions, 439
Temperate zones, 778, 1031
Temperature, 77, 329, 515, 778, 1047, 1058, 1059, 1060, 1076
Temperature gradient, 345
Temperature models, 81
Temperature rise, 820
Temperature stress, 844
Temperature threshold, 130
Tennis, 755
Tennis courts, 736
Tensiometer, 745
Tenure systems, 1146
Terminal FLower (TFL), 107
Terminalia pruniodes, 720
Termites, 1033
Terpenes, 212
Terpenoids, 212, 650, 656, 967, 979
Terra Madre, 1261
Terrarium, 766
Terra, the Compleat Gardener, 1279
Territorial Approach to Rural Agro-enterprise Development, 1160
Terroir, 200, 234
Teruel and Kuroda, 1147
Testes, 984
Tetranychus urticae, 781
Texas, 80, 81, 84–88, 797
Texture, 408
Thailand, 8, 159, 264, 269, 270, 418, 436, 781, 1159
Thanet Earth, 615
Thanksgiving Day, 411
The Access Chain, 960
The Agriculture Course, 862
The Annunciation, 412
The Arboricultural Association (AA), 1133
Theatrum Orbis Terrarum, 1220
The British Library, 1249
The Carbon Trust, 512
The Chrysanthemum Throne, 409
The Dumbarton Oaks, 1249
The Fruit Seller, 1214
The Garden History Museum, 1249
The Golden Ass, 1212
The greatest show on Earth, 2
The Great Glasshouse, 763
The Harber-Bosch process, 507
The Living Soil, 863
The Lloyd Library, 1249
The Netherlands, 331, 337, 346, 347, 414–417, 419, 421, 424, 425, 427, 428–431, 436, 685, 781, 782, 1013, 1127, 1128, 1176, 1188
Theobroma cacao, 267
The Odyssey, 1233
Theophrastus, 4, 160, 310, 1232, 1233, 1257, 1266, 1267
Theoprastus, 1265

Theorie der Gartenkunst, 679
Theory of Gaia, 2
Theory of garden art, 679
Theory of mutations, 1287
Therapeutic gardens, 1015
Therapeutic horticulture, 1053, 1119
Therapeutic intervention, 954
Therapeutic landscape, 1011
Therapy, 13, 798
Therapy Gardens, 1015
The Renaissance, 1270
Thermal comfort, 778
Thermal cooling, 765
Thermal degradation, 333
Thermal dissipation, 366
Thermal energy, 348
Thermal screens, 347
Thermocouple psychrometers, 220
Thermo-stability, 88
The Soil Association, 863
Thespesia populnea, 720
The Water Footprint Assessment Manual Setting the Global Standard, 511
The Weathered Company, 417
The Winters Tale, 1246, 1247
Thiazols, 103
Thigmomorphogenesis, 365
Thiocyanates, 975, 976, 984
Thioglucosidase, 974
Thiols, 212
Thiosulfinates, 977
Thomas and Hangula 2011, 1158
Thomas Malthus, 794
Three-phase Clustering Framework, 1162
Threshold temperatures, 833
Thrips, 306, 637
Thuja, 625
Thunberg's barberry, 446
Thymus vulgaris, 451
Tiarella, 455
Tiberius, 416
Ticks, 751
Tiergarten, 679
Tigris and Euphrates Rivers, 1205, 1301
Tigris-Euphrates, 3
Tiliqua scincoides scincoides, 1031
Tillage, 747, 750
Tillering, 739
Timbers, 295, 719, 720
Timer-based, 373
Tipburn, 352
Tipburn of lettuce, 339
Tissue culture, 444, 621, 622, 626, 782, 1177, 1184, 1235
Titratable acidity, 210, 211, 375
Tits, 1035
Tobacco, 264, 362, 721
To erosion, 519
To global warming, 520
Togo, 715
Tokaido trail, 672
Tokyo, 685
Toluene, 773
Tomatoes, 77–79, 85–87, 335–337, 339–341, 344, 350–353, 355, 356, 361–365, 367, 368, 375–378, 381–390, 823, 827, 843, 846, 847, 984, 1208, 1230, 1235
Tomato leaf curl virus, 847
Tomato (Solanum esculentum), 85
Tonga, 269
Topdressing, 748, 750
Topiarius, 673
Topiary, 638, 1204
Toronja, 168
Total acidity, 175
Total acidity (TA), 189
Total packaged oxygen, 252
Total soluble fruit solids, 375
Total soluble salts (TST), 457
Total soluble solids, 175, 375, 382, 383, 476
Total soluble sugars (TSS), 351
Tourism, 13, 800, 806, 807, 1061, 1071, 1073, 1074
Tourism Authority of Thailand (TAT), 808
Tourist attractions, 849
Toxicity, 510
T oxygen quenching capacity, 980
T. patula, 414
Traceability systems, 1152
Trace elements, 244
Trade, 1171, 1172, 1174, 1176, 1179, 1181
Trade volume, 1174, 1177
Traffic calming, 1055–1057
Trailing plants, 777
Training, 516
Training and visit (T and V), 1124
Training system, 388
Train the trainer, 1124
Traité du Citrus, 1258, 1282, 1284
Transgenic papayas, 136, 137
Transgenic technology, 450
Transition phase, 105, 107
Transit of Venus, 1172
Transmission, 332

Transmittance, 330, 331, 332
Transparent foils, 328
Transpiration, 364, 368, 369, 385
Transpirational cooling, 355
Transpiration models, 367
Transpiration rates, 355, 356, 367, 376
Transplant, 79, 1061
Transplanting, 78, 640, 641
Transport, 466, 467, 504, 519
Transportation, 420
Transport hubs, 1176
Transporting, 507
Transporting horticultural produce, 520
Transport life, 276
Trans- stereoisomers, 979
Trattato dell'agricoltura, 1269
Travel, 507
Treatise on agriculture, 1269
Treatise on tree growing, 1295
Treaty of Tordesillas, 649
Tree, 1226, 1274, 1275, 1281, 1285, 1296, 1297, 1303
Tree care, 701, 706
Tree cover, 5
Tree crops, 1157
Tree culture, 1285
Tree dimension, 714
Tree ecophysiology, 697, 698
Tree establishment, 698
Tree fruit, 821
Tree grower, 1297
Tree guards, 726
Tree health, 725, 726
Tree maintenance and management, 705
Tree management, 695, 702, 703, 706, 708, 1276
Tree nuts, 289
Tree pathogens, 708, 829
Tree populations, 694
Tree protection, 726
Trees, 764, 797, 803, 849, 1132, 1198, 1199, 1226, 1233, 1236, 1262, 1267, 1274, 1275, 1278, 1279, 1281, 1285–1287, 1296, 1297, 1299, 1303
Trees and Design Action Group, 708
Trees and ornamentals, 1185
Trees and shrubs, 620, 1122, 1172, 1186, 1236
Tree selection, 697, 698, 707, 708
Trees fruits, 1285
Tree strategy, 703, 704
Trellis designs, 223
Trellising, 234, 235
Trellis system, 389
Tresco Abbey Gardens, 716
Trialeurodes vaporariorum (glasshouse whitefly), 781
Trichoglossus haematodus, 1032
Trichoplusia ni, 1029
Trickle, 80
Trickle irrigation, 194
Trifoliate orange, 172
Trinidad, 285
Tripartite model, 1156
Tripeptide, 249
Triploids, 112, 128
Triploidy, 111, 133
Triterpene, 981
Triterpene saponins, 981
Tropical, 329, 717, 725, 727
Tropical Africa, 715
Tropical America, 716
Tropical foliage, 13
Tropical forests, 801
Tropical fruits, 1235
Tropical grasslands, 716
Tropical Palm House, 766
Tropical plants, 453
Tropical rain forest, 437–439, 442
Tropical storms, 844
Tropics, 98, 716
Tropisms, 1233
Trueness meter, 753
Trunk, 203
T. spinosa, 720
Tuber rotting pathogens, 846
Tubers, 843, 1144
Tuber yield, 846
Tuff, 378
Tulameen, 308
Tulip, 410, 412, 413, 430, 839, 1029
Tulipa, 766, 1029
Tundra, 437
Tunisia, 1209
Tupy, 313
Turbulence, 360
Turbulent velocity, 358
Turdus merula, 1032
Turf, 456, 734, 736, 737, 739, 742, 744, 750, 751, 803, 807, 1198
Turf aesthetics, 752
Turf grass, 752, 807
Turfgrass, 11, 13, 740, 742, 744, 751, 1129, 1132, 1239
Turf grasses, 733, 739, 751, 1226

Turfgrass management, 1232
Turf industry, 752
Turf laying, 737
Turf maintenance, 732
Turf management, 734, 806
Turf pests, 751
Turf production, 1119
Turf quality, 807
Turf recovery, 751
Turf sciencc, 734, 752
Turf-stripping, 737
Turf vigour, 752
Turf weeds, 749
Turgor, 364, 382
Turgor pressure, 364
Turkey, 8, 77, 84, 164, 165, 304, 328, 411, 413, 723, 801, 1209, 1218
Turmeric (Cucurma longa), 267, 842
Turnover, 508
Two-spotted spider mite (Tetranychus urticae), 637
Type 2 diabetes (T2D), 985, 986
Type II diabetes, 1007, 1013
Typically glasshouses with heating and generally a high content of technology, 506
Tyrosine, 970

U
Uganda, 270, 419, 715, 841
UK, 284, 315, 429, 605, 615, 779, 781, 782, 795, 797, 799, 805, 807, 834, 863, 868, 1006, 1007, 1036, 1174, 1185, 1186
UK Biodiversity Action Plan, 806
UKCIP98 Medium High, 830
UK hardy nursery industry, 1177
Ukraine, 722, 1141
UK soft fruit, 842
Ule rubber, 270
Ulisse Aldrovandi, 1270
Ulmus spp., 1063
Ultra low oxygen, 477
Ultraviolet-stable, 345
Ulysses, 1241, 1248
Ulysses Prentiss Hedrick, 1291
Umami, 383
Unani, 648
Unani medicine, 648
Uncinula necator, 228
Uniform stand, 77
United Fruit Company (UFC), 275
United Fruits Plantation, 285
United Kingdom, 447, 453, 659, 672, 716, 718, 724, 725, 795, 798, 1027, 1119, 1121, 1132, 1238
United Kingdom fraise mowing, 748
United Nations (UN), 1179
United States, 78, 318, 426, 439, 453, 620, 630, 651, 659, 661, 663–665, 682, 733, 797, 827, 830, 868, 1186, 1249, 1288, 1293
United States Department of Agriculture, 863
United States Golf Association, 736
United States of America, 97, 200, 678, 680, 681, 737, 744, 749, 864
University of Arkansas, 311
University of British Columbia Plant introduction Scheme, 447
University of California, 304
Unsaturated lipids, 967
Unsustainable, 805
U.P. Hedrick, 1290
Upland Marketing Foundation Incorporated (UMFI), 1149
Urban Agriculture, 962, 1048, 1067–1070, 1119
Urban and peri-urban, 1119
Urban boulevards, 678
Urban built environments, 1009
Urban climate change, 708
Urban communities, 14
Urban cooling, 694
Urban density, 784
Urban design, 707, 1132
Urban development, 1027
Urban Environmental Health, 1058
Urban environments, 696–698, 701, 705, 1047, 1048, 1058, 1059, 1062, 1069, 1071, 1076, 1077
Urban food production, 1069, 1070
Urban forestry, 694, 702, 705, 706, 708, 1077, 1131
Urban forests, 703–705, 708, 1002
Urban forest/tree strategy, 703
Urban green deficits, 685
Urban greening, 672, 673, 685, 1132
Urban green open, 790
Urban green open space, 797
Urban green spaces, 783, 790, 796, 1008, 1010, 1122
Urban health, 1123, 1134
Urban heat island, 1059
Urban heat island effect, 696
Urban Horticulture, 127, 1047–1050, 1068–1070, 1072, 1077, 1119, 1132

Urban Horticulture Institute, 1077
Urbanisation, 505, 1143, 1237
Urbanised food production, 17
Urbanization, 16, 685, 796, 804, 810, 1047, 1049, 1062, 1065, 1066, 1069, 1076, 1077
Urbanized landscape, 685
Urban landscape management, 1122
Urban landscapes, 707, 803, 1122, 1131
Urban landscaping, 1237
Urban lifestyle, 783
Urban open space, 672
Urban parks, 13, 1009
Urban parks and gardens, 1122
Urban parks movement, 1120, 1122
Urban planner, 679
Urban planning, 702, 1201
Urban populations, 8, 808, 1047, 1048, 1061, 1065, 1067–1069, 1071, 1074, 1076–1078, 1143
Urban public space, 734
Urban settings, 1009
Urban societies, 693
Urban spaces, 678, 685, 707
Urban trees, 697, 699, 701, 702, 705, 707, 708
Urban vegetation, 1123
Urbs in horto, 680
Ureide synthesis, 187
Uric acid, 982
Urocissa whiteheadi, 1032
Uruguay, 164, 165, 166, 168, 170, 172
Uruk, 672
Uryuk, 1206
U.S., 80, 274, 378, 471, 620–622
US, 80, 89, 90, 136, 274, 309, 310, 315, 377, 446, 449, 454, 471, 620–622, 683, 721, 863–865, 1145, 1173, 1178, 1185, 1186, 1192, 1238
USA, 8, 76, 84, 136, 139, 148, 161, 164–168, 170, 172, 199, 200, 232, 264, 268, 270, 304, 306, 309, 317, 318, 409, 411, 413, 415–419, 421, 426, 436, 439, 442, 446, 456, 722, 734, 772, 781, 782, 796, 799, 806, 808, 821, 822, 826, 862, 864, 1006, 1013, 1029, 1030, 1032, 1035, 1036, 1041, 1119, 1121, 1125, 1128, 1141, 1173, 1174, 1178, 1181, 1183, 1186–1188, 1190, 1238, 1288, 1289, 1291, 1293, 1294, 1301, 1303
USA National Health and Nutrition Examination Survey (NHANES), 985
USDA, 317
USDA Forest Service, 704
USDA Forest Service USA, 1003
U.S. Department of Agriculture-Agricultural Research Service (USDA-ARS), 311
User groups, 1008
US Fish and Wildlife Service, 806
US fruit industry, 1291
Ustilago tritici, 1300
Utility Patents, 421
Utopia, 674
Utz Certified, 511
UV radiation, 329, 333

V

Vacciniums, 317, 987
Vacuoles, 977
Valencia, 162, 163
Valkila (2009), 1159
Valuation of biodiversity, 808
Value added chain, 16
Value chain, 329, 479, 481, 1124, 1135, 1146, 1154
Value chain model, 479
Value chains, 1152, 1155, 1156
Value chain systems, 479
Value supply chain, 17
Vancouver Botanical Gardens, 447
Vanda, 418
Vandalism, 800
V. angustifolium, 317
Vanilla, 267
Vanilla planifolia, 267
Vanillylamine, 981
Van Mons, 1297
Van Mons, J.B., 1288
Van Niel, C.B., 1301
Vanuatu, 269, 717
Vapor pressure, 366
Vapor Pressure Deficit (VPD), 354, 355
Vapor pressure-temperature curve, 366
Vapour Pressure Deficit, 221
Variation of Animals and Plants under Domestication, 1258
Variegation, 636
Varro, 1270
Varroa, 1030
Varro (Marcus Terentius Varro), 1265
Vasco da Gamma, 649
Vase life, 356, 421
V. ashei, 317

Vatican Museum, 411
V. corymbosum, 317
Vectors, 832, 847
Vegetable agribusiness system, 1152
Vegetable and fruit gardens, 1272
Vegetable Consumption, 89
Vegetable crops, 823
Vegetable gardens, 1257, 1266, 1274
Vegetable irrigation, 80
Vegetable oil, 269
Vegetable production, 1234
Vegetables, 11, 12, 76–78, 81, 83–85, 87–91, 329, 335–337, 348–353, 356, 362, 365, 373–375, 380, 382, 387, 388, 390, 451, 466, 467, 470–475, 477–481, 606, 796, 842, 843, 846, 850, 860, 966, 1119, 1157, 1161, 1199, 1212, 1218, 1226, 1231, 1235, 1236, 1239, 1248, 1256, 1266, 1274, 1276, 1278, 1281, 1291, 1299
Vegetables and fruits, 1143
Vegetable transplants, 78
Vegetation management, 1119, 1123
Vegetative phase, 104
Vegetative phase l, 823
Vegetative propagation, 737
Veihmeyer, F.J., 1303
Veitchia merrillii, 764
Veldt, 439
Velocity, 838
Venezuela, 280, 285, 438
Ventilated, 367, 368
Ventilated tunnel, 328
Ventilation, 329, 343, 345, 357–359, 361
Ventilation rate, 359, 360, 368
Venus, 1205
Veraison, 118, 203, 204, 208, 209–211, 215, 218, 222, 223, 226, 234
Verbena hybrids, 451
Verbenas, 454
Verdelli, 169
Vergil, 672, 673
Vermeulen and Cotula 2010, 1155, 1156, 1162
Vermiculite, 378
Vernalisation, 820, 821, 846
Vernalization, 178
Versailles, 674, 1235, 1257, 1276, 1278, 1279
Vertebrate pests, 836
Vertical bounce test, 753
Vertical garden walls, 436
Vertical turbulent flux, 368
Verticillium, 306
Verticillium longisporum, 828
Verticordia spp., 442
Vesuvius, 1212
VET, 1129
Vetch, 1264
Veteran trees, 718
Viburnum, 1185
Victoria, 229, 719, 839
Victorian, 411
Victorian era, 412
Victor Loret, 160
Vienna, 678
Vienna codex, 1218
Vietnam, 270, 280, 725
Villa Farnesina, 1212
Ville Contemporaine, 683
Villeggiatura, 674
Ville Radieuse, 683
Vincent Van Gogh, 1215
Vincenzo Campi, 1214
Vine crops, 127
Vine development, 214, 216
Vine nutrition, 216
Vine pull scheme, 198
Vines, 115, 198, 201, 203, 209, 621, 822, 1178, 1257, 1260, 1262, 1269, 1270, 1280, 1281, 1290, 1293, 1297, 1300
Vine weevil, 834
Vineyard budgets, 225
Vineyard establishment, 234, 235
Vineyard hygiene, 230, 231, 232
Vineyard Management, 213, 214, 216, 223, 235, 236
Vineyard managers, 223, 226, 233, 235
Vineyard mechanisation, 236
Vineyard productivity, 215
Vineyards, 212, 215, 216, 219, 222, 225, 229, 233, 605, 1243, 1260, 1265, 1300
Vinitor, 673
Vintage, 234
Violas, 419, 443, 766
Viola tricolour, 455
Violaxanthin, 119
Violence, 17
Violet, 410
Vireya, 439
Virgil, 160, 1257, 1259, 1261, 1262, 1265, 1267, 1288
Virgil (Publius Vergilius Maro), 1261
Virgin forests, 606

Virginiana glauca, 1280
Virtual water, 6, 519
Virus diseases, 847
Viruses, 333, 1030, 1175, 1179
Virus incidence, 848
Virus pathogens, 832, 848
Virus vectors, 846, 847
Visible impacts, 615
Visitor experience, 961
Visual appearance, 476
Visual interest, 1198
Vitamin A, 383, 715
Vitamin C, 139, 173, 176, 190, 315, 353, 383, 416, 480, 715, 982
Vitamin C *see* Ascorbic acid, 126, 967
Vitamin E, 982
Vitamins, 76, 125, 126, 238, 244, 480, 627, 966, 967, 1241
Vitellaria paradoxa, 269, 715
Viticultural practices, 235, 1243
Viticulture, 12, 234, 252, 1226, 1231
Vitis, 199, 625
Vitis sylvestris, 198
Vitis vinifera, 199
Vitis vinifera sylvestris, 110
Vocational and professional skills, 1119
Vocational education and training (VET), 1129
Volatile acidity, 215
volatile organic compounds (VOCs), 773
Volcanic porous rock, 378
Voles, 1035
Volkamer lemon, 172
Von Bertalanffy 1968, 1152
von Humboldt, Alexander, 1172
Von Liebig, 1299, 1300
Von Liebig, J., 1299
Vorley, B., 1150, 1155–1157, 1161
Vulpes vulpes, 1028, 1033
V. vinifera ssp sylvestris, 199
V. virgatum, 317

W

Wagner's Greenhouses, 419
Wales, 830
Walkable communities, 1050, 1052
Walnuts, 99, 109, 289, 822, 1259, 1270
Walte Palm (Wettinia maynensis), 720
Waratah, 440
Warblers, 1035
Wardian Case, 766, 1172
Warm season grasses, 734
Warm temperate, 440, 454
Warm temperate forest, 438
Warner, Kahan, and Lehel 2008, 1148
Wart disease, 848
Washington, 416
Washington, DC, 414, 678
Washington State University, 1121
Washington State University Master Garden program, 1121
Wasps, 751
Waste, 467, 472, 473, 475–479, 481
Waste disposal, 470
Wasteland, 716
Waste management, 802
Waste water, 514, 515
Water, 513–515, 606–608, 630, 632
Water allocations, 218
Water and labour, 510
Water availability, 310
Water balance, 354
Water budgeting, 218
Water budgets, 218
Water conservation, 1121
Water consumption patterns, 320
Watercourses, 514, 714, 719, 726
Water cycling, 802
Water deficits, 201, 234, 375, 376, 1303
Water demand, 218
Water extraction, 518
Water features, 1029
Water footprints, 6, 511, 607, 608
Water Footprint Network, 607, 608
Water holding capacity, 217, 450, 630, 745
Water-holding capacity (WHC), 456
Water hyacinth, 519
Water infiltration, 747, 840
Water(irrigation), 633
Water loss, 467, 468, 480
Water management, 605, 608, 797
Watermelons, 84, 353, 375, 384, 1210, 1218, 1239, 1266
Water molds, 635
Water percolation, 747
Water pollution, 513, 802
Water potential, 77, 85, 219–221, 364
Water production, 808
Water purification, 723, 724
Water quality, 505, 807, 1047, 1059, 1061, 1076
Water resource management, 519
Water resources, 606, 1142

Water Resources Management Authority in Kenya, 519
Water scarcity, 5, 18, 80, 366, 505
Water security, 505
Waterside locations, 1009
Water snails, 1033
Water-soluble pectins, 187
Water status, 210, 775
Water stress, 85, 208, 216, 220–223, 352, 364, 375, 376
Water stress survival mechanisms, 514
Water supply, 363, 373
Water uptake, 364
Water use, 606, 608, 615
Water use efficiency, 219, 368, 607, 837, 850
Water use efficiency (WUE), 80, 81, 329, 371
Water vapor, 341, 364, 368
Water vapour, 820
Waterways, 724
Water withdrawal, 513
Wattle (Acacia spp), 716
Wattlebird, 1032
Wax, 721
Waxes, 466
Waxflower, 450
Wealth creation, 18
Weather, 1027, 1043
Weather index, 848
Weather patterns, 845, 849
Weather windows, 837
Web-based technologies, 1128
Weed colonization, 750
Weed competition, 836
Weed control, 630, 631
Weed management, 630–632, 1239
Weeds, 836, 1059, 1063, 1064, 1076
Weevils, 306, 751, 834
Welfare, 11
Well-being, 18, 772, 1038
Wellbeing and therapy, 1226
West Africa, 135, 269, 280, 281, 283, 438
Westchester County, 683
Western Asia, 101
Western Australia, 442, 454, 456, 719, 839
Western dewberry, 311
West, F.L., 1302
West Indian, 143–145
West Indies, 268, 441
Westonbirt, 725
West Pomerania, 675
Wetlands, 514, 515, 801
Wetland system, 514
Wet pad, 329, 357, 368
Wet pad cooling, 368
Wet tropical regions, 453
Wheat, 274, 805, 826, 839, 1178, 1300
Wheat bunt, 1178
Wheat (Triticum spp.), 648
Whinham's Industry, 315
White apples (Malus alba), 1271
Whiteflies, 847
Whitefly, 847
White ginger, H. coronatrium, 803
White grapes, 238
White juices, 240, 241, 244
White Leadtree, 716
White pine blister rust, 315–317, 829
White pines, 829
Whitesbog (NJ), 317
White wines, 219, 236, 250
Wholesale, 408
Wholesale markets, 468, 475, 1156
Wholesalers, 1152
Wild areas, 801
Wilderness, 13, 1007, 1009
Wilderness Therapy, 1015
Wild flower meadows, 1002, 1035
Wild grasses, 1029
Wild harvesting, 667
Wild lands, 13
Wildlife, 839, 841, 849, 1028, 1059, 1062
Wildlife Botanic Gardens at Bush Prairie, Washington USA., 1041
Wildlife festivals, 1042
Wildlife garden, 1026
Wildlife gardener, 1043
Wildlife gardening, 1026
Wildlife habitat, 790
Wild life sanctuary, 798
Wildlife television, 1040
Wildlife Trusts, 1035, 1036
Wildlife values, 1026
Wildlife watching, 1027, 1039
Wilhelm Miller, 1233
William Bartram, 1172
William Shakespeare, 1245
Willows, 724, 1202, 1269
Willow (Salix spp), 639
Wilting, 467
Wilting coefficient, 1303
Wind, 328, 329, 345, 357, 359, 360, 457, 723, 838, 844

Wind and water, 504
Windbreak, 839
Windbreaks, 716, 723, 724, 838, 839
Windbreak trees, 839
Wind damage, 136
Wind erosion, 504, 723, 839
Wind flow, 797, 839
Wind frequency, 838
Wind injuries, 176
Window pruning, 150
Wind pollinated, 408
Winds, 716, 839
Wind speed, 838, 839
Wine, 198, 211, 212, 214, 242, 970, 973, 1210, 1243, 1260, 1291
Wine composition, 215
Winegrape canopy ideotype, 225
Wine industry, 200
Winemakers, 239, 246
Winemaking, 198, 200, 215, 236, 239, 243, 247
Wineries, 236, 240, 248
Winery, 215
Wine science, 252
Wine styles, 236
Wine tannin, 212
Winkles, 1033
Winter chill, 131
Winter chilling, 778, 821
Winter freeze, 206
Winter Gardens in Sheffield, 764
Winter hardiness, 317
Winter injury, 822
Wisley, 1040
Woerlitz Park, 672, 675, 676
Wolffia columbiana, 408
Wolfhart, C., 1271
Wollemia nobilis, 450
Wollemi pine, 450
Wood-borers, 1174
Wood fibers, 378
Woodlands, 713, 714, 719, 723–725
Woodpecker, 1033
Woodpigeon, 836, 1032
Wood products, 273
Wood pulp, 273
Woody, 627
Woody florals, 639
Woody horsetail, 409
Woody ornamentals, 620–623, 625, 626, 628, 630, 632–635, 637, 638, 640, 641
Woody perennial, 819
Woody plants, 127, 820
Woody shrubs, 408, 1204
Woody storage tissues, 204
Woolly bush (Adenanthos sericea), 457
Worker productivity, 800
Worker welfare, 504
Working environments, 773
Workplace productivity, 773
Work productivity, 799
Work satisfaction, 771
World Bank (WB), 849, 1140
World Cancer Research Fund, 966, 987
World economy, 1174
World Health Organization, 651, 655, 656, 659, 966
World Intellectual Property Organisation (WIPO), 459
World's population, 17, 18, 804, 1066
World trade, 1179
World Trade Organisation (WTO), 459, 1173
World Trade Organisation (WTO) and the World Bank, 508
Worms, 1033
Worshipful Company of Fruiterers, 4
Wounding, 974
WTO, 1179, 1181
WTO Agreement on the Application of Sanitary and Phytosanitary measures (SPS Agreement), 1179
WuYi, 266

X
Xanthomonas axonopodis, 172
Xanthomonas campestris pv. campestris, 831
Xanthophylls, 119
X. campestris pv. campestris, 831
Xenobiotic detoxification, 978
Xenobiotic enzymes, 987
Xenobiotics, 983
Xenophon, 1276
Xeriscaping, 1059
Xerophytic, 438
Xiaolan, 414
Xylella fastidiosa, 232
Xylene, 773
Xyloglucan endotransglicosidases (Xet), 118

Y
Yam, 981
Yarrow, 409
Year-round production, 419
Yeast autolysis, 247

Yeast-derived, 248
Yeast lees, 241, 247
Yeasts, 211, 238, 240–245, 247–249, 282
Yeast species, 244
Yellow ginger, 842
Yield, 83, 104
Yield control, 225
Yin-Yang, 648
Yucca elephantipes, 764, 768
Yunnan, 266, 674

Z
Zagros Mountains, 198
Zambia, 419, 840
Zamio, 450
Zamioculcas zamiifolia, 450, 455
Zanzibar rubber vine, 270
Zarskoje Selo, 676
Zeller, M., 1159
Zenaida aurita, 1033
Zenaida macroura, 1032
Zeolite, 378
Zimbabwe, 84, 419
Zinc, 210, 634, 967
Zingiberales, 128
Zingiber officinale, 267
Zinnia, 449
Zizyphus jujuba, 721
Zizyphus mauratiana, 717
Zizyphus sativus, 721
Zonal geranium, 382
Zoosporic pathogens, 513
Zoysia japonica, 734, 739
Zoysia matrella, 451
Zucchini, 390
Zucchini squash, 384
Zuckerman's Inventory of Personal Emotional Reactions Score, 770
Zygomorphic, 408